AF361708

Supramolecular Chemistry in the 3rd Millennium

Supramolecular Chemistry in the 3rd Millennium

Editor

Catherine E. Housecroft

MDPI • Basel • Beijing • Wuhan • Barcelona • Belgrade • Manchester • Tokyo • Cluj • Tianjin

Editor
Catherine E. Housecroft
University of Basel
Switzerland

Editorial Office
MDPI
St. Alban-Anlage 66
4052 Basel, Switzerland

This is a reprint of articles from the Special Issue published online in the open access journal *Chemistry* (ISSN 2624-8549) (available at: https://www.mdpi.com/journal/chemistry/special_issues/supramol_chem_2020).

For citation purposes, cite each article independently as indicated on the article page online and as indicated below:

LastName, A.A.; LastName, B.B.; LastName, C.C. Article Title. *Journal Name* **Year**, *Volume Number*, Page Range.

ISBN 978-3-0365-1463-5 (Hbk)
ISBN 978-3-0365-1464-2 (PDF)

Contents

About the Editor

Catherine E. Housecroft is Professor of Chemistry in the Department of Chemistry at the University of Basel, Switzerland. Her research group, run jointly with Professor Edwin Constable, is engaged in aspects of materials, coordination, structural and supramolecular chemistries. Much of the emphasis is upon the design and development of materials which will contribute to mankind achieving the 17 Sustainable Development Goals designated in 2015 by the United Nations General Assembly. Compounds of interest range from coordination compounds which can absorb or emit light to coordination networks which can act as hosts for small molecules.

 *chemistry*

MDPI

Editorial

Supramolecular Chemistry in the 3rd Millennium

Catherine E. Housecroft

Department of Chemistry, University of Basel, Building 1096, Mattenstrasse 24a, CH-4058 Basel, Switzerland; catherine.housecroft@unibas.ch

Citation: Housecroft, C.E. Supramolecular Chemistry in the 3rd Millennium. *Chemistry* **2021**, *3*, 509–510. https://doi.org/10.3390/chemistry3020035

Received: 19 February 2021
Accepted: 24 March 2021
Published: 8 April 2021

Publisher's Note: MDPI stays neutral with regard to jurisdictional claims in published maps and institutional affiliations.

The description of supramolecular chemistry as "chemistry beyond the molecule" (Jean-Marie Lehn, 1987 Nobel Lecture and Gautam R. Desiraju, *Nature*, **2001**, *412*, 397) encompasses the wide variety of weak, non-covalent interactions that are the basis for the assembly of supramolecular architectures, molecular receptors and molecular recognition, programed molecular systems, dynamic combinatorial libraries, coordination networks and functional supramolecular materials. For this issue of *Chemistry*, the theme of "Supramolecular Chemistry in the 3rd Millennium" attracted eighteen contributions that cover a broad spectrum of supramolecular assemblies and exemplify the unity of contemporary multidisciplinary science, in which organic, inorganic, physical and theoretical chemists work together with molecular biologists and physicists to develop a systems-level understanding of molecular interactions.

The issue features two reviews which focus, respectively, on supramolecular metalla-assemblies combining coordination and hydrogen bonds (Therrien) [1] and uranyl ion coordination compounds of polycarboxylates (Harrowfield and Thuéry) [2].

Supramolecular interactions are critical to selective binding within receptor molecules and host–guest chemistry, and several papers illustrate these principles using hydroquinone-based anion receptors (Gale and coworkers) [3], cucurbit[7]uril (Redshaw and coworkers) [4] and self-assembled *n*-alkyl-benzoureido-15-crown-5-ethers as selective ion-channels for K^+ cations (Barboiu and coworkers) [5]. Dalgarno and coworkers contribute with a fascinating investigation of cage assembly using *p*-^{t}Bu-calix[4]arene building blocks [6]. Catalysis carried out within the confines of molecular or supramolecular cages is a topical area, and Ward and his coauthors describe a beautiful example exploiting a cubic M_8L_{12} coordination cage [7]. The assembly of highly symmetric metal–cyclo-tricatechylene cages supported within a three-dimensional cubic hydrogen-bonded network is described by Abrahams and coworkers [8].

The introduction of sulfur atoms into ligands often leads to interesting supramolecular interactions. For example, close S...S contacts. Carballo, Belén Lago and coworkers illustrate the different supramolecular interactions that predominate in the structures of the copper(II) coordination compounds of two flexible bis-tetrazole organosulfur ligands [9]. Copper(II) coordination compounds containing Schiff base ligands have been designed by Marques Netto and coworkers as model systems to realize allosteric behavior by regulating the equilibrium between monomeric and dimeric species [10].

Supramolecular interactions play a vital role in the assembly of molecular helicates, and it is therefore fitting that this area of chemistry is represented in this themed issue – the assembly of enantiopure M_4L_4 helicates is described in a study from Turner and coworkers [11]. Moving from multinuclear helicates to coordination polymers takes us on to contributions that describe supramolecular assemblies containing trinuclear copper(II)–pyrazolato units (Raptis, Boudalis, Herchel and coworkers) [12] and chloro-substituted pyrazin-2-aminocopper(I) assemblies featuring hydrogen bond and halogen bond interactions (Mailman, Rissanen and coworkers) [13]. Halogen bonds are a relatively new addition to the array of supramolecular interactions and also feature in the assembly of architectures comprising tetrakis(4-(iodoethynyl)phenyl)methane and 1,3,5,7-tetrakis(4-(iodoethynyl)phenyl)adamantane building blocks (Aakeröy and coworkers) [14]. This

1

work nicely illustrates the potential for halogen bonding in the assembly of porous molecular solids. Bourne and coworkers report a series of cobalt- or zinc-based metal organic frameworks (MOFs) containing pyridylbenzoate linkers [15]. The three-dimensional assemblies comprise non-interpenetrated frameworks that retain their structure upon activation under vacuum, and the study extends to the sorption capacity of the assemblies and their selectivity for volatile organic compounds (VOCs).

Cocrystallization and polymorphism, respectively, are the topics of articles from Merz and coworkers [16] and from Baisch, Vella-Zarb and coworker [17]. This latter article presents an interesting holistic crystallographic study of the antiviral ganciclovir. Baisch and Vella-Zarb also present the crystal structure of N,N',N'',N'''-tetraisopropylpyrophosphoramide and compare the supramolecular interactions with those found in the solid-state structures of other pyrophosphoramides [18].

The range of topics in this themed issue of *Chemistry* illustrates the diverse nature of the research areas which depend upon supramolecular interactions, and I am grateful to all the authors who contributed to this issue.

Funding: This research received no external funding.

Conflicts of Interest: The author declares no conflict of interest.

References

1. Therrien, B. Combining Coordination and Hydrogen Bonds to Develop Discrete Supramolecular Metalla-Assemblies. *Chemistry* **2020**, *2*, 565–576. [CrossRef]
2. Harrowfield, J.; Thuéry, P. Uranyl Ion Complexes of Polycarboxylates: Steps towards Isolated Photoactive Cavities. *Chemistry* **2020**, *2*, 63–79. [CrossRef]
3. McNaughton, D.; Fu, X.; Lewis, W.; D'Alessandro, D.; Gale, P. Hydroquinone-Based Anion Receptors for Redox-Switchable Chloride Binding. *Chemistry* **2019**, *1*, 80–88. [CrossRef]
4. Xu, W.; Zhu, X.; Bian, B.; Xiao, X.; Tao, Z.; Redshaw, C. A Study of the Interaction between Cucurbit[7]uril and Alkyl Substituted 4-Pyrrolidinopyridinium Salts. *Chemistry* **2020**, *2*, 262–273. [CrossRef]
5. Li, Y.; Zheng, S.; Wang, D.; Barboiu, M. Selective Proton-Mediated Transport by Electrogenic K^+-Binding Macrocycles. *Chemistry* **2020**, *2*, 11–21. [CrossRef]
6. Coletta, M.; Palacios, M.; Brechin, E.; Dalgarno, S. A Brucite-Like Mixed-Valent Cluster Capped by $[Mn^{III}p$-^tBu-calix[4]arene$]^-$ Moieties. *Chemistry* **2020**, *2*, 253–261. [CrossRef]
7. Mozaceanu, C.; Taylor, C.; Piper, J.; Argent, S.; Ward, M. Catalysis of an Aldol Condensation Using a Coordination Cage. *Chemistry* **2020**, *2*, 22–32. [CrossRef]
8. Holmes, J.; Russell, S.; Abrahams, B.; Hudson, T.; White, K. Complementarity in Cyclotricatechylene Assemblies: Symmetric Cages Linked within 3D Cubic Hydrogen Bonded Networks. *Chemistry* **2020**, *2*, 577–590. [CrossRef]
9. Gómez-Paz, O.; Carballo, R.; Lago, A.; Vázquez-López, E. Metallosupramolecular Compounds Based on Copper(II/I) Chloride and Two Bis-Tetrazole Organosulfur Ligands. *Chemistry* **2020**, *2*, 33–49. [CrossRef]
10. Castro, C.; Silveira, R.; Colombari, F.; de Moura, A.; Nascimento, O.; Marques Netto, C. Solvent Effect on the Regulation of Urea Hydrolysis Reactions by Copper Complexes. *Chemistry* **2020**, *2*, 525–544. [CrossRef]
11. Boer, S.; Cao, W.; Glascott, B.; Turner, D. Towards a Generalized Synthetic Strategy for Variable Sized Enantiopure M_4L_4 Helicates. *Chemistry* **2020**, *2*, 613–625. [CrossRef]
12. Shi, K.; Mathivathanan, L.; Herchel, R.; Boudalis, A.; Raptis, R. Supramolecular Assemblies of Trinuclear Copper(II)-Pyrazolato Units: A Structural, Magnetic and EPR Study. *Chemistry* **2020**, *2*, 626–644. [CrossRef]
13. Mailman, A.; Puttreddy, R.; Lahtinen, M.; Svahn, N.; Rissanen, K. Hydrogen and Halogen Bond Mediated Coordination Polymers of Chloro-Substituted Pyrazin-2-Amine Copper(I) Bromide Complexes. *Chemistry* **2020**, *2*, 700–713. [CrossRef]
14. Gunawardana, C.; Sinha, A.; Reinheimer, E.; Aakeröy, C. From Frustrated Packing to Tecton-Driven Porous Molecular Solids. *Chemistry* **2020**, *2*, 179–192. [CrossRef]
15. Ndamyabera, C.; Zacharias, S.; Oliver, C.; Bourne, S. Solvatochromism and Selective Sorption of Volatile Organic Solvents in Pyridylbenzoate Metal-Organic Frameworks. *Chemistry* **2019**, *1*, 111–125. [CrossRef]
16. Komisarek, D.; Schauerte, C.; Merz, K. Controlled Stepwise Synthesis and Characterization of a Ternary Multicomponent Crystal with 2-Methylresorcinol. *Chemistry* **2020**, *2*, 93–100. [CrossRef]
17. Spiteri, L.; Baisch, U.; Vella-Zarb, L. Understanding Conformational Polymorphism in Ganciclovir: A Holistic Approach. *Chemistry* **2021**, *3*, 126–137. [CrossRef]
18. Micallef, D.; Vella-Zarb, L.; Baisch, U. Exploring the Structural Chemistry of Pyrophosphoramides: N,N',N'',N'''-Tetraisopropylpyrophosphoramide. *Chemistry* **2021**, *3*, 149–163. [CrossRef]

 *chemistry*

Review

Combining Coordination and Hydrogen Bonds to Develop Discrete Supramolecular Metalla-Assemblies

Bruno Therrien

Institute of Chemistry, University of Neuchatel, Avenue de Bellevaux 51, CH-2000 Neuchatel, Switzerland; bruno.therrien@unine.ch; Tel.: +41-32-718-2499

Received: 6 April 2020; Accepted: 18 May 2020; Published: 8 June 2020

Abstract: In Nature, metal ions play critical roles at different levels, and they are often found in proteins. Therefore, metal ions are naturally incorporated in hydrogen-bonded systems. In addition, the combination of metal coordination and hydrogen bonds have been used extensively to develop supramolecular materials. However, despite this win-win combination between coordination and hydrogen bonds in many supramolecular systems, the same combination remains scarce in the field of coordination-driven self-assemblies. Indeed, as illustrated in this mini-review, only a few discrete supramolecular metalla-assemblies combining coordination and hydrogen bonds can be found in the literature, but that figure might change rapidly.

Keywords: coordination chemistry; hydrogen bonds; supramolecular chemistry; metalla-assemblies; coordination-driven self-assembly; orthogonality; ligands; metal ions

---

1. Introduction

The preparation and characterization of discrete metal-based assemblies have been the focus of several research groups. Such metalla-assemblies are obtained by combining metal ions and multidentate ligands in a pre-designed and controlled manner [1–5]. These supramolecular metalla-assemblies can be used as sensors [6–8], anticancer agents [9,10], hosts for guest molecules [11,12], drug carriers [13], mesogens [14,15], or molecular flasks [16,17]. About 40 years ago, the first coordination-driven metal-based squares (Figure 1), composed of linear diphosphine ligands and tetracarbonyl metal ions (Cr, Mo, W), were synthesized [18].

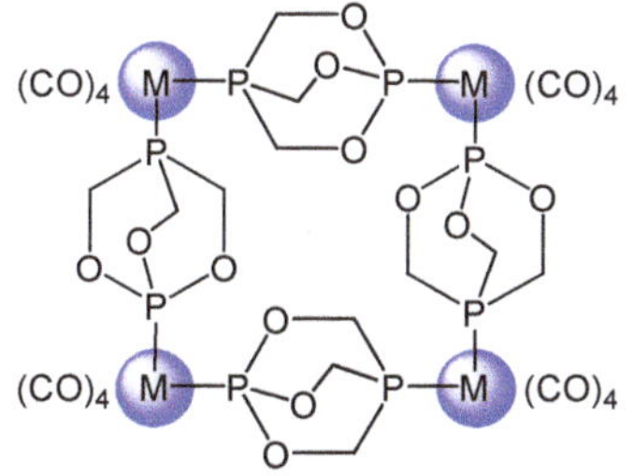

Figure 1. Molecular structure of the first coordination-linked metalla-squares [18].

A few years later, the field really took off with the introduction of 90° square-planar palladium ions, which are versatile building blocks in supramolecular chemistry [19]. Nowadays, all kind of transition metal ions with different coordination geometries have been introduced in metalla-assemblies, and the field is flourishing.

Like the field of coordination-driven self-assemblies, hydrogen-bonded self-assembled systems have received for many years a great deal of attention [20–25]. The directionality, stability, reversibility, and biological importance of hydrogen bonds have encouraged research groups to use hydrogen-bonded motifs to construct supramolecular assemblies. An appropriate selection of donor and acceptor groups, in a pre-organized fashion, can control the strength and geometry of the designed supramolecular structures, and accordingly, allow the formation of two-dimensional and three-dimensional assemblies. Nowadays, the degree of sophistication has reached an incredible level, far beyond the simple DNA helices, and beautiful examples are published at a regular pace with different applications in mind [20–25].

In the field of materials science, coordination and hydrogen bonds have been joint to generate polymers, dendrimers, and other supramolecular assemblies [26–32]. An early example of such materials comes from the group of Reinhoudt, in which a barbituric acid entity was coupled to palladium-based metallo-dendrons to generate metallo-dendrimers [33]. In these systems, the barbituric acid residue forms a rosette type structure via hydrogen bonds, while the supramolecular network is further extended by the dendritic arms: The two functions are linked together by coordination chemistry. Following this pioneer report, similar combinations have been used to develop coordination and hydrogen-bonded materials [26–35].

Surprisingly, despite this relative popularity, the combination of coordination and hydrogen bonds to form discrete supramolecular metalla-assemblies remains scarce. Most examples are limited to cyclic and planar entities (one and two dimensions), and only recently, systems showing cavities and cage-like structures (three dimensions) have appeared in the literature. These hybrid self-assembled systems involving coordination chemistry and hydrogen-bonded interactions to form discrete entities are presented and discussed in this short review, thus showing the great potential of combining coordination and hydrogen bonds to develop new supramolecular metalla-assemblies.

2. Planar and Macrocyclic Assemblies Exploiting Coordination and Hydrogen Bonds

In crystal engineering, the combination of metal ions and hydrogen bonds has been extensively explored [36], and the first examples of discrete coordination and hydrogen-bonded systems were probably inspired by solid-state chemistry. Joining several metal-based chromophores is needed for the preparation of light-harvesting systems, however, to better understand the electronic pathway and metal-metal communications involved in such systems, having a dinuclear compound can be more appropriate. With that in mind, the groups of Ward and Barigelletti have studied the electronic energy transfer process between metal-polypyridyl complexes linked by complementary hydrogen-bonded groups [37,38]. Bispyridyl ligands functionalized with nucleobases were synthesized and used to connect two metal ions, see Figure 2.

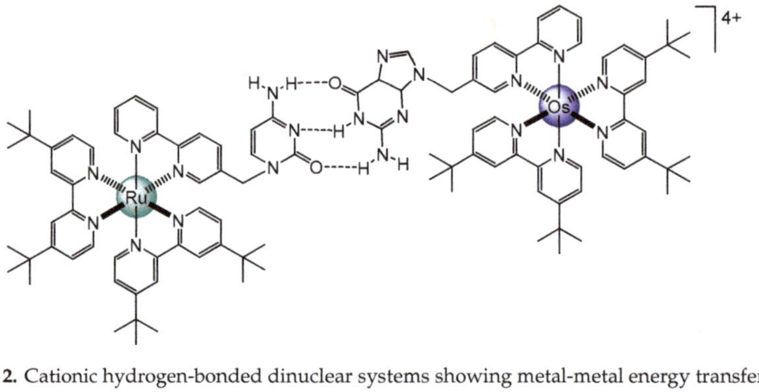

Figure 2. Cationic hydrogen-bonded dinuclear systems showing metal-metal energy transfer [37].

The relatively high binding constant of the nucleotide base pairs and the nature of the dinuclear systems have suggested that the energy transfer occurs, even in solution (CH_2Cl_2), via the hydrogen-bonded interface.

A similar bis-rhodium complex has been synthesized [39], and a single crystal X-ray structure analysis has confirmed the dimeric nature of the system (Figure 3). Interestingly, upon coordination to the rhodium pentamethylcyclopentadienyl unit, the hydrogen bond pairing between two 7-diphenylphosphino-1*H*-quinolin-2-one ligands is not disturbed. Diffusion-ordered NMR spectroscopy in CD_2Cl_2 shows that the dimeric structure is stable in aprotic solvents.

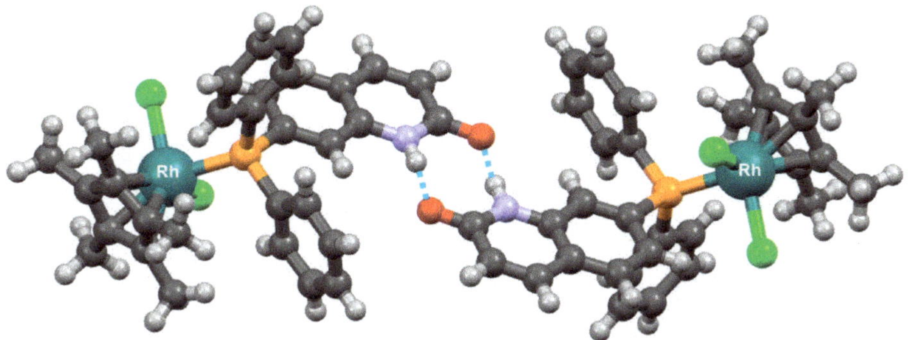

Figure 3. Dimeric structure of a hydrogen-bonded bis-rhodium complex [39].

Other dinuclear systems have been prepared, in which a combination of coordination and hydrogen-bonded interactions were used. For instance, a platinum-based dimer has been prepared in view to synthesize higher nuclearity systems [40]. Unfortunately, the self-complementary quinolone hydrogen bonds were too weak compared to the π-stacking interactions of the ligands, thus forming in solution a coordination macrocycle instead of a hydrogen-bonded tetranuclear system (Scheme 1).

Scheme 1. Self-assembly of a coordination macrocycle (**top**) over a tetranuclear hydrogen-bonded system (**bottom**) [40].

Quinolone-based ligands were also used with octahedral metal center. Indeed, a dinuclear rhodium-based complex was obtained by reacting [Cp*RhCl$_2$]$_2$ (Cp* = pentamethylcyclopentadienyl) with 7-diphenylphosphino-1*H*-quinolin-2-one in a 1:2 stoichiometry [39]. The cationic dinuclear complex (Figure 4) is stable in solution (CD_2Cl_2), and NMR studies suggest that no dynamic behavior (assembly-disassembly) is occurring at room temperature in aprotic solvent. As emphasized in Figure 4, strong π-π stacking interactions take place, which increases the stability of the macrocyclic structure. Analogous dinuclear systems were obtained by reacting [(*para*-cymene)RuCl$_2$]$_2$ with 1-(4-oxo-6-undecyl-1,4-dihydropyrimidin-2-yl)-3-(pyridine-4-ylethyl)urea (UPy-L) in a 1:2 fashion [41]. The neutral complex (Figure 5) is stable under ESI-MS (electro-spray ionization–mass spectrometry) conditions. The dinuclear complexes were also incorporated in tetranuclear systems, in which the UPy-L units were parallel to each other to generate metalla-rectangles [42].

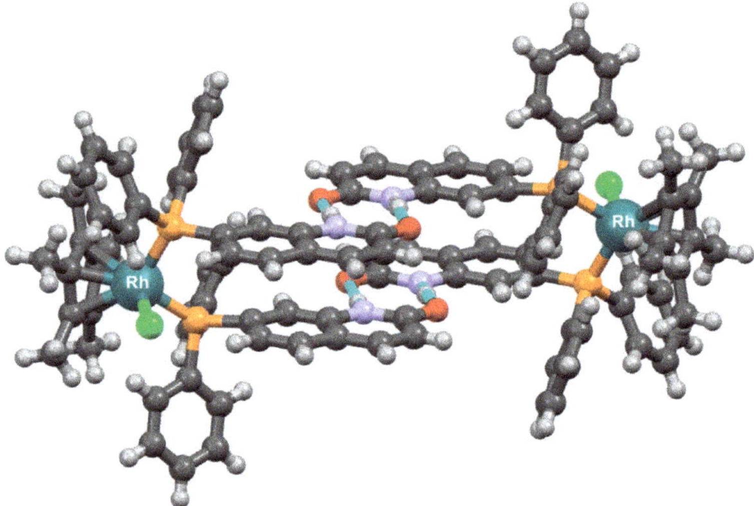

Figure 4. Dinuclear rhodium-based metallacycle [39].

Figure 5. A dinuclear complex incorporating piano-stool complexes and derived-ligands allowing hydrogen-bonded assemblies [41].

Two-dimensional assemblies with more than two metal ions have also been synthesized, using for example square-planar complexes. Tetranuclear and hexanuclear platinum-based metalla-cycles were prepared by Rendina and his coworkers [43]. The nicotinic acid pair acts as a 120° bridging ligand, and upon coordination to *trans*-bis(diphenylphosphine)platinum units, it forms a dinuclear sub-unit that can be coupled to other bidentate ligands. In combination with a 180° bidentate spacer (4,4′-biphenyl), a hexanuclear metalla-cycle is obtained (Figure 6A), while in combination with a 120° bidentate ligand (4,4′-benzophenone), a tetranuclear metalla-cycle is isolated (Figure 6B).

When *iso*-nicotinic acid is used instead, oligomeric and polymeric species are formed, demonstrating the importance of ligands and metal ions geometry for the preparation of discrete metalla-assemblies.

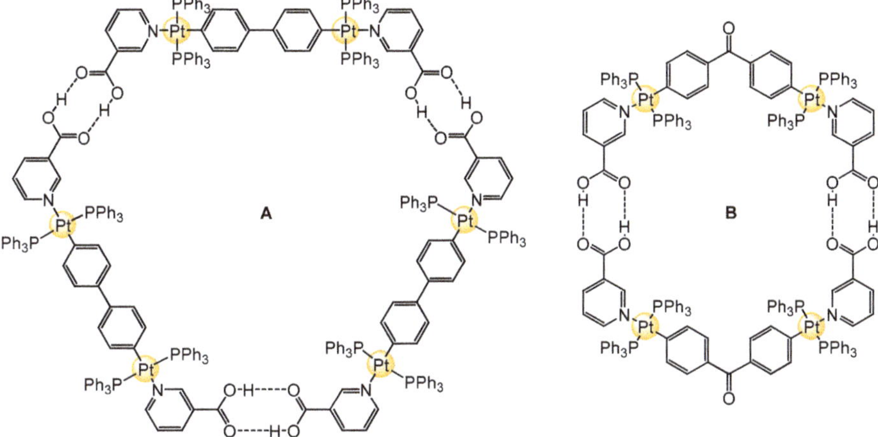

Figure 6. Hexanuclear (**A**) and tetranuclear (**B**) metalla-cycles built from nicotinic acid and square-planar platinum ions [43].

Hydrogen-bonded dimers of *para*-pyridyl-substituted 2-ureido-4-1*H*-pyrimidinone and *cis*-coordinated palladium complexes have been combined to afford a tetranuclear metalla-cycle [44]. In solution (CDCl$_3$), a mixture of a metalla-square (Figure 7) and a metalla-triangle was observed. At low concentrations (1 mM), the triangular assembly is favored, while at higher concentrations, the amount of the square-like structure is increasing significantly. This study confirms that the nature of metalla-cycles can be controlled by steric factors, by the solubility of the final entity, and by the geometry of the different building blocks.

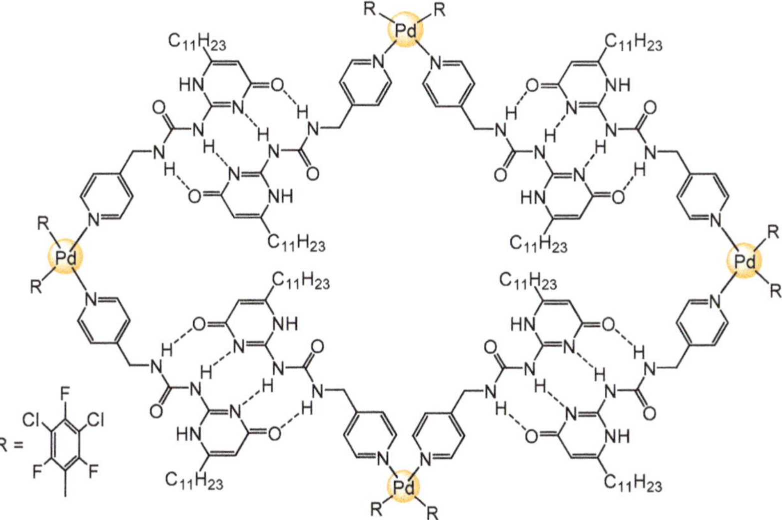

Figure 7. Molecular structure of a palladium-based tetranuclear assembly [44].

The same *para*-pyridyl-substituted 2-ureido-4-1*H*-pyrimidinone hydrogen-bonded dimer was used to construct tetranuclear arene ruthenium metalla-rectangles [42]. NMR spectroscopy and

DFT calculations showed that the formation of the hydrogen-bonded assembly results in an energy gain of $\Delta E = -146.8$ kJ mol^{-1}, thus confirming the stability in solution of these multiple hydrogen-bonded assemblies.

The melamine–cyanuric acid (barbituric acid) pairing is among the most studied hydrogen-bonded system [45–48]. Rosette-type and tapelike structures can be achieved by the controlled functionalization of the sub-units [49,50]. Steric groups will favored the formation of discrete rosette-type structures, while small and highly soluble groups will increase tapelike structures. Therefore, in view to obtain discrete metal-coordinated rosette-type systems, a series of pyridyl-functionalized cyanuric acid [51] and melamine [52] derivatives were synthesized. Coordination of arene ruthenium complexes to the pyridyl groups has generated trinuclear (Figure 8A) and hexanuclear (Figure 8B) rosette-type assemblies.

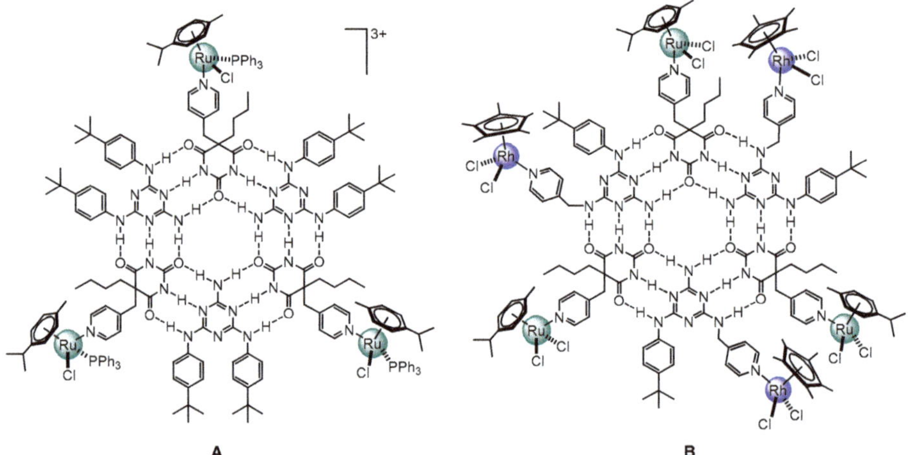

A   **B**

Figure 8. Examples of trinuclear (**A**) and hexanuclear (**B**) rosette-type metalla-assemblies [51,52].

The use of metalated nucleobases has been another approach in supramolecular chemistry that combines coordination chemistry and hydrogen bonds [53–59]. In such systems, the natural pairing of nucleic acids is replaced by metal-mediated base pairs, which allows the generation of hybrid DNA structures. Various applications have been foreseen for these derivatives (sensing, expending the genetic code, forming nanoclusters or nanowires, DNA technology) and they have been the subject of many reviews [53–59]. Therefore, this abundant literature will not be covered here, and the readers who are interested in that particular area are encouraged to refer to these reviews to complete the discussion.

3. Cage-Like Assemblies Exploiting Coordination and Hydrogen Bonds

Several supramolecular capsules built by a combination of two functionalized C_2 symmetrical calix[4]arene cavitands have been synthesized by Yamanaka and his coworkers [60–62]. In these systems, the two capsules are linked by two metal ions and two pairs of hydrogen bonds (Figure 9). The size of the calix[4]arene and the length of the functional groups (hydrogen-bonded derivatives and pyridyl groups) dictate the size of the cavity. In some cases, the cavity is filled by an anion, while in other systems, a guest molecule is trapped. The guest exchange dynamics are linked to the nature of the anions used, and their ability to generate conformational changes by disrupting the intramolecular hydrogen-bonded system.

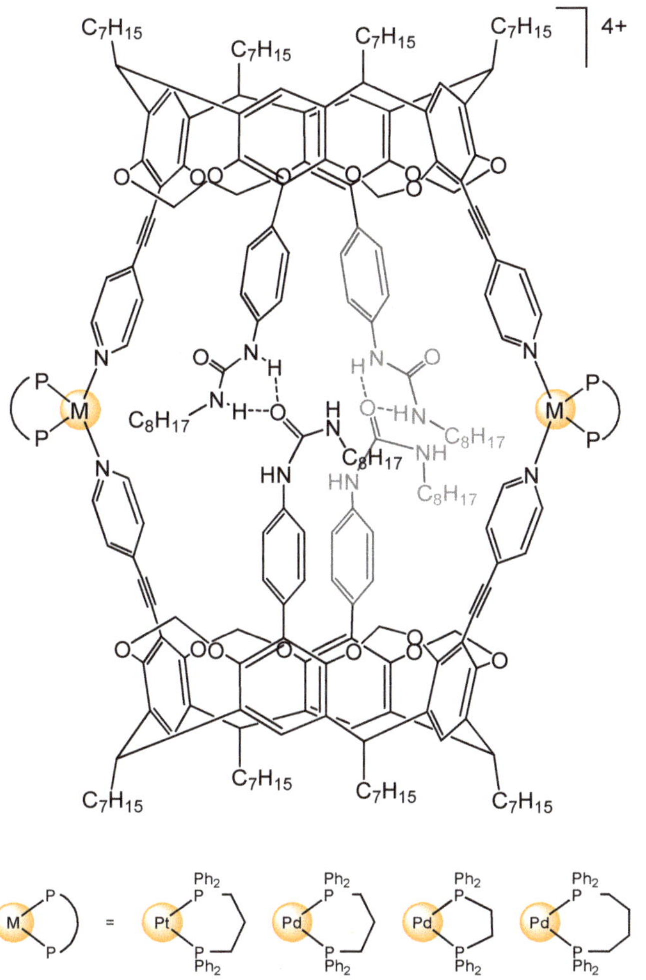

Figure 9. Hybrid capsules built from two functionalized calix[4]arene cavitands [62].

The conical-shape of calix[4]arene was also used to generate giant uranyl-based cages [63]. In these systems, hydrogen bond interactions are exploited to ensure that the calixarene carboxylate ligands adopt a stable and symmetrical conformation, which ultimately forces the carboxylate anion to coordinate to the uranyl cation (UO_2^{2+}) in a controlled manner. This strategy has allowed the formation of several discrete icosahedral cage-like structure (Figure 10), in which the metals are not located at the corners or edges of the assemblies, and for which the cavity of the large anionic capsule is relatively well shielded.

The cooperative action of coordination bonds and quadruple hydrogen-bonded interactions has allowed the synthesis of tetrahedron cage-like structures [64]. The symmetry, size, and nature of the assembly are linked to the flexibility of the ligands, the choice of the metal ions (Hg^{2+}, Fe^{2+}, Zn^{2+}) and the conditions used (solvent polarity, concentration, anion, temperature). In the iron(II) derivatives, the quadruple hydrogen-bonded units are linked to 2,2′-bipyridyl group, to produce a tetrahedron cage-like structure (Figure 11). Stability studies have showed that protic solvents (DMSO, H_2O) initiate the disassembly of the cage-like structure. However, some derivatives show remarkable stability in polar solvents, and even in the presence of coordinating competing agents, thus suggesting that these capsules can be used as reactors for catalytic reactions.

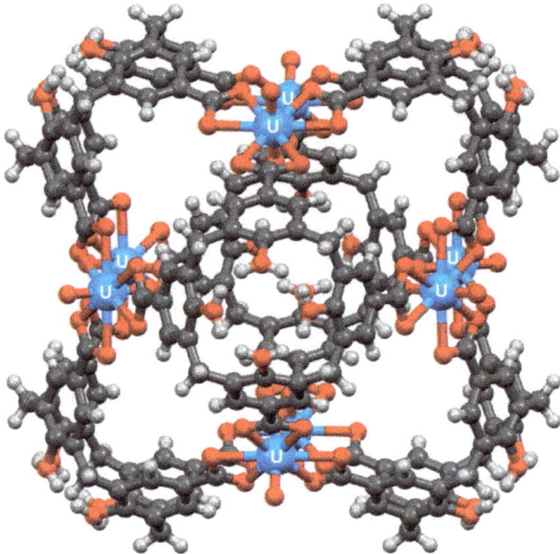

Figure 10. Molecular structure of a hexameric cage-like structure built from six calix[4]arene carboxylates and eight uranyl cations [63].

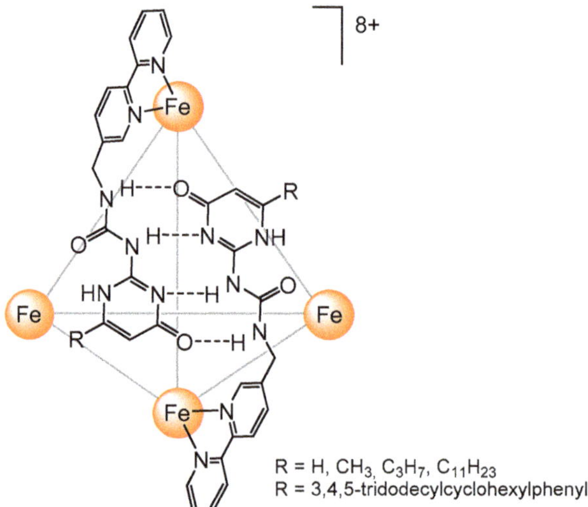

Figure 11. Molecular structure of a hydrogen-bonded tetrahedron cage-like system [64].

The nature of the metal ion was also a critical point when dealing with these quadruple hydrogen-bonded units linked to 2,2′-bipyridyl ligand [64]. Replacing Zn^{2+} or Fe^{2+} with Hg^{2+} not only modified the stability, but also the overall geometry. The large ionic radius of Hg^{2+} provides a wider separation of the coordinated 2,2′-bipyridyl ligands, thus allowing the quadruple hydrogen-bonded units to stack on top of each other and to form a triple decker system (Figure 12). The helicate structure is less stable than the tetrahedron systems, as the coordination energy of 2,2′-bipyridyl to Hg^{2+} remains relatively weak compared to Fe^{2+}.

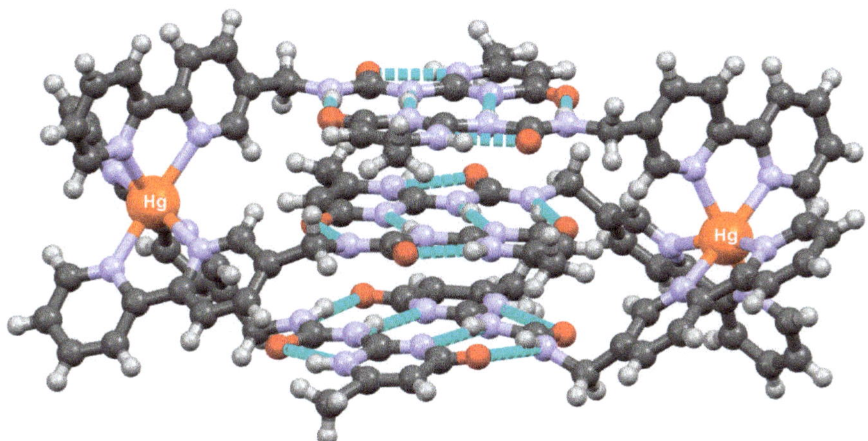

Figure 12. Dinuclear mercury-based hydrogen-bonded helicate [64].

4. Conclusions

As pointed out here, as well as in several reviews and publications [26–35], allowing two or more non-covalent interactions to take place simultaneously can be quite challenging (orthogonality concept) [30]. To be successful, a compatibility between the hydrogen-bonded and coordination interactions is essential, as individually they show different solubility, different flexibility and different stability. Moreover, to form a discrete supramolecular metalla-assembly, the ligands and the hydrogen bonded units should not compete for the metal ions, and they should cooperate. Therefore, it is not so surprising that so far the number of discrete metalla-assemblies combining coordination and hydrogen bonds remains limited. Nevertheless, we can assume that considering recent progress in the understanding on how such orthogonal concepts can be applied to supramolecular systems, and how innovative strategies have recently emerged in the field, that soon, we will see more of these discrete supramolecular metalla-assemblies.

Funding: This research received no external funding.

Acknowledgments: The author would like to thank past and present members of his group.

Conflicts of Interest: The authors declare no conflict of interest.

References

1.   Chen, L.-J.; Yang, H.-B. Construction of stimuli-responsive functional materials via hierarchical self-assembly involving coordination interactions. *Acc. Chem. Res.* **2018**, *51*, 2699–2710. [CrossRef] [PubMed]

2.   Cook, T.R.; Stang, P.J. Recent developments in the preparation and chemistry of metallacycles and metallacages via coordination. *Chem. Rev.* **2015**, *115*, 7001–7045. [CrossRef] [PubMed]

3.   Datta, S.; Saha, M.L.; Stang, P.J. Hierarchical assemblies of supramolecular coordination complexes. *Acc. Chem. Res.* **2018**, *51*, 2047–2063. [CrossRef] [PubMed]

4.   Stíbal, D.; Therrien, B. Organometallic chemistry in directed assembly. In *Comprehensive Supramolecular Chemistry II*; Atwood, J.L., Ed.; Elsevier: Oxford, UK, 2017; Volume 6, pp. 305–329.

5.   Hosseini, M.W. Molecular tectonics: From simple tectons to complex molecular networks. *Acc. Chem. Res.* **2005**, *38*, 313–323. [CrossRef] [PubMed]

6.   Li, M.; Chen, L.-J.; Zhang, Z.; Luo, Q.; Yang, H.-B.; Tian, H.; Zhu, W.-H. Conformer-dependent self-assembled metallacycles with photo-reversible response. *Chem. Sci.* **2019**, *10*, 4896–4904. [CrossRef] [PubMed]

7.   Percástegui, E.G.; Jancik, V. Coordination-driven assemblies based on *meso*-substituted porphyrins: Metal-organic cages and a new type of *meso*-metallaporphyrin macrocycles. *Coord. Chem. Rev.* **2020**, *407*, 213165. [CrossRef]

8. Barry, N.P.E.; Therrien, B. Pyrene: The guest of honor. In *Organic Nanoreactors*; Sadjadi, S., Ed.; Elsevier: London, UK, 2016; pp. 421–461.

9. Pöthig, A.; Casini, A. Recent developments of supramolecular metal-based structures for applications in cancer therapy and imaging. *Theranostics* **2019**, *9*, 93150–93169. [CrossRef]

10. Sepehrpour, H.; Fu, W.; Sun, Y.; Stang, P.J. Biomedically relevant self-assembled metallacycles and metallacages. *J. Am. Chem. Soc.* **2019**, *141*, 14005–14020. [CrossRef]

11. Amouri, H.; Desmarets, C.; Moussa, J. Confined nanospaces in metallocages: Guest molecules, weakly encapsulated anions, and catalyst sequestration. *Chem. Rev.* **2012**, *112*, 2015–2041. [CrossRef]

12. Wang, G.-L.; Lin, Y.-J.; Berke, H.; Jin, G.-X. Two-step assembly of multinuclear metallacycles with half-sandwich Ir, Rh, and Ru fragments for counteranion encapsulation. *Inorg. Chem.* **2010**, *49*, 2193–2201. [CrossRef]

13. Therrien, B. Biologically relevant arene ruthenium metalla-assemblies. *CrystEngComm* **2015**, *17*, 484–491. [CrossRef]

14. Chen, L.; Chen, C.; Sun, Y.; Lu, S.; Huo, H.; Tan, T.; Li, A.; Li, X.; Ungar, G.; Liu, F.; et al. Luminescent metallacycle-cored liquid crystals induced by metal coordination. *Angew. Chem. Int. Ed.* **2020**, *59*. [CrossRef]

15. Therrien, B. Thermotropic liquid-crystalline materials based on supramolecular coordination complex. *Inorganics* **2020**, *8*, 2. [CrossRef]

16. Cullen, W.; Misuraca, M.C.; Hunter, C.A.; Williams, N.H.; Ward, M.D. Highly efficient catalysis of the Kemp elimination in the cavity of a cubic coordination cage. *Nat. Chem.* **2016**, *8*, 231–236. [CrossRef] [PubMed]

17. Yoshizawa, M.; Klosterman, J.K.; Fujita, M. Functional molecular flasks: New properties and reactions within discrete, self-assembled hosts. *Angew. Chem. Int. Ed.* **2009**, *48*, 3418–3438. [CrossRef]

18. Stricklen, P.M.; Volcko, E.J.; Verkade, J.G. Novel homo- and heterometallic coordination macrocycles. *J. Am. Chem. Soc.* **1983**, *105*, 2494–2495. [CrossRef]

19. Fujita, M.; Tominaga, M.; Hori, A.; Therrien, B. Coordination assemblies from a Pd (II)-cornered square complex. *Acc. Chem. Res.* **2005**, *38*, 371–380. [CrossRef]

20. Conn, M.M.; Rebek, J., Jr. Self-assembling capsules. *Chem. Rev.* **1997**, *97*, 1647–1668. [CrossRef]

21. Adriaenssens, L.; Ballester, P. Hydrogen bonded supramolecular capsules with functionalized interiors: The controlled orientation of included guests. *Chem. Soc. Rev.* **2013**, *42*, 3261–3277. [CrossRef]

22. Sigel, R.K.O.; Freisinger, E.; Metzger, S.; Lippert, B. Metalated nucleobase quartets: Dimerization of a metal-modified guanine, cytosine pair of *trans*-$(NH_3)_2Pt^{II}$ and formation of CH···N hydrogen bonds. *J. Am. Chem. Soc.* **1998**, *120*, 12000–12007. [CrossRef]

23. Young, M.C.; Holloway, L.R.; Johnson, A.M.; Hooley, R.J. A supramolecular sorting hat: Stereocontrol in metal-ligand self-assembly by complementary hydrogen bonding. *Angew. Chem. Int. Ed.* **2014**, *53*, 9832–9836. [CrossRef] [PubMed]

24. McLaughlin, C.K.; Hamblin, G.D.; Sleiman, H.F. Supramolecular DNA assembly. *Chem. Soc. Rev.* **2011**, *40*, 5647–5656. [CrossRef] [PubMed]

25. Philp, D.; Stoddart, J.F. Self-assembly in natural and unnatural systems. *Angew. Chem. Int. Ed.* **1996**, *35*, 1154–1196. [CrossRef]

26. Krische, M.J.; Lehn, J.-M. The utilization of persistent H-bonding motifs in the self-assembly of supramolecular architectures. *Struct. Bond.* **2000**, *96*, 3–29.

27. Sijbesma, R.P.; Meijer, E.W. Quadruple hydrogen bonded systems. *Chem. Commun.* **2003**, 5–16. [CrossRef] [PubMed]

28. Van Manen, H.-J.; Paraschiv, V.; García-López, J.J.; Schönherr, H.; Zapotoczny, S.; Vancso, G.J.; Crego-Calama, M.; Reinhoudt, D.N. Hydrogen-bonded assemblies as a scaffold for metal-containing nanostructures: From zero to two dimensions. *Nano Lett.* **2004**, *4*, 441–446. [CrossRef]

29. Lehn, J.-M. Perspectives in Chemistry—Steps towards complex matter. *Angew. Chem. Int. Ed.* **2013**, *52*, 2836–2850. [CrossRef]

30. Wong, C.-H.; Zimmerman, S.C. Orthogonality in organic, polymer, and supramolecular chemistry: From Merrifield to click chemistry. *Chem. Commun.* **2013**, *49*, 1679–1695. [CrossRef]

31. Krieg, E.; Bastings, M.M.C.; Besenius, P.; Rybtchinski, B. Supramolecular polymers in aqueous media. *Chem. Rev.* **2016**, *116*, 2414–2477. [CrossRef]

32. Ayme, J.-F.; Lehn, J.-M. From coordination chemistry to adaptive chemistry. *Adv. Inorg. Chem.* **2018**, *71*, 3–78.

33. Huck, W.T.S.; Hulst, R.; Timmerman, P.; van Veggel, F.C.J.M.; Reinhoudt, D.N. Noncovalent synthesis of nanostructures: Combining coordination chemistry and hydrogen bonding. *Angew. Chem. Int. Ed.* **1997**, *36*, 1006–1008. [CrossRef]

34. Yan, X.; Li, S.; Pollock, J.B.; Cook, T.R.; Chen, J.; Zhang, Y.; Ji, X.; Yu, Y.; Huang, F.; Stang, P.J. Supramolecular polymers with tunable topologies via hierarchical coordination-driven self-assembly and hydrogen bonding interfaces. *Proc. Nat. Acad. Sci. USA* **2013**, *110*, 15585–15590. [CrossRef] [PubMed]

35. Zhang, Q.; Li, T.; Duan, A.; Dong, S.; Zhao, W.; Stang, P.J. Formation of a supramolecular polymeric adhesive via water-participant hydrogen bond formation. *J. Am. Chem. Soc.* **2019**, *141*, 8058–8063. [CrossRef] [PubMed]

36. Burrows, A.D.; Chan, C.-W.; Chowdhry, M.M.; McGrady, J.E.; Mingos, D.M.P. Multidimensional crystal engineering of bifunctional metal complexes containing complementary triple hydrogen bonds. *Chem. Soc. Rev.* **1995**, *24*, 329–339. [CrossRef]

37. Armaroli, N.; Barigelletti, F.; Calogero, G.; Flamigni, L.; White, C.M.; Ward, M.D. Electronic energy transfer between ruthenium (II) and osmium (II) polypyridyl luminophores in a hydrogen-bonded supramolecular assembly. *Chem. Commun.* **1997**, 2181–2182. [CrossRef]

38. White, C.M.; Fernandez Gonzalez, M.; Bardwell, D.A.; Rees, L.H.; Jeffery, J.C.; Ward, M.D.; Armaroli, N.; Calogero, G.; Barigelletti, F. Derivatives of luminescent metal-polypyridyl complexes with pendant adenine or thymine groups: Building blocks for supramolecular assemblies based on hydrogen bonding. *J. Chem. Soc. Dalton Trans.* **1997**, 727–735. [CrossRef]

39. Sommer, S.K.; Zakharov, L.N.; Pluth, M.D. Design, synthesis, and characterization of hybrid metal-ligand hydrogen-bonded (MLHB) supramolecular architectures. *Inorg. Chem.* **2015**, *54*, 1912–1918. [CrossRef]

40. Sommer, S.K.; Henle, E.A.; Zakharov, L.N.; Pluth, M.D. Selection for a single self-assembled macrocycle from a hybrid metal-ligand hydrogen-bonded (MLHB) ligand subunit. *Inorg. Chem.* **2015**, *54*, 6910–6916. [CrossRef]

41. Appavoo, D.; Carnevale, D.; Deschenaux, R.; Therrien, B. Combining coordination and hydrogen-bonds to form arene ruthenium metalla-assemblies. *J. Organomet. Chem.* **2016**, *824*, 80–87. [CrossRef]

42. Appavoo, D.; Raja, N.; Deschenaux, R.; Therrien, B.; Carnevale, D. NMR spectroscopy and DFT calculations of a self-assembled arene ruthenium rectangle obtained from a combination of coordination and hydrogen bonds. *Dalton Trans.* **2016**, *45*, 1410–1421. [CrossRef]

43. Gianneschi, N.C.; Tiekink, E.R.T.; Rendina, L.M. Dinuclear platinum complexes with hydrogen-bonding functionality: Noncovalent assembly of nanoscale cyclic arrays. *J. Am. Chem. Soc.* **2000**, *122*, 8474–8479. [CrossRef]

44. Marshall, L.J.; de Mendoza, J. Self-assembled squares and triangles by simultaneous hydrogen bonding and metal coordination. *Org. Lett.* **2013**, *15*, 1548–1551. [CrossRef] [PubMed]

45. Yagai, S.; Mahesh, S.; Kikkawa, Y.; Unoike, K.; Karatsu, T.; Kitamura, A.; Ajayaghosh, A. Toroidal Nanoobjects from rosette assemblies of melamine-linked oligo(*p*-phenyleneethynylene)s and cyanurates. *Angew. Chem. Int. Ed.* **2008**, *47*, 4691–4694. [CrossRef]

46. Petelski, A.N.; Peruchena, N.M.; Sosa, G.L. Evolution of the hydrogen-bonding motif in the melamine-cyanuric acid co-crystal: A topological study. *J. Mol. Model.* **2016**, *22*, 202. [CrossRef]

47. Adhikari, B.; Lin, X.; Yamauchi, M.; Ouchi, H.; Aratsu, K.; Yagai, S. Hydrogen-bonded rosettes comprising π-conjugated systems as building blocks for functional one-dimensional assemblies. *Chem. Commun.* **2017**, *53*, 9663–9683. [CrossRef] [PubMed]

48. Ferigán, B.; Folcia, C.L.; Termine, R.; Golemme, A.; Granadino-Roldán, J.M.; Navarro, A.; Serrano, J.L.; Giménez, R.; Sierra, T. Inspecting the electronic architecture and semiconducting properties of a rosette-like supramolecular columnar liquid crystal. *Chem. Eur. J.* **2018**, *24*, 17459–17463. [CrossRef]

49. Igci, C.; Karaman, O.; Fan, Y.; Gonzales III, A.A.; Fenniri, H.; Gumbas, G. Synthesis of N-bridged pyrido[4,3-*d*]pyrimidines and self-assembly into twin rosette cages and nanotubes in organic media. *Sci. Rep.* **2018**, *8*, 15949. [CrossRef]

50. Whitesides, G.M.; Simanek, E.E.; Mathias, J.P.; Seto, C.T.; Chin, D.N.; Mammen, M.; Gordon, D.M. Noncovalent synthesis: Using physical-organic chemistry to make aggregates. *Acc. Chem. Res.* **1995**, *28*, 37–44. [CrossRef]

51. Zhang, F.; Therrien, B. Coordination of piano-stool complexes to a hydrogen-bonded rosette-type assembly. *Eur. J. Inorg. Chem.* **2017**, 3214–3221. [CrossRef]

52. Zhang, F.; Therrien, B. Using hydrogen-bonded rosette-type scaffold to generate heteronuclear metalla-assemblies. *Eur. J. Inorg. Chem.* **2018**, 2399–2407. [CrossRef]

53. Müller, J. Nucleic acid duplexes with metal-mediated base pairs and their structures. *Coord. Chem. Rev.* **2019**, *393*, 37–47. [CrossRef]

54. Park, K.S.; Jung, C.; Park, H.G. "Illusionary" polymerase activity triggered by metal ions: Use for molecular logic-gate operations. *Angew. Chem. Int. Ed.* **2010**, *49*, 9757–9760. [CrossRef]

55. Ono, A.; Togashi, H. Highly selective oligonucleotide-based sensor for mercury (II) in aqueous solutions. *Angew. Chem. Int. Ed.* **2004**, *43*, 4300–4302. [CrossRef]

56. Kobayashi, T.; Takezawa, Y.; Sakamoto, A.; Shionoya, M. Enzymatic synthesis of ligand-bearing DNAs for metal-mediated base pairing utilizing a template-independent polymerase. *Chem. Commun.* **2016**, *52*, 3762–3765. [CrossRef] [PubMed]

57. Hirao, I.; Kimoto, M.; Yamashige, R. Natural versus artificial creation of base pairs in DNA: Origin of nucleobases from the perspectives of unnatural base pair studies. *Acc. Chem. Res.* **2012**, *45*, 2055–2065. [CrossRef] [PubMed]

58. Takezawa, Y.; Shionoya, M. Metal-mediated DNA base pairing: Alternatives to hydrogen-bonded Watson-Crick base pairs. *Acc. Chem. Res.* **2012**, *45*, 2066–2076. [CrossRef] [PubMed]

59. Scharf, P.; Müller, J. Nucleic acids with metal-mediated base pairs and their applications. *ChemPlusChem* **2013**, *78*, 20–34. [CrossRef]

60. Yamanaka, M.; Toyoda, N.; Kobayashi, K. Hybrid cavitand capsule with hydrogen bonds and metal-ligand coordination bonds: Guest encapsulation with anion assistance. *J. Am. Chem. Soc.* **2009**, *131*, 9880–9881. [CrossRef] [PubMed]

61. Yamanaka, M.; Kawaharada, M.; Nito, Y.; Takaya, H.; Kobayashi, K. Structural alteration of hybrid supramolecular capsule induced by guest encapsulation. *J. Am. Chem. Soc.* **2011**, *133*, 16650–16656. [CrossRef] [PubMed]

62. Nito, Y.; Adachi, H.; Toyoda, N.; Takaya, H.; Kobayashi, K.; Yamanaka, M. Hydrogen-bond and metal-ligand coordination bond hybrid supramolecular capsules: Identification of hemicapsular intermediate and dual control of guest exchange dynamics. *Chem. Asian J.* **2014**, *9*, 1076–1082. [CrossRef]

63. Pasquale, S.; Sattin, S.; Escudero-Adán, E.C.; Martínez-Belmonte, M.; de Mendoza, J. Giant regular polyhedral from calixarene carboxylates and uranyl. *Nat. Commun.* **2012**, *3*, 785. [CrossRef] [PubMed]

64. Shi, Q.; Zhou, X.; Yuan, W.; Su, X.; Neniškis, A.; Wei, X.; Taujenis, L.; Snarskis, G.; Ward, J.S.; Rissanen, K.; et al. Selective formation of S_4- and T-symmetric supramolecular tetrahedral cages and helicates in polar media assembled via cooperative action of coordination and hydrogen bonds. *J. Am. Chem. Soc.* **2020**, *142*, 3658–3670. [CrossRef] [PubMed]

 *chemistry* 

Review

Uranyl Ion Complexes of Polycarboxylates: Steps towards Isolated Photoactive Cavities

Jack Harrowfield [1,*] and Pierre Thuéry [2,*]

[1] Institut de Science et d'Ingénierie Supramoléculaires, Université de Strasbourg, 8 allée Gaspard Monge, 67083 Strasbourg, France

[2] Université Paris-Saclay, CEA, CNRS, NIMBE, 91191 Gif-sur-Yvette, France

[*] Correspondence: harrowfield@unistra.fr (J.H.); pierre.thuery@cea.fr (P.T.)

Received: 29 January 2020; Accepted: 18 February 2020; Published: 20 February 2020

Abstract: Consideration of the extensive family of known uranyl ion complexes of polycarboxylate ligands shows that there are quite numerous examples of crystalline solids containing capsular, closed oligomeric species with the potential for use as selective heterogeneous photo-oxidation catalysts. None of them have yet been assessed for this purpose, and some have obvious deficiencies, although related framework species have been shown to have the necessary luminescence, porosity and, to some degree, selectivity. Aspects of ligand design and complex composition necessary for the synthesis of uranyl ion cages with appropriate luminescence and chemical properties for use in selective photo-oxidation catalysis have been analysed in relation to the characteristics of known capsules.

Keywords: uranium(VI); carboxylates; capsules; structure; luminescence

---

1. Introduction

Intensive research over the past few decades has been devoted to the synthesis of crystalline cavity-containing, framework, and coordination polymer species of a porous nature suited to the storage, immobilisation, sensing, or reaction of a wide variety of substrates of environmental and economic importance [1–15]. (The references cited here are a somewhat eclectic selection intended to illustrate the range of chemistry involved, rather than to be comprehensive, which is far from the case.) This has resulted not only in real advances towards practical objectives in gas storage [1,5,16,17] but also in unanticipated developments such as that of the "crystalline sponge" method [18,19] of determining the molecular structures of molecules otherwise difficult to crystallise. While metal ions clearly have a fundamental role in determining the structure of these materials, of equal importance is that they endow the solids with functionality specific to the given metal ion. One such function is that of photoactivity, a property which may have various manifestations [2,8], but which in the case of uranium(VI) as uranyl ion, UO_2^{2+}, derivatives, is anticipated to be that of photo-oxidation catalysis, long known in their solution chemistry [20–22].

Although photocatalysis by metal-organic framework (MOF) systems in particular could be described in 2017 as a "largely unexplored field" [23], it has rapidly become a popular area of study [24]. Investigations of hetereogeneous photocatalysis by uranyl-containing solids [25–35], however, have remained largely limited to those of oxidative destruction of environmental pollutants or to basic mechanistic work, although water splitting has been frequently cited as a possible application. Selectivity of these reactions has not been a major focus and in some early instances [34] would be expected to have been determined by the nature of the preformed support upon which uranyl centres were immobilised. Given that the use of a radioactive material would pose problems in any large-scale application for environmental remediation or water splitting, an alternative, more appealing prospect is that of selective photochemical synthesis within cavities of a porous uranyl complex crystal, a prospect which parallels what has already been realised for synthesis in general with other metal ion derivatives [2,36–38] and which is rendered worthy of wider investigation by the observation of selective incorporation of various materials into the cavities of some known uranyl ion coordination polymers [39–45].

In general, polycarboxylates, often in the company of aza-aromatic species, are the most important class of ligands giving rise to coordination polymers, metal-organic frameworks, and closed metallo-clusters [1–14,46]. This is particularly true of uranyl ion containing systems [30,47–51], and it is for this reason that the present report is focussed upon uranyl polycarboxylates, though this is not to say that less-investigated species such as, for example, those based on polyphosphonates [52–55] are not of equal potential interest. We do note, however, that while uranyl ion photocatalysed oxidation of carboxylic acids is a long known reaction [20,21], it is slow and there is little evidence that the synthesis of uranyl carboxylates [49] is significantly influenced by it, so that the extraordinary variety of known carboxylate systems is open to exploration. With the particular objective of defining possibly more efficient pathways to photoactive closed uranyl-polycarboxylate oligomers, expected to be the most stringent form of receptor, we present an analysis of both positive and negative aspects of the crystal structures and composition of currently known system.

2. Discussion

The first closed uranyl polycarboxylate oligomer to be structurally characterised [56] was that formed by a monoester derivative of the *cis,trans* stereoisomer of 1,3,5-trimethylcyclohexane-1,3,5-tricarboxylic acid in its dianionic form (L^{2-}) and with the composition $(HNEt_3)_8[(UO_2)_8(L)_8(O_2)_4] \cdot 5CHCl_3 \cdot 16H_2O \cdot 6CH_3OH$ (**A**, CSD refcode GOPVUC). The box-like, octa-anionic oligomer found in this structure (Figure 1) defines a cavity large enough to accommodate two triethylammonium cations and (partly) two chloroform molecules, indicating that small molecule reactions within the cavity could be possible provided the cations could be replaced by reactive species. It also has features found in many other uranyl complexes in that the carboxylate groups are bound as $\kappa^2 O,O'$ chelates and the peroxide ligands act as bridges to produce convergent $U(O_2)U$ units. The adventitious presence of peroxide in the complex is not an unusual observation in uranyl ion coordination chemistry and detailed studies [57,58] have led to its rationalisation as a result of photochemical reduction of uranyl ion by water or organic substrates (such as methanol) to give U(V), which subsequently reacts with atmospheric oxygen to give peroxide. The bent form of the $U(O_2)U$ unit is favourable for the formation of a closed species and this effect is spectacularly exemplified in the extraordinary family of cages formed by uranyl ion in the presence of peroxide ion and various co-ligands such as oxide, hydroxide, nitrate, phosphate and other simple oxyanions, a family known to extend up to a multi-compartmental cage built from 124 uranyl units [59,60]. The presence of bound peroxide on uranyl ion, however, has the unfortunate consequence that uranyl ion emission, with its characteristic multiple vibronic components [21,61], is quenched, though ligand-centred emission is observed in some cases [58,62].

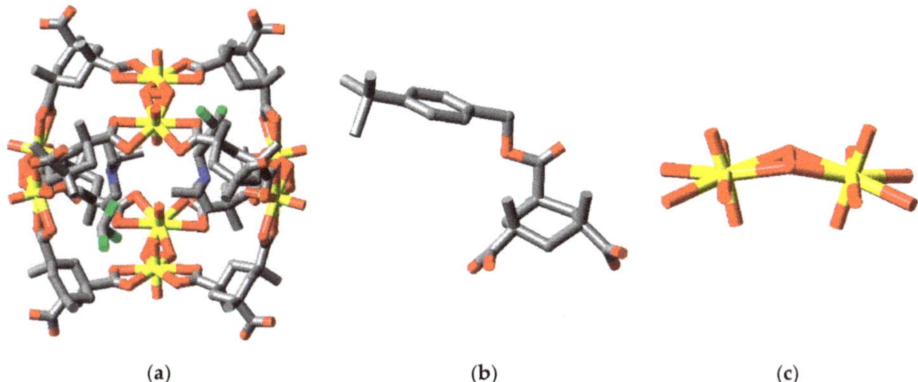

(a)   (b)   (c)

Figure 1. (**a**) A perspective view of the octa-uranate cage present in the crystal of complex **A**, showing the included triethylammonium ions and chloroform molecules but not the ester groups on the ligand; (**b**) The complete ligand, with its ester group; (**c**) The bent $O_6U(O_2)UO_6$ unit present in the complex. H-atoms were not included in the structure. (Colour code: grey = C, blue = N, red = O, green = Cl, yellow = U.)

Thus, the $[(UO_2)_8(L)_8(O_2)_4]^{8-}$ cavity must be considered unsuitable for photocatalysed oxidation reactions involving the excited UO_2^{2+} ion. The same conclusion must be drawn in relation to the octanuclear uranyl cage (Figure 2), found in the complex of composition $(HNEt_3)_8[(UO_2)_8(H_2bcat)_4(O_2)_8]\cdot22H_2O$, (**B**, CSD refcode QAGCOR) [63], obtained with a *bis*-catechol ligand in its doubly deprotonated form (H_2bcat^{2-}). The origin of the peroxide ligands is presumably the same as that of peroxide in complex **A**, although any loss of uranyl emission (not actually demonstrated) here could be due to the phenoxide ligands, similar highly coloured but non-emissive complexes being well known for the calixarenes [64]. $(UO_2)_4(O_2)_4$ units form two bowl-shaped entities that provide caps to the cage. That the estimated internal volume [63] of the cage in **B** is less than that of the cavity in **A** may explain why only water molecules are found within and the triethylammonium counter cations are located externally, being involved in H-bonding to peroxo-O atoms.

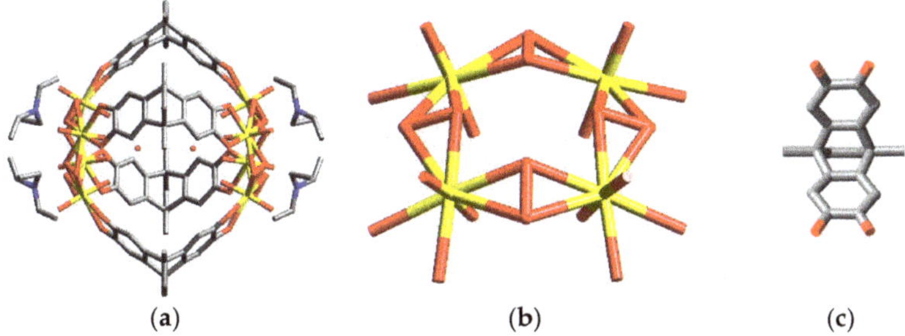

(a)   (b)   (c)

Figure 2. Perspective views of (**a**) the octanuclear cage found in the crystal of complex **B**, showing the oxygen atoms of the two included water molecules (H-atoms not located) and the external location of triethylammonium ions; (**b**) the capping unit of the cage formed by 4 uranyl ions bridged by 4 peroxide ions and (**c**) the bis(catecholate) ligand linking the capping units. H-atoms on C are not shown.

The same limitation to its utility must be applied again to the more recently described cavity-containing complex obtained through reaction of uranyl nitrate with a dicarboxylate derivative of calix[4]pyrrole ($cpdc^{2-}$) to give a product of composition $[(UO_2)_4(cpdc)_4(O_2)_2](pyH)_4\cdot4dmf$, (**C**, CSD refcode IDOKIY; dmf = dimethylformamide) (Figure 3) [65]. Interestingly, the capsular form of the $[(UO_2)_4(cpdc)_4(O_2)_2]^{4-}$ anion present cannot here be attributed to the convergent nature of the $U(O_2)U$ units, since these are not centred on a common point, and must instead be a consequence of the convergent array of the carboxylate substituents of $cpdc^{2-}$. While the dimensions of the cavity in **C** are similar to those of that in **A**, with four small molecules/ions found in each cavity, in **C** the occupying species are all neutral, disordered dmf, with the pyridinium counter cations being confined to the exterior of the cavity by insertion into the calixpyrrole cups. A means of controlling the species entering an anionic cavity, differing from that seen in complex **B**, is therefore evident. In this regard, it should also be noted that the anionic capsules in **C** form stacks parallel to the *a* axis so as to define a narrow channel, a structure which could be regarded as suitable for the insertion through several capsules of a long molecular chain species or simply as a pathway for small molecule entry into the cavities. Another significant aspect of the synthesis of complex **C** is that it appears to form via the intermediacy of a photoactive $[(UO_2)_2(cpdc)_3]^{2-}$ anion, also considered likely to be capsular, though not characterised as such crystallographically. Such a capsule might be too small to be useful as a molecular flask for photo-oxidation of molecules larger than water (see ahead).

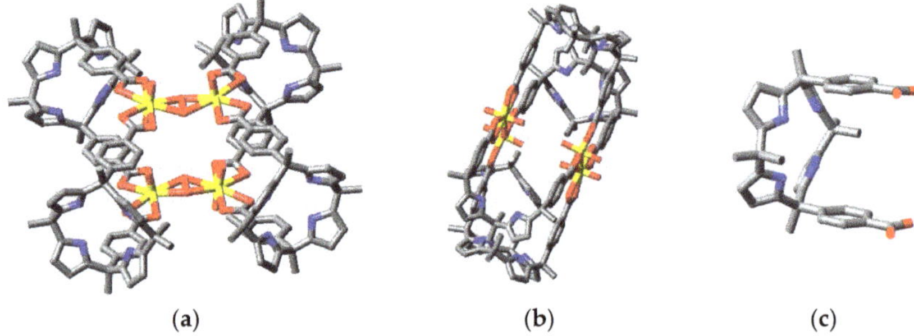

(a)   (b)   (c)

Figure 3. (**a**) A perspective view of the tetranuclear cage present in complex **C** (4 disordered dmf molecules included are not shown) showing the opposed bending of the $U(O_2)U$ units; (**b**) A view down one column of cages showing the rather constricted channel formed. (Dmf molecules, which do not block the central region of the channel, are again not shown.); (**c**) perspective view of the disubstituted calixpyrrole ligand.

The structure of complex **C** provides far from the first example of a uranyl ion complex where there are enclosed channels which might engender porosity in the crystal, a simple early example being that of the chiral tubes (Figure S1) defined by helical polymer chains in $[UO_2(dipic)(OH_2)]$ [66] (CSD refcode PYDCUO; dipic = dipicolinate = pyridine-2,6-dicarboxylate), although here it seems that the inner space of the tube is too small even to include water molecules and it is unoccupied. Many other "nanotubular" species (not all based on carboxylates) have since been characterised [47,67–76], their significance lying not only in their possible suitability as reaction vessels for the synthesis or oxidation of long, linear molecules but also, relating to the focus of the present discussion, as channels which might be used to link and provide access to capsular reaction vessels. The objective here would be the creation of uranyl ion-based structures analogous to those of zeolites and mesoporous silicas, an objective, which despite an early success [77] (discussed ahead), has been attained in but a few instances [42–44]. An intriguing comparative consideration here is that of the relationships between graphite, fullerenes, and carbon nanotubes, since a common feature of the crystal structures of anionic uranyl ion complexes of dicarboxylates is the presence of diperiodic honeycomb layers with a hexagonal form similar to that of graphite [42,47–49,78]. While the appropriate choice of ligand has certainly enabled this tendency to be overcome, the actual outcome has proved difficult to predict, as is well illustrated by various investigations of the dianion of camphoric acid as a ligand for uranyl ion [79–82].

(1*R*,3*S*)-Camphoric acid (H$_2$cam) is a readily available, chiral dicarboxylic acid with the desirable feature that it can provide two carboxylate groups oriented such that although they are too far apart to simply chelate a single metal ion, they can be convergently arranged so as to favour closed oligomeric complex units. Thus, a U(camphorate)U unit can adopt a form equivalent to that of the $U(O_2)U$ unit considered above, although the equivalence is inexact in that the carboxylate groups need not necessarily adopt $\kappa^2 O,O'$ chelation and rotation about the $C-CO_2-$ bonds can occur. In the neutral (1:1 uranyl:dicarboxylate) complex $[UO_2(cam)(py)_2]\cdot py$ (py = pyridine), (**D**, CSD refcode PENFIY) [82], the uranium is 8-coordinate with the pyridine ligands in *trans* positions and although the cam^{2-} ligands bind as bis($\kappa^2 O,O'$) chelates and have a convergent form, they are only present in sufficient number to link uranyl centres into chains or rings. Thus, what is found is that the complex is a 1D zig-zag polymer (Figure 4a) rather than a metallacycle, perhaps as a consequence of the pyridine ligands forcing the carboxylate units to be as remote as possible in the uranium coordination sphere. With methanol as the co-ligand rather than pyridine in $[UO_2(cam)(CH_3OH)]\cdot CH_3OH$, (**E**, CSD refcode PENFOE), the uranium is now 7-coordinate and the crystal contains diperiodic sheets involving fused 8- and 32-membered metallacyclic units where each cam^{2-} binds to three uranium centres with one carboxylate forming a $\kappa^2 O,O'$ chelate and the other forming a μ^2-$\kappa^1 O,\kappa^1 O'$ bridge (Figure 4b). Since the cam^{2-} conformation is very similar in both complexes, it is apparent that this cannot be the only factor controlling the structures. One obvious additional influence is the coordination mode of the carboxylate units, as well-exemplified in the 1:1 complexes of uranyl ion with cyclobutane-1,1-dicarboxylate, where both 4- and 6-membered chelate rings form part of a simple binuclear species (CSD refcode PENFEU) [82], and with (2R,3R,4S,5S)-tetrahydrofurantetracarboxylate, where 7-membered chelate rings are found in metallamacrocyclic oligomers possibly of sufficient depth to accommodate small molecules (CSD refcodes IZOHEK and IZOHIO) [83].

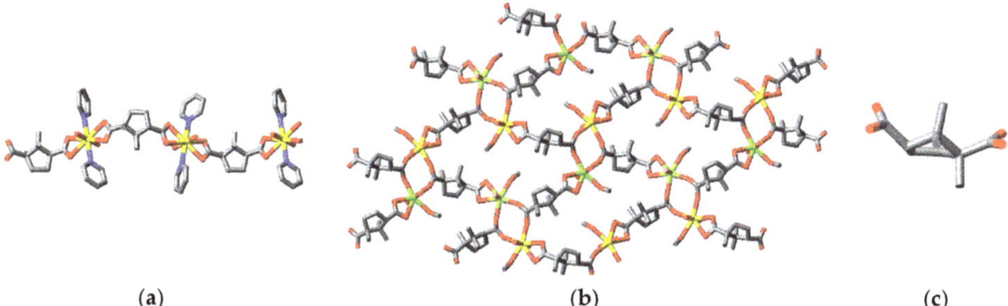

(a)   (b)   (c)

Figure 4. (a) Section of the monoperiodic chains found in the crystal of complex **D**; (b) partial view of the diperiodic sheet found in the crystal of complex **E**; (c) perspective view of the camphorate ligand present in both.

When uranyl ion and $H_2 cam$ are reacted in the presence of 1,4-diazabicyclo[2.2.2]octane (DABCO), however, the bent arrangement does appear to have the desired effect in that in the crystal of $[(UO_2)_8\{(cam)_{12}H_8\}]\cdot 12H_2O$, (**F**, CSD refcode MUNKOW), an octanuclear cage species (Figure 5a) with both carboxylate groups bound in the 4-membered, $\kappa^2 O,O'$ chelate mode, is found [81]. This chiral cage has quite large portals and its packing in the crystal results in facing arrays which define channels indicating it might well have 3-dimensional porosity, although this property has not been established. As a neutral species, the cage might be expected to be able to encapsulate neutral small molecules but the resolved water molecules of the structure are found either on the faces of the cages or in between cages, where H-bond acceptor sites are most abundant. That the cage has significant stability is indicated by the fact that it can be crystallised in its fully deprotonated form as Ba(II) [81] and K(I) [80] derivatives (CSD refcodes MUNKUC and LIYRAO), although here the channels are now blocked by the counter cations. This could mean, nonetheless, that the complexes might be used as ion-exchange materials but the fact that in the presence of NH_4^+ and $CH_3PPh_3^+$ cations, camphoric acid and uranyl ion react [79] to give crystals of composition $[CH_3PPh_3]_3[NH_4]_3[(UO_2)_6(cam)_9]$, (**G**) (CSD refcode

JIVBOI), containing a hexanuclear cage complex (Figure 5b) in which a phosphonium cation occupies the cage, indicates that any such capacity would be limited. It also must be noted that although luminescence measurements have not been made on all these complexes, where they have [79,80], uranyl ion emission appears to be largely, if not completely quenched, indicating a limited potential for photo-oxidation catalysis. In some instances, while uranyl ion emission is not observed or is weak, broad-band emission of obscure origin is observed, providing yet another indication that the photophysics of uranyl ion complexes in the solid state is yet to be fully understood [61,84,85].

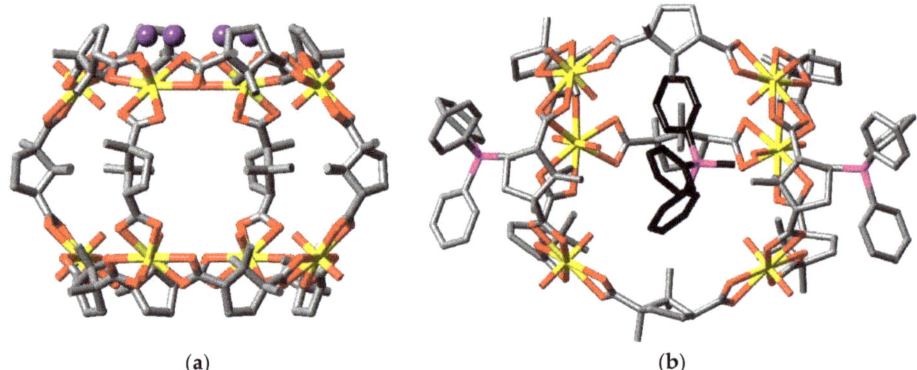

(a)   (b)

Figure 5. Perspective views of (**a**) the octanuclear cage found in the crystal of complex **F**, with the oxygen atoms of water molecules (H-atoms not located) associated with a cage portal shown as violet spheres); (**b**) the hexanuclear cage, with associated [MePPh$_3$]$^+$ cations, found in the crystal of complex **G** (P atoms in violet; carbon atoms of the included cation are shown in black).

Structural characterisation [86–89] of uranyl ion complexes of dicarboxylate ligands rather closely related to camphorate, adamantane-1,3-dicarboxylate (adc^{2-}) and adamantane-1,3-diacetate (ada^{2-}) has further exposed the variety of influences determining their nature in the solid state. Thus, the ligand with the closer similarity to camphorate, adc^{2-}, reacts with uranyl ion in a 1.5:1 (ligand:metal) ratio to give not a closed, cage complex but a triperiodic network, of composition [H$_2$NMe$_2$]$_2$[(UO$_2$)$_2$(adc)$_3$]·1.5H$_2$O, (**H**, CSD refcode ZOZCIC), in which channels (Figure S2) are occupied by dimethylammonium cations (formed by hydrolysis of the dimethylformamide cosolvent), even though the ligand is again bound in a *bis*($\kappa^2 O,O'$) mode [89]. When Cu(II) replaces the dimethylammonium cations, a triperiodic network is again formed but it is one involving diperiodic polymers of [(UO$_2$)$_2$(adc)$_3$]$^{2-}$ units linked by Cu(II) bridges involving Cu–O(carboxylate) bonding which disrupts the uranyl-carboxylate interactions, so that the ligands function only as *bis*($\kappa^1 O$) donors to uranium (CSD refcode ZOZDID). The effects of other metal ions on uranyl-carboxylate complex structures are so varied as to require separate analysis but one untoward effect which must be noted here is that it is common to find that the presence of the hetero-metal ion leads to quenching of uranyl ion emission [61,85,90,91], as in fact is complete in the present instance probably because of the close proximity of the Cu and U centres. With ada^{2-}, a complex of similar stoichiometry to **H**, of composition [H$_2$NMe$_2$]$_2$[(UO$_2$)$_2$(ada)$_3$]·1.5H$_2$O, (**I**, CSD refcode IHOGIX) [88], can be isolated in which sheets of rather convoluted diperiodic polymer (Figure S3) are present, with the conformational freedom resulting from the presence of the CH$_2$–CO$_2$– bonds seemingly allowing the carboxylate units of one ligand unit to adopt more divergent relative orientations than those in the adc^{2-} units of **H**, even though the ligand is bound as a *bis*($\kappa^2 O,O'$) species. In the complex [H$_2$NMe$_2$][PPh$_3$Me][(UO$_2$)$_2$(ada)$_3$]·H$_2$O, (**J**, CSD refcode YEXDIR) [87], where methyltriphenylphosphonium has replaced one dimethylammonium cation of **I**, one of the 3 inequivalent units adopts a completely divergent arrangement of its carboxylate groups and diperiodic polymer sheets with a distorted honeycomb topology (Figure S4) are formed, all three ligand units still being bound in the *bis*(κ^2O,O') mode. When the (formal) ion exchange is complete for the 1:1.5 U:adc system as in $[NH_4]_2[PPh_4]_2[(UO_2)_4(ada)_6]\cdot H_2O$, (**K**, CSD refcode YEXDAJ), and $[NH_4]_2[PPh_3Me]_2[(UO_2)_4(ada)_6]\cdot H_2O$, (**L**, CSD refcode YEXDEN), an essentially identical tetranuclear, metallatricyclic cage species is now found in both (Figure 6). Of the six ada^{2-} ligands in a cage, four have a convergent array of carboxylates and two a divergent array, although all six behave as *bis*(κ^2O,O') chelates. While the ammonium ions are H-bonded to the exterior of the cage, the phosphonium cations and particularly the $[PPh_3Me]^+$ species partly occupy the interior through CH···O interactions and it is not evident that the cage could accommodate other molecules or that any exchange could occur without transformation of the cage. The dependence of the structure of complexes based on 1:1.5 uranyl:ligand units on the counter cation is further illustrated in the structures [86] of $[PPh_4]_2[(UO_2)_2(adc)_3]\cdot 2H_2O$, (**M**) (CSD refcode GOTPAJ), and $[PPh_4]_2[(UO_2)_2(ada)_3]$, (**N**) (CSD refcode GOTPIR), where very similar 1D, trough-like polymers (Figure S5) are present. The cavities defined by these troughs are occupied by the counter cations, so that once again, although these complexes are like other monometallic uranyl complexes of both adc^{2-} and ada^{2-} in showing uranyl ion luminescence, they do not offer any obvious prospect of being useful for photo-oxidation catalysis.

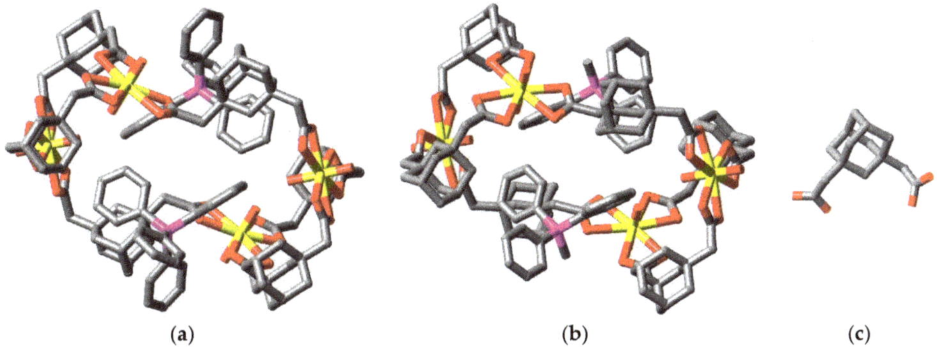

(a)   (b)   (c)

Figure 6. The near identical macrotricyclic, tetranuclear cages, along with their nearest phosphonium cations, found in the crystals of (**a**) complex **K** and (**b**) complex **L**; (**c**) one conformation of the adc^{2-} ligand found in these complexes.

Where conformational restrictions are somewhat diminished compared to camphor or adamantane derivatives in the *cis* and *trans* isomers of 1,2-cyclohexane dicarboxylates ccdc^{2-} and tcdc^{2-}), tetrahedral cage species based on $UO_2(\kappa^2O,O'$-carboxylate)$_3$ apices have been obtained for the *trans* isomer and an octanuclear cage for the *cis* (complexes **O** and **P**, Figure 7, CSD refcodes WANKAA and LICNIX, respectively) [92–94]. Broader investigations [95–97] of the uranyl ion complexes of these ligands have shown that these particular results are due to the choice of counter cation for the anionic oligomers, although the range is quite wide for the tetranuclear cages from the *trans* isomer and it has been suggested that the cage may be the favoured form for the stoichiometry 1:1.5 U:ligand [92]. These tetranuclear cages are luminescent and uranyl-O atoms, potentially sites for photoreaction [21], are directed towards the interior, but the internal space of the cage is too small to accommodate any molecule of real interest. The octanuclear cage derived from the *cis* isomer has a near-cubic array of U centres with one oxygen on each directed towards the interior and four involved in H-bonds to an encapsulated ammonium ion, and could be expected to be suitable for the inclusion of small molecules, although unfortunately it shows very weak luminescence.

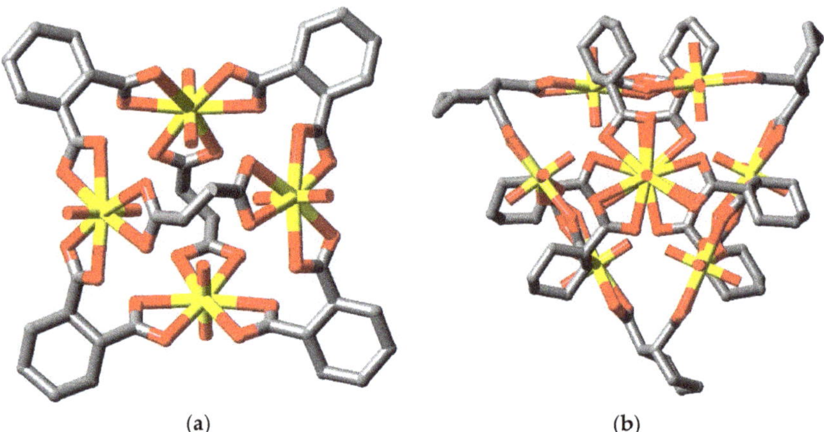

Figure 7. Views (**a**) of the tetranuclear, tetrahedral cluster found in the structure of [NH$_4$]$_4$[(UO$_2$)$_4$(tcdc)$_6$], **O** and (**b**) the octanuclear, near-cubic cage found in the structure of [NH$_4$][PPh$_4$][(UO$_2$)$_8$(ccdc)$_9$(H$_2$O)$_6$]·3H$_2$O, **P**.

Although the examples given above show that conformational restrictions in dicarboxylate ligands do have some influence on the structure of their uranyl ion complexes, it is worthy of note that even conformationally highly flexible aliphatic α,ω-dicarboxylates [98,99], which typically give diperiodic coordination polymers [98], can be induced to form anionic cage oligomers of helical form (helicates) in the presence of particular counter cations [99]. Thus, in the presence of [Co(bipy)$_3$]$^{2+}$ or [Ni(bipy)$_3$]$^{2+}$ (bipy = 2,2′-bipyridine), uranyl ion and 1,7-heptanedicarboxylic acid (H$_2$C9) react to give isomorphous crystals of composition [M(bipy)$_3$][(UO$_2$)$_2$(C9)$_3$] (M = Co and Ni, CSD refcodes DACGIA and DACGOG, respectively), while with [Mn(phen)$_3$]$^{2+}$ or [Co(phen)$_3$]$^{2+}$ (phen = 1,10-phenanthroline) and 1,10-decanedicarboxylic acid (H$_2$C12), isomorphous [M(phen)$_3$][(UO$_2$)$_2$(C12)$_3$] crystals result (**Q**, CSD refcodes DACGUM and DACHAT for Mn and Co, respectively). The anionic capsules present (Figure 8) are small, with an internal space partly occupied by the aza-aromatic ligands on the counter cations and with little space for any guest. As the U···U separations (~7.5 Å) in these species are very close to that in complex **C** described above, this is taken as an indication that the supposed [(UO$_2$)$_2$(cpdc)$_3$]$^{2-}$ precursor to **C** would also lack the capacity to act as a reaction vessel.

Recognition of the fact that four carboxylate groups disposed on a scaffold such that they are tetrahedrally oriented provide two orthogonal, bent dicarboxylate entities leads to the expectation that an appropriate such ligand could be used to provide linked cavities within a triperiodic framework polymer. This expectation was first realised (fortuitously) in the synthesis of the uranyl ion complex of the *trans,trans,trans* isomer of 1,2,3,4-cyclobutanetetracarboxylate (cbtc^{4-}) formed from its *cis,trans,cis* isomer under solvothermal conditions [77]. Thus, the structure of the complex [H$_3$O]$_2$[(UO$_2$)$_5$(cbtc)$_3$(H$_2$O)$_6$], (**R**, CSD refcode GOJFAN), contains octanuclear boxes (Figure 9), where each uranyl centre is bound by three κ^2*O,O′* carboxylate units, linked by tetranuclear metallacycles where each uranyl centre is bound to two κ^2*O,O′* carboxylate units and two water molecules (in *trans* positions). Neither the luminescence nor porosity of complex **R**, nor of the more recently isolated framework complex [H$_2$NMe$_2$]$_4$[(UO$_2$)$_4$(cbtc)$_3$] (similar but of a different topological type, CSD refcode TOJJAG) [100], have yet been studied but certainly the porosity of the uranyl ion complexes of *tetrakis*(4-carboxyphenyl)methane, where a single atom is the source of the tetrahedral orientation, has been demonstrated [42]. It is of course not essential that dicarboxylate units be orthogonally directed in order to generate triperiodic structures of linked cavities, as is seen in the formation of such structures with *cis,trans,cis*-cyclobutanetetracarboxylate [77] and in the more recently studied structures of uranyl ion complexes of porphyrin-derived tetracarboxylates [39] (where photoreactivity is associated with the porphyrin centres).

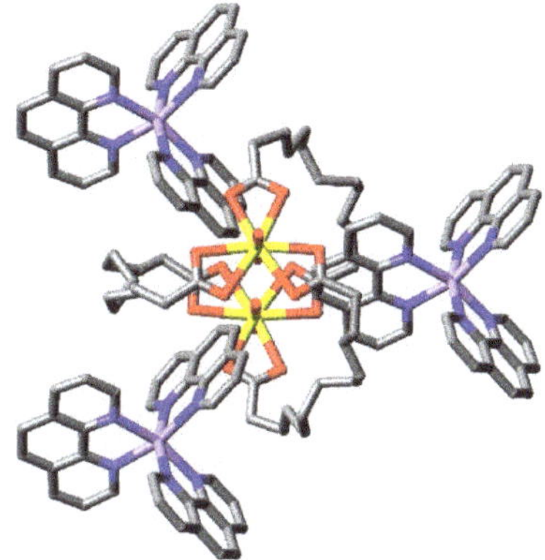

Figure 8. Perspective view of the binuclear, triple-stranded, anionic helicate and associated cations found in the crystal of [Mn(phen)$_3$][(UO$_2$)$_2$(C12)$_3$], one of the isomorphous complexes **Q**. (Violet = Mn; C12 = 1,10-dodecanedicarboxylate; H-atoms and partial disorder of the polymethylene chains are not shown.)

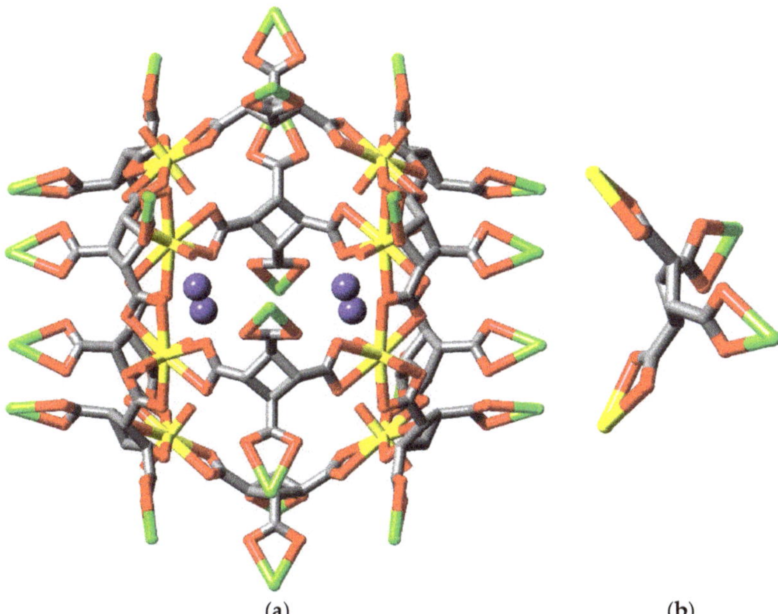

(a)   (b)

Figure 9. (a) Perspective view of the box-like unit, defined by the uranium atoms shown in yellow, and linked to others through tetranuclear metallacyclic units (not shown) involving the uranium atoms shown in green, found in the crystal of complex **R**. Water molecule oxygen atoms within the box are shown in violet; (b) the tetrahedral array produced by the *trans,trans,trans* conformation of the ligand.

That convergent polycarboxylates can be used as well to generate capsular structures is beautifully demonstrated by the structures of the complexes formed by calix[4]- and calix[5]-arene carboxylates (e.g., structures **S**, Figure 10, CSD refcodes YANGUR and YANHAY, respectively) [101]. The cavities formed here are large and can accommodate species as big as tetraprotonated cyclen (1,4,7,10-tetra-azacyclododecane) in the case of the calix[4]arene tetracarboxylate or several pyridinium and pyridine species in the calix[5]arene pentacarboxylate derivative, where the estimated effective volume of the cavity is 7000 Å^3. Luminescence measurements are not available for these complexes and a concern is that the presence of the calixarene units may lead to quenching (see above). Since the capsules are anionic, it is unsurprising that they include cations but this may be a barrier to the inclusion of neutral potential substrate molecules.

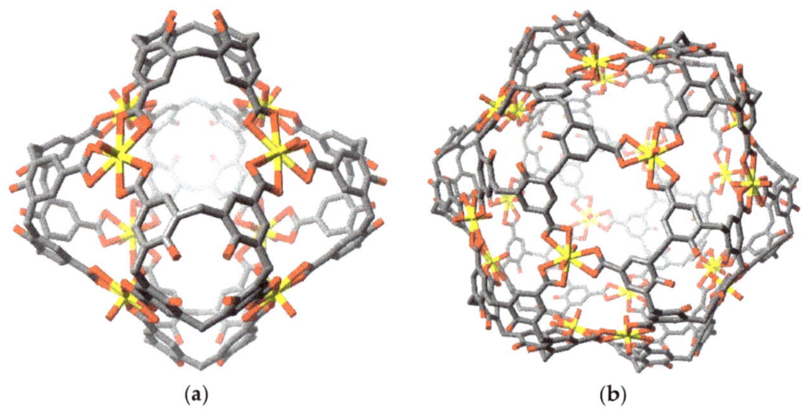

(a)   (b)

Figure 10. The (**a**) octanuclear (**S1**) and (**b**) icosanuclear (**S2**) cages formed from complexation of uranyl ion by calix[4]arene tetracarboxylate and calix[5]arene pentacarboxylate, respectively. (Figures shown with depth fading of the atoms.).

Just as bent dicarboxylate units can be seen as possible struts to a uranyl ion capsule, tripodal tricarboxylates can be seen as possible caps and an obvious candidate for this role is the trianion of Kemp's triacid, *cis,cis*-1,3,5-trimethyl-1,3,5-cyclohexane tricarboxylic acid (H$_3$kta) [102], in its chair conformation where all three carboxylate groups are axially disposed. In the complex [Ni(bipy)(OH$_2$)$_4$][(UO$_2$)$_8$(kta)$_6$(OH$_2$)$_6$], (**T**, CSD refcode POGZIW) [103], the kta^{3-} ligands do indeed sit upon the six faces of a near-cubic octanuclear anion, bridging four uranium centres as a result of two carboxylates forming κ^1O,κ^1O' bridges and one forming a κ^2O,O' chelate. Once again, the internal volume of the cage is not great and but a single water molecule, H-bonded to uranyl-O appears to be encapsulated (Figure 11). Luminescence measurements were not reported but the presence of Ni(II) in the counter cation raises the possibility that uranyl emission would be quenched, as would be expected in the case of several heterometallic compounds, some involving the same cage as just described and other larger aggregates (described in detail elsewhere [47]), characterised [104] in extension of the initial study. In a further extension [105], however, where luminescence (but not quantum yield) measurements were conducted on some similar heterometallic complexes of *cis,cis*-1,3,5-cyclohexanetricarboxylate (ctc^{2-}), quenching there was clearly not complete, although a factor here may have been an apparent preference for the triequatorial disposition of the carboxylates leading to the predominant formation of honeycomb-like diperiodic polymer sheets, just as indeed observed for kta^{3-} complexes involving counter cations other than [Ni(bipy)(OH$_2$)$_4$]$^{2+}$ [104,105]. Both ligands have been shown to form tubular species, with all carboxylate groups axial and additional Ni(II) cations in the case of Kemp's triacid, and with all carboxylate groups equatorial in the case of Hctc$^-$, the latter exemplifying the introduction of curvature into a honeycomb sheet precursor (CSD refcodes POHPEJ and RORROH, respectively). The trianion (kta^{3-}) is even found in a boat conformation in M[UO$_2$(kta)] complexes (M = H$_2$NMe$_2$ (CSD refcode QUKLAL) or Cs) [106,107], rendering the synthesis of complexes of its triaxial chair form more a matter of chance rather than design. Nonetheless, that a tricarboxylate constrained to a discoidal form with all carboxylates oriented so as to favour a planar form of their complex does not necessarily generate diperiodic honeycomb species and in fact gives a triperiodic complex with multiple large, linked cavities is seen in the remarkable structure of the uranyl ion complex of 1,3,5-trimethyl-2,4,6-*tris*(4'-carboxyphenyl)benzene (CSD refcode UNUNEY) [43,44].

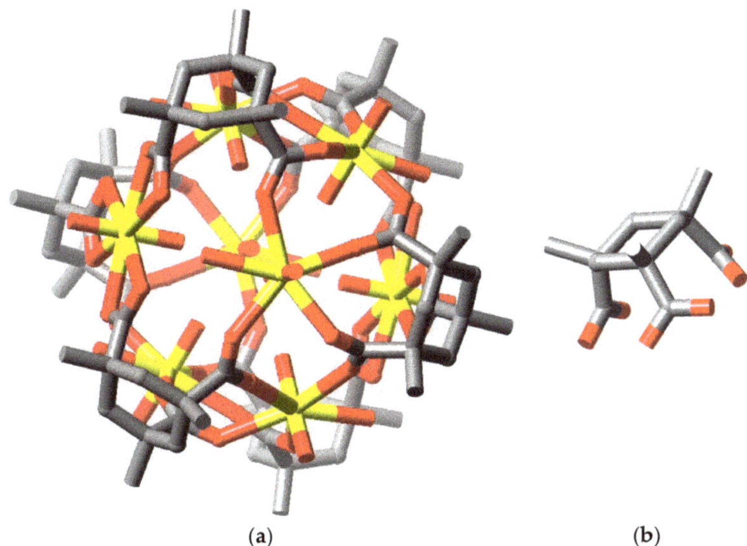

(a)   (b)

Figure 11. (a) The octanuclear cage (again shown with depth fading) found in the crystal of complex **T**; (b) perspective view of the convergent triaxial conformation of the ligand found in the complex.

3. Conclusions

While the lack of complete luminescence and porosity measurements for known capsular oligomers of uranyl carboxylates creates some uncertainty as to their potential value as photo-oxidation catalysts, there seems little likelihood that any would be selective due to their capacity to encapsulate substrates of moderate molecular size. Practical aspects of application, such as stability under reaction conditions are also completely unexplored. Geometrical analysis analogous to that applied to other metallacapsule design [108,109] could of course be used along with the "extended ligand" approach [110] to prepare ligands suited to the formation of larger cavities, although this involves the danger of generating interpenetration in the structures, already seen in numerous uranyl coordination polymers [111]. Another potential drawback with most known capsules is that they are anionic and thus favour interaction with cations, so that one objective of continuing efforts of synthesis would be to couple a neutral bridging ligand, such as a bis(naphthyridine), with two carboxylate units on every uranium. The focus of this brief review has been on solid materials containing capsular species, in part because known examples are all solids of low solubility in any solvent and in part because product separation is more straightforward with heterogeneous catalysis but soluble capsular species would also be of interest. The attraction of a capsular species as a reaction vessel is that any selectivity depends on the structure of the capsule itself and not upon the environment in which it is found and capsular species which align in crystals so as to define channels, as found in various instances described herein, could offer heterogeneous catalysts of this type. Here, tubular complexes as found with selenates [71,72] and phosphonates [52–55] as well as with polycarboxylates such as tricarballylate [112], iminodiacetate [69] and phenylenediacetates [68], would also be of interest, especially if a better understanding of metal ion quenching of uranyl ion luminescence in solids could be attained, since many tubular systems are those involving heterometallic species [47]. As solvothermal synthesis [113,114] is widely applied for the isolation of crystalline uranyl ion complexes, another need is for more data concerning kinetics and equilibria of complex formation under conditions of high temperature and pressure, particularly in mixed solvents. Finally, it is essential to note that the crystal structures of known uranyl ion complexes are frequently seen [115–118] to be sensitive to a wide range of weak interactions, so that the supramolecular behaviour of a bound ligand is a crucial aspect of its design but one yet to be mastered.

Supplementary Materials: The following are available online at http://www.mdpi.com/2624-8549/2/1/7/s1, Figure S1: Perspective view of one helical tube within the crystal of $[UO_2(dipic)(OH_2)]$; Figure S2: Views, down a, of the triperiodic structure of $[H_2NMe_2]_2[(UO_2)_2(adc)_3]\cdot1.5H_2O$; Figure S3: Views of one of the diperiodic sheets found in the crystal of $[H_2NMe_2]_2[(UO_2)_2(ada)_3]\cdot1.5H_2O$; Figure S4: Views of the diperiodic anionic polymer sheets (counter cations not shown) in the crystal of $[H_2NMe_2][PPh_3Me][(UO_2)_2(ada)_3]\cdot H_2O$; Figure S5: Views of the trough-like anionic monoperiodic polymers and their closest cations found in the crystals of (a) $[PPh_4]_2[(UO_2)_2(adc)_3]\cdot2H_2O$ and (b) $[PPh_4]_2[(UO_2)_2(ada)_3]$.

Funding: This research received no external funding.

Conflicts of Interest: The authors declare no conflict of interest.

References and Notes

1. Kaneko, K.; Rodríguez-Reinoso, F. (Eds.) *Nanoporous Materials for Gas Storage. Green Energy and Technology;* Springer: Singapore, 2019.

2. Fang, Y.; Powell, J.A.; Li, E.; Wang, Q.; Perry, Z.; Kirchon, A.; Yang, X.; Xiao, Z.; Zhu, C.; Zhang, L.; et al. Catalytic reactions within the cavity of coordination cages. *Chem. Soc. Rev.* **2019**, *48*, 4707–4730. [CrossRef] [PubMed]

3. Wang, Y.; Wöll, C. Chemical Reactions at Isolated Single-Sites inside Metal-Organic Frameworks. *Catal. Lett.* **2018**, *48*, 2201–2222. [CrossRef]

4. Hooley, R.J. Rings and Things: The Magic of Building Self-Assembled Cages and Macrocycles. *Inorg. Chem.* **2018**, *57*, 3497–3499, and following articles. [CrossRef] [PubMed]

5. Ma, S.; Perman, J.A. (Eds.) *Elaboration and Applications of Metal-Organic Frameworks;* World Scientific: Singapore, 2017.

6. Lin, Y.; Kong, C.; Zhang, Q.; Chen, L. Metal-Organic Frameworks for Carbon Dioxide Capture and Methane Storage. *Adv. Energy Mater.* **2017**, *7*, 1601296. [CrossRef]

7. Rogge, S.M.J.; Bavykina, A.; Hajek, J.; Garcia, H.; Olivos-Suarez, A.I.; Sepulveda-Escribano, A.; Vimont, A.; Clet, G.; Bazin, P.; Kapteijn, F.; et al. Metal-organic and covalent organic frameworks as single-site catalysts. *Chem. Soc. Rev.* **2017**, *46*, 3134–3184. [CrossRef]

8. Gao, C.; Wang, J.; Xu, H.X.; Xiong, Y.J. Coordination chemistry in the design of heterogeneous photocatalysts. *Chem. Soc. Rev.* **2017**, *46*, 2799–2823. [CrossRef]

9. Huang, Y.B.; Liang, J.; Wang, X.S.; Cao, R. Multifunctional metal-organic framework catalysts: Synergistic catalysis and tandem reactions. *Chem. Soc. Rev.* **2017**, *46*, 126–157. [CrossRef]

10. Zhu, L.; Liu, X.Q.; Jiang, H.L.; Sun, L.B. Metal-Organic Frameworks for Heterogeneous Basic Catalysis. *Chem. Rev.* **2017**, *117*, 8129–8176. [CrossRef]

11. Zhuang, J.L.; Terfort, A.; Wöll, C. Formation of oriented and patterned films of metal-organic frameworks by liquid phase epitaxy: A review. *Coord. Chem. Rev.* **2016**, *307*, 391–424. [CrossRef]

12. Liu, J.W.; Chen, L.F.; Cui, H.; Zhang, J.Y.; Zhang, L.; Su, C.Y. Applications of metal-organic frameworks in heterogeneous supramolecular catalysis. *Chem. Soc. Rev.* **2014**, *43*, 6011–6061. [CrossRef]

13. Cook, T.R.; Zheng, Y.-R.; Stang, P.J. Metal-Organic Frameworks and Self-Assembled Supramolecular Coordination Complexes: Comparing and Contrasting the Design, Synthesis and Functionality of Metal-Organic Materials. *Chem. Rev.* **2013**, *113*, 734–777. [CrossRef] [PubMed]

14. Suh, M.P.; Park, H.J.; Prasad, T.K.; Lim, D.-W. Hydrogen Storage in Metal-Organic Frameworks. *Chem. Rev.* **2012**, *112*, 782–835. [CrossRef] [PubMed]

15. Inokuma, Y.; Kawano, M.; Fujita, M. Crystalline Molecular Flasks. *Nat. Chem.* **2011**, *3*, 349–358. [CrossRef] [PubMed]

16. Kapelewski, M.T.; Runčevski, T.; Tarver, J.D.; Jiang, H.Z.H.; Hurst, K.E.; Parilla, P.A.; Ayala, A.; Gennett, T.; Fitzgerald, S.A.; Brown, C.M.; et al. Record High Hydrogen Storage Capacity in the Metal-Organic Framework Ni$_2$(*m*-dobc) at Near-Ambient Temperature. *Chem. Mater.* **2018**, *30*, 8179–8189. [CrossRef]

17. Sumida, K.; Rogow, D.L.; Mason, J.A.; McDonald, T.M.; Bloch, E.D.; Herm, Z.R.; Bae, T.-H.; Long, J.R. Carbon Dioxide Capture in Metal-Organic Frameworks. *Chem. Rev.* **2012**, *112*, 724–781. [CrossRef]

18. Hoshino, M.; Khutia, A.; Xing, H.; Inokuma, Y.; Fujita, M. The crystalline sponge method updated. *IUCr J.* **2016**, *3*, 139–151. [CrossRef]

19. Gee, W.J. The growing importance of crystalline molecular flasks and the crystalline sponge method. *Dalton. Trans.* **2017**, *46*, 15979–15986. [CrossRef]

20. Sarakha, M.; Bolte, M.; Burrows, H.D. Electron-Transfer Oxidation of Chlorophenols by Uranyl Ion Excited State in Aqueous Solution. Steady-State and Nanosecond Flash Photolysis Studies. *J. Phys. Chem. A* **2000**, *104*, 3142–3149. [CrossRef]

21. Burrows, H.D.; Kemp, T.J. The Photochemistry of the Uranyl Ion. *Chem. Soc. Rev.* **1974**, *3*, 139–165. [CrossRef]

22. Yusov, A.B.; Shilov, V.P. Reduction of the photo-excited uranyl ion by water. *Russ. Chem. Bull.* **2000**, *49*, 285–290. [CrossRef]

23. Deng, X.; Li, Z.; Garcia, H. Visible Light Induced Organic Tranformations over Metal-Organic-Frameworks (MOFs). *Chem. Eur. J.* **2017**, *28*, 11189–11209. [CrossRef] [PubMed]

24. Wang, Q.; Gao, Q.; El-Anizi, A.M.; Nafady, A.; Ma, S. Recent advances in MOF-based photocatalysis: Environmental remediation under visible light. *Inorg. Chem. Front.* **2020**, *7*, 300–339. [CrossRef]

25. Khandan, F.M.; Afzali, D.; Sargazi, G.; Gordan, M. Novel uranyl-curcumin-MOF photocatalysts with high-performance photocatalytic activity toward the degradation of phenol red from aqueous solution: Effective synthesis route, design and a controllable systematic study. *J. Mater. Sci. Mater. Electron.* **2018**, *29*, 18600–18613.

26. Li, H.-H.; Zeng, X.-H.; Wu, H.-Y.; Jie, X.; Zheng, S.-T.; Chen, Z.-R. Incorporating Guest Molecules into Honeycomb Structures Constructed from Uranium(VI) Polycarboxylates: Structural Diversity and Photocatalytic Activity for the Degradation of Organic Dyes. *Cryst. Growth Des.* **2015**, *15*, 10–13. [CrossRef]

27. Yang, W.; Tiang, W.-G.; Liu, X.-X.; Wang, L.; Sun, Z.-M. Syntheses, Structures, Luminescence and Photocatalytic Properties of a Series of Uranyl Coordination Polymers. *Cryst. Growth Des.* **2014**, *14*, 5904–5911. [CrossRef]

28. Hou, Y.-N.; Xu, X.-T.; Xing, N.; Bai, F.-Y.; Duan, S.-B.; Sun, Q.; Wei, S.-Y.; Shi, Z.; Zhang, H.-Z.; Xing, Y.-H. Photocatalytic Application of 4f-5f Inorganic-Organic Frameworks: Influence of the Lanthanide Contraction on the Structure and Functional Properies of Uranyl-Lanthanide Complexes. *ChemPlusChem* **2014**, *79*, 1304–1315. [CrossRef]

29. Kolinko, P.A.; Fillipov, T.N.; Kozlov, D.V.; Parmon, V.N. Ethanol vapour photocatalytic oxidation with uranyl-modified titania under visible light: Comparison with silica and alumina. *J. Photochem. Photobiol. A Chem.* **2012**, *250*, 72–77. [CrossRef]

30. Wang, K.-X.; Chen, J.-S. Extended Structures and Physicochemical Properties of Uranyl-Organic Compounds. *Acc. Chem. Res.* **2011**, *44*, 531–540. [CrossRef]

31. Krishna, V.; Kamble, V.S.; Gupta, N.M.; Selvam, P. Uranyl-Anchored MCM-41 as a Highly Efficient Photocatalyst in the Oxidative Destruction of Short-Chain Linear Alkanes: An in-situ FTIR Study. *J. Phys. Chem. C* **2008**, *112*, 15832–15843. [CrossRef]

32. Liao, Z.-L.; Li, G.-D.; Bi, M.-H.; Chen, J.-S. Preparation, Structures and Photocatalytic Properies of Three New Uranyl-Organic Assembly Compounds. *Inorg. Chem.* **2008**, *47*, 4844–4853. [CrossRef]

33. Yu, Z.-T.; Liao, Z.-L.; Jiang, Y.-S.; Li, G.-H.; Chen, J.-S. Water Insoluble Ag-U-Organic Assemblies with Photocatalytic Activity. *Chem. Eur. J.* **2005**, *11*, 2642–2650. [CrossRef] [PubMed]

34. Nieweg, J.A.; Lemma, K.; Trewyn, B.G.; Lin, V.S.-Y.; Bakac, A. Mesoporous Silica-Supported Uranyl: Synthesis and Photoreactivity. *Inorg. Chem.* **2005**, *44*, 5641–5648, and references therein. [CrossRef] [PubMed]

35. Burrows, H.D.; da G. Miguel, M. Applications and limitations of uranyl ion as a photophysical probe. *Adv. Coll. Interf. Sci.* **2001**, *89–90*, 485–496, and references therein. [CrossRef]

36. Zhang, D.; Ronson, T.K.; Nitschke, J.R. Functional Capsules via Subcomponent Self-Assembly. *Acc. Chem. Res.* **2018**, *51*, 2423–2436. [CrossRef]

37. Yoshizawa, M.; Tamura, M.; Fujita, M. Diels-Alder in Aqueous Molecular Hosts: Unusual Regioselectivity and Efficient Catalysis. *Science* **2006**, *312*, 251–254. [CrossRef]

38. Fiedler, D.; Leung, D.H.; Bergman, R.G.; Raymond, K.N. Selective Molecular Recognition, C–H Bond Activation and Catalysis in Nanocale Reaction Vessels. *Acc. Chem. Res.* **2005**, *38*, 351–360. [CrossRef]

39. Shao, L.; Zhai, F.; Wang, Y.; Yue, G.; Li, Y.; Chu, M.; Wang, S. Assembly of porphyrin-based uranium organic frameworks with (3,4)-connected *pto* and *tbo* topologies. *Dalton. Trans.* **2019**, *48*, 1595–1598. [CrossRef]

40. Hu, K.-Q.; Huang, Z.-W.; Zhang, Z.-H.; Mei, L.; Qian, B.-B.; Yu, J.-P.; Chai, Z.-F.; Shi, W.-P. Actinide-based Porphyrinic MOF as Dehydrogenation Catalyst. *Chem. Eur. J.* **2018**, *24*, 16766–16769. [CrossRef]

41. Hu, F.; Di, Z.; Lin, P.; Huang, P.; Wu, M.; Jiang, F.; Hong, M. An Anionic Urnaium-Based Metal-Organic Framework with Ultralarge Nanocages for Selective Dye Adsorption. *Cryst. Growth Des.* **2018**, *18*, 576–580. [CrossRef]

42. Hu, K.-Q.; Jiang, X.; Wang, C.-Z.; Mei, L.; Xie, Z.-N.; Tao, W.-Q.; Zhang, X.-L.; Chai, Z.-F.; Shi, W.-Q. Solvent-dependent Synthesis of Porous Anionic Uranyl-organic Frameworks Featuring Highly Symmetrical (3,4)-connected *ctn* or *bor* Topology for Selective Dye Adsorption. *Chem. Eur. J.* **2017**, *23*, 529–532. [CrossRef]

43. Li, P.; Vermeulen, N.A.; Malliakas, C.D.; Gomez-Gualdron, D.A.; Howarth, A.J.; Mehdi, B.L.; Dohnalkova, A.; Browning, N.D.; O'Keeffe, M.; Farha, O.A. Bottom-up construction of a superstructure in a porous uranium-organic crystal. *Science* **2017**, *356*, 624–627. [CrossRef] [PubMed]

44. Li, P.; Vermeulen, N.A.; Gong, X.R.; Malliakas, C.D.; Stoddart, J.F.; Hupp, J.T.; Farha, O.K. Design and synthesis of a water-stable anionic uranium-based metal–organic framework (MOF) with ultra large pores. *Angew. Chem. Int. Ed.* **2016**, *128*, 10514–10518. [CrossRef]

45. Wang, Y.; Liu, Z.; Li, Y.; Bai, Z.; Liu, W.; Wang, Y.; Xu, X.; Xiao, C.; Sheng, D.; Diwu, J.; et al. Umbellate Distortions of the Uranyl Coordination Environment Result in a Stable and Porous Polycatenated Framework that can Effectively Remove Cesium from Aqueous Solutions. *J. Am. Chem. Soc.* **2015**, *137*, 6144–6147. [CrossRef] [PubMed]

46. Ahmad, N.; Chugtai, A.H.; Younus, H.A.; Verpoort, F. Discrete metal-carboxylate self-assembled cages: Design, synthesis and applications. *Coord. Chem. Rev.* **2014**, *280*, 1–27. [CrossRef]

47. Thuéry, P.; Harrowfield, J. Recent advances in structural studies of heterometallic uranyl-containing coordination polymers and polynuclear closed species. *Dalton. Trans.* **2017**, *46*, 13660–13667. [CrossRef] [PubMed]

48. Su, J.; Chen, J.-S. MOFs of Uranium and the Actinides. *Struct. Bond. (Berlin)* **2015**, *163*, 265–296.

49. Loiseau, T.; Mihalcea, I.; Henry, N.; Volkringer, C. The crystal chemistry of uranium carboxylates. *Coord. Chem. Rev.* **2014**, *266–267*, 69–109. [CrossRef]

50. Andrews, M.B.; Cahill, C.L. Uranyl-Bearing Hybrid Materials: Synthesis, Speciation and Solid State Structures. *Chem. Rev.* **2013**, *113*, 1121–1136. [CrossRef]

51. Giesting, P.A.; Burns, P.C. Uranyl-organic complexes: Structure symbols, classification of carboxylates, and uranyl polyhedral geometries. *Crystallogr. Rev.* **2006**, *12*, 205–255. [CrossRef]

52. Zheng, T.; Wu, Q.-Y.; Gao, Y.; Gui, D.; Qiu, S.; Chen, L.; Sheng, D.; Diwu, J.; Shi, W.-Q.; Chai, Z.; et al. Probing the Influence of Phosphonate Bonding Modes to Uranium(VI) on Structural Topology and Stability: A Complementary Experimental and Computational Investigation. *Inorg. Chem.* **2015**, *54*, 3864–3874. [CrossRef]

53. Yang, W.; Yi, F.-Y.; Tian, T.; Tian, W.-G.; Sun, Z.-M. Structural Variation within Heterometallic Uranyl Hybrids based on Flexible Alkyldiphosphonate Ligands. *Cryst. Growth Des.* **2014**, *14*, 1366–1374. [CrossRef]

54. Parker, T.G.; Cross, J.M.; Polinski, M.J.; Liu, J.; Albrecht-Schmitt, T.E. Ionothermal and Hydrothermal Flux Syntheses of Five New Uranyl Phosphonates. *Cryst. Growth Des.* **2014**, *14*, 228–235. [CrossRef]

55. Tian, T.; Yang, W.; Wang, H.; Dang, S.; Sun, Z.-M. Flexible Diphosphonic Acids for the Isolation of Uranyl Hybrids with Heterometallic U(VI)=O–Zn(II) Cation-Cation Interactions. *Inorg. Chem.* **2013**, *52*, 8288–8290. [CrossRef] [PubMed]

56. Thuéry, P.; Nierlich, M.; Baldwin, B.W.; Komatsuzaki, N.; Hirose, T. A metal-organic molecular box obtained from self assembling around uranyl ions. *J. Chem. Soc. Dalton. Trans.* **1999**, 1047–1048. [CrossRef]

57. McGrail, B.T.; Pianowski, L.S.; Burns, P.C. Photochemical Water Oxidation and Origin of Nonaqueous Uranyl Peroxide Complexes. *J. Am. Chem. Soc.* **2014**, *136*, 4797–4800. [CrossRef]

58. Thangavelu, S.G.; Cahill, C.L. Uranyl-Promoted Peroxide Generation: Synthesis and Characterisation of Three Uranyl Peroxo $[(UO_2)_2(O_2)]$ Complexes. *Inorg. Chem.* **2015**, *54*, 4208–4221. [CrossRef]

59. Qiu, J.; Burns, P.C. Clusters of Actinides with Oxide, Peroxide or Hydroxide Bridges. *Chem. Rev.* **2013**, *113*, 1097–1120. [CrossRef]

60. Qiu, J.; Ling, J.; Jouffret, L.; Thomas, R.; Szymanowski, J.E.S.; Burns, P.C. Water soluble multi-cage super tetrahedral uranyl peroxide phosphate clusters. *Chem. Sci.* **2014**, *5*, 303–310. [CrossRef]

61. Natrajan, L.S. Developments in the Photophysics and Photochemistry of Actinide Ions and their Coordination Compounds. *Coord. Chem. Rev.* **2012**, *256*, 1583–1603. [CrossRef]

62. Thuéry, P.; Harrowfield, J. Anchoring Flexible Uranyl Dicarboxylate Chains through Stacking Interactions of Ancillary Ligands on the Chiral U(VI) Centres. *CrystEngComm* **2016**, *18*, 3905–3918. [CrossRef]

63. Thuéry, P.; Masci, B. Self Assembly of an Octa-uranate Cage Complex with a Rigid Bis-Catechol Ligand. *Supramol. Chem.* **2003**, *15*, 95–99. [CrossRef]

64. Thuéry, P.; Nierlich, M.; Harrowfield, J.; Ogden, M. Phenoxide complexes of f-elements. In *Calixarenes 2001*; Asfari, Z., Böhmer, V., Harrowfield, J., Vicens, J., Eds.; Kluwer Academic Publishers: Dordrtecht, The Netherlands, 2001; Chapter 30; pp. 561–582.

65. Lee, J.; Brewster, J.T., II; Song, B.; Lynch, V.M.; Hwang, I.; Li, X.; Sessler, J.L. Uranyl dication mediated photoswitching of a calix[4]pyrrole-based metal coordination cage. *Chem. Commun.* **2018**, *54*, 9422–9425. [CrossRef] [PubMed]

66. Immirzi, A.; Bombieri, G.; Degetto, S.; Marangoni, G. The crystal and molecular structure of pyridine-2,6-dicarboxylatodioxouranium(VI) monohydrate. *Acta Crystallogr. Sect. B Struct. Crystallogr. Cryst. Chem.* **1975**, *31*, 1023–1028. [CrossRef]

67. Jayasinghe, A.S.; Unruh, D.K.; Kral, A.; Libo, A.; Forbes, T.Z. Structural Features in Metal–Organic Nanotube Crystals That Influence Stability and Solvent Uptake. *Cryst. Growth Des.* **2015**, *15*, 4062–4070. [CrossRef]

68. Thuéry, P.; Atoini, Y.; Harrowfield, J. Tubelike Uranyl–Phenylenediacetate Assemblies from Screening of Ligand Isomers and Structure-Directing Counterions. *Inorg. Chem.* **2019**, *58*, 6550–6564. [CrossRef] [PubMed]

69. Unruh, D.K.; Gojdas, K.; Libo, A.; Forbes, T.Z. Development of Metal-Organic Nanotubes Exhibiting Low Temperature, Reversible Exchange of Confined "Ice Channels". *J. Am. Chem. Soc.* **2013**, *135*, 7398–7401. [CrossRef]

70. Mihalcea, I.; Henry, N.; Loiseau, T. Revisiting the Uranyl-phthalate System: Isolation and Crystal Structures of Two Types of Uranyl-Organic Frameworks (UOF). *Cryst. Growth Des.* **2011**, *11*, 1940–1947. [CrossRef]

71. Krivovichev, S.V.; Kahlenberg, V.; Tananaev, I.G.; Kaindl, R.; Mersdorf, E.; Myasoedov, B.F. Highly Porous Uranyl Selenate Nanotubes. *J. Am. Chem. Soc.* **2005**, *127*, 1072–1073. [CrossRef]

72. Krivovichev, S.V.; Kahlenberg, V.; Kaindl, R.; Mersdorf, E.; Tananaev, I.G.; Myasoedov, B.F. Nanoscale tubules in uranyl selenates. *Angew. Chem. Int. Ed.* **2005**, *44*, 1134–1136. [CrossRef]

73. Aranda, M.A.G.; Cobeza, A.; Bruque, S.; Poojary, D.M.; Clearfield, A. Polymorphism and Phase Transition in Nanotubular Uranyl Phenylphosphonate: $(UO_2)_3(HO_3PC_6H_5)_2(O_3PC_6H_5)_2$.$H_2O$. *Inorg. Chem.* **1998**, *37*, 1827–1832. [CrossRef]

74. Grohol, D.; Clearfield, A. Alkali-Ion-Catalysed Transformation of Two Linear Uranyl Phosphonates into a Tubular One. *J. Am. Chem. Soc.* **1997**, *119*, 9301–9302. [CrossRef]

75. Poojary, D.M.; Cabeza, A.; Aranda, M.A.G.; Bruque, S.; Clearfield, A. Structure Determination of a Complex Tubular Uranyl Phenylphosphonate, $(UO_2)_3(HO_3PC_6H_5)_2(O_3PC_6H_5)_2$.$H_2O$, from Conventional X-ray Powder Diffraction Data. *Inorg. Chem.* **1996**, *35*, 1468–1473. [CrossRef] [PubMed]

76. Poojary, D.M.; Grohol, D.; Clearfield, A. Synthesis and X-ray Powder Structure of a Novel Porous Uranyl Phenylphosphonate Containing Unidimensional Channels Flanked by Hydrophobic Regions. *Angew. Chem. Int. Ed.* **1995**, *34*, 1508–1510. [CrossRef]

77. Thuéry, P.; Masci, B. Uranyl-Organic Frameworks with 1,2,3,4-Butanetetracarboxylate and 1,2,3,4-Cyclobutanetetracarboxylate Ligands. *Cryst. Growth Des.* **2008**, *8*, 3430–3436. [CrossRef]

78. Cahill, C.L.; Borkowski, L.A. *Structural Chemistry of Inorganic Actinide Compounds*; Krivovichev, S.V., Burns, P.C., Tananaev, I.G., Eds.; Elsevier: Amsterdam, The Netherlands, 2007; Chapter 11.

79. Thuéry, P.; Atoini, Y.; Harrowfield, J. Chiral Discrete and Polymeric Uranyl Ion Complexes with (1*R*,3*S*)-(+)-Camphorate Ligands: Counterion-Dependent Formation of a Hexanuclear Cage. *Inorg. Chem.* **2019**, *58*, 870–880. [CrossRef]

80. Thuéry, P.; Harrowfield, J.M. Chiral One- to Three-dimensional Uranyl-organic Assemblies from (1*R*,3*S*)-(+)-Camphoric acid. *CrystEngComm* **2014**, *16*, 2996–3004. [CrossRef]

81. Thuéry, P. A Nanosized Uranyl Camphorate Cage and its Use as a Building Unit in a Metal–Organic Framework. *Cryst. Growth Des.* **2009**, *9*, 4592–4594. [CrossRef]

82. Thuéry, P. Solvothermal Synthesis and Crystal Structure of Uranyl Complexes with 1,1-Cyclobutanedicarboxylic and (1*R*,3*S*)-(+)-Camphoric Acids—Novel Chiral Uranyl–Organic Frameworks. *Eur. J. Inorg. Chem.* **2006**, 3646–3651. [CrossRef]

83. Thuéry, P.; Villiers, C.; Jaud, J.; Ephritikhine, M.; Masci, B. Uranyl-Based Metallamacrocycles: Tri- and Tetranuclear Complexes with (2*R*,3*R*,4*S*,5*S*)-Tetrahydrofurantetracarboxylic Acid. *J. Am. Chem. Soc.* **2004**, *126*, 6838–6839. [CrossRef]

84. Thuéry, P.; Atoini, Y.; Harrowfield, J. Functionalized Aromatic Dicarboxylate Ligands in Uranyl-Organic Assemblies: The Cases of Carboxycinnamate and 1,2-/1,3-Phenylenedioxydiacetate. *Inorg. Chem.* **2020**, *59*. in press, and references therein. [CrossRef]

85. Adelani, P.O.; Albrecht-Schmitt, T.E. Heterobimetallic Copper(II) Uranyl Carboxyphenylphosphonates. *Cryst. Growth Des.* **2011**, *11*, 4676–4683. [CrossRef]

86. Thuéry, P.; Atoini, Y.; Harrowfield, J. 1,3-Adamantanedicarboxylate and 1,3-Adamantanediacetate as Uranyl Ion Linkers: Effect of Counterions, Solvents and Differences in Flexibility. *Eur. J. Inorg. Chem.* **2019**, 4440–4449. [CrossRef]

87. Thuéry, P.; Atoini, Y.; Harrowfield, J. Closed Uranyl-Dicarboxylate Oligomers: A Tetranuclear Metallatricycle with Uranyl Bridgeheads and 1,3-Adamantanedicarboxylate Linkers. *Inorg. Chem.* **2018**, *57*, 7932–7939. [CrossRef] [PubMed]

88. Thuéry, P.; Harrowfield, J. Solvent effects in solvo-hydrothermal synthesis of uranyl ion complexes with 1,3-adamantanediacetate. *CrystEngComm* **2015**, *17*, 4006–4018. [CrossRef]

89. Thuéry, P.; Rivière, E.; Harrowfield, J. Uranyl- and Uranyl-3d-Block-Cation Complexes with 1,3-adamantanedicarboxylate: Crystal Structures, Luminescence and Magnetic Properties. *Inorg. Chem.* **2015**, *54*, 2838–2850. [CrossRef]

90. Heine, J.; Müller-Buschbaum, K. Engineering metal-based luminescence in coordination polymers and metal-organic frameworks. *Chem. Soc. Rev.* **2014**, *42*, 9232–9242. [CrossRef]

91. Thuéry, P.; Atoini, Y.; Harrowfield, J. Zero-, mono- and diperiodic uranyl ion complexes with the diphenate dianion: Influences of transition metal ion coordination and differential U^{VI} chelation. *Dalton. Trans.* **2020**, *49*, 817–828. [CrossRef]

92. Thuéry, P.; Atoini, Y.; Harrowfield, J. Counterion-Controlled Formation of an Octanuclear Uranyl Cage with *cis*-1,2-Cyclohexanedicarboxylate Ligands. *Inorg. Chem.* **2018**, *57*, 6283–6288. [CrossRef]

93. Thuéry, P.; Harrowfield, J. Tetrahedral and cuboidal clusters in Complexes of Uranyl and Alkali or Alkaline-Earth Metal Ions with *rac*- and (1*R*,2*R*)-*trans*-1,2-Cyclohexanedicarboxylate. *Cryst. Growth Des.* **2017**, *17*, 2881–2892. [CrossRef]

94. Thuéry, P.; Harrowfield, J. Coordination Polymers and Cage-Containing Frameworks in Uranyl Ion Complexes with rac- and (1*R*,2*R*)-trans-1,2-Cyclohexanedicarboxylates: Consequences of Chirality. *Inorg. Chem.* **2017**, *56*, 1455–1469. [CrossRef]

95. Thuéry, P.; Harrowfield, J. [Ni(cyclam)]$^{2+}$ and [Ni(*R,S*-Me$_6$cyclam)]$^{2+}$ as Linkers or Counterions in Uranyl–Organic Species with *cis*- and *trans*-1,2-Cyclohexanedicarboxylate Ligands. *Cryst. Growth Des.* **2018**, *18*, 5512–5520. [CrossRef]

96. Thuéry, P.; Atoini, Y.; Harrowfield, J. Crown Ethers and their Alkali Metal Ion Complexes as Assembler Groups in Uranyl–Organic Coordination Polymers with *cis*-1,3-, *cis*-1,2- and *trans*-1,2-Cyclohexanedicarboxylates. *Cryst. Growth Des.* **2018**, *18*, 3167–3177. [CrossRef]

97. Thuéry, P.; Atoini, Y.; Harrowfield, J. Uranyl–Organic Coordination Polymers with *trans*-1,2-, *trans*-1,4- and *cis*-1,4-Cyclohexanedicarboxylates: Effects of Bulky PPh$_4$$^+$ and PPh$_3$Me$^+$ Counterions. *Cryst. Growth Des.* **2018**, *18*, 2609–2619. [CrossRef]

98. Borkowski, L.A.; Cahill, C.L. Crystal Engineering with the Uranyl Cation I. Aliphatic Carboxylate Coordination Polymers: Synthesis, Crystal Structures, and Fluorescent Properties. *Cryst. Growth Des.* **2006**, *6*, 2241–2247. [CrossRef]

99. Thuéry, P.; Harrowfield, J. A New Form of Triple-stranded Helicate found in Uranyl Complexes of Aliphatic α,ω-Dicarboxylates. *Inorg. Chem.* **2015**, *54*, 10539–10541. [CrossRef]

100. Thuéry, P.; Atoini, Y.; Harrowfield, J. Favoring Framework Formation through Structure-Directing Effects in Uranyl Ion Complexes with 1,2,3,4-(Cyclo)butanetetracarboxylate Ligands. *Cryst. Growth Des.* **2019**, *19*, 4109–4120. [CrossRef]

101. Pasquale, S.; Sattin, S.; Escudero-Adan, E.C.; Martinez-Belmonte, M.; De Mendoza, J. Giant regular polyhedra from calixarene carboxylates and uranyl. *Nature Commun.* **2012**, *3*, 785–791. [CrossRef]

102. Kemp, D.S.; Petrakis, K.S. Synthesis and Conformational Analysis of *cis,cis*-1,3,5-Trimethylcyclohexane-1,3,5-tricarboxylic Acid. *J. Org. Chem.* **1981**, *46*, 5140–5143. [CrossRef]

103. Thuéry, P. A Highly Adjustable Coordination System: Nanotubular and Molecular Cage Species in Uranyl Ion Complexes with Kemp's Triacid. *Cryst. Growth Des.* **2014**, *14*, 901–904. [CrossRef]

104. Thuéry, P. Increasing Complexity in the Uranyl Ion–Kemp's Triacid System: From One- and Two-Dimensional Polymers to Uranyl–Copper(II) Dodeca- and Hexadecanuclear Species. *Cryst. Growth Des.* **2014**, *14*, 2665–2676. [CrossRef]

105. Thuéry, P.; Harrowfield, J. Uranyl Ion Complexes with all-cis 1,3,5-Cyclohexanetricarboxylate: Unexpected Framework and Nanotubular Assemblies. *Cryst. Growth Des.* **2014**, *14*, 4214–4225. [CrossRef]

106. Thuéry, P.; Harrowfield, J. Two-dimensional assemblies in f-element ion (UO_2^{2+}, Yb^{3+}) complexes with two cyclohexyl-based polycarboxylates. *Polyhedron* **2015**, *2015 98*, 5–11. [CrossRef]

107. Harrowfield, J.; Thuéry, P. Dipodal, Tripodal, and Discoidal Coordination Modes of Kemp's Triacid Anions. *Eur. J. Inorg. Chem.* **2020**, in press. [CrossRef]

108. Caulder, D.; Raymond, K.N. Supermolecules by Design. *Acc. Chem. Res.* **1999**, *32*, 975–982. [CrossRef]

109. Jansze, S.M.; Cecot, G.; Wise, M.D.; Zhurov, K.O.; Ronson, T.K.; Castilla, A.M.; Finelli, A.; Pattison, P.; Solari, E.; Scopelliti, R.; et al. Ligand Aspect Ratio as a Decisive Factor for the Self-Assembly of Coordination Cages. *J. Am. Chem. Soc.* **2016**, *138*, 2046–2054. [CrossRef] [PubMed]

110. Constable, E.C. Expanded ligands—An assembly principle for supramolecular chemistry. *Coord. Chem. Rev.* **2008**, *252*, 842–855. [CrossRef]

111. Thuéry, P.; Harrowfield, J. Uranyl Ion-Containing Polymeric Assemblies with cis/trans Isomers of 1,2-, 1,3- and 1,4-Cyclohexanedicarboxylates, Including a Helical Chain and a Sixfold-Interpenetrated Framework. *Cryst. Growth Des.* **2020**, *20*, 262–273, and references therein. [CrossRef]

112. Thuéry, P.; Harrowfield, J. Variations on the Honeycomb Topology: From Triangular- and Square-Grooved Networks to Tubular Assemblies in Uranyl Tricarballylate Complexes. *Cryst. Growth Des.* **2017**, *17*, 963–966. [CrossRef]

113. Demazeau, G. Solvothermal Processes: Definition, Key Factors Governing the Involved Chemical Reactions and New Trends. *Z. Naturforsch.* **2010**, *65b*, 999–1006. [CrossRef]

114. Demazeau, G. Solvothermal reactions: An original route for the synthesis of novel materials. *J. Mater. Sci.* **2008**, *43*, 2104–2114. [CrossRef]

115. Thuéry, P.; Atoini, Y.; Harrowfield, J. Structure-directing effects of coordinating solvents, ammonium and phosphonium counterions in uranyl ion complexes with 1,2-, 1,3- and 1,4-phenylenediacetate. *Inorg. Chem.* **2020**, *59*, 2503–2518. [CrossRef] [PubMed]

116. Carter, K.P.; Kalaj, M.; Kerridge, A.; Cahill, C.L. Probing hydrogen and halogen-oxo interactions in uranyl coordination polymers: A combined crystallographic and computational study. *CrystEngComm* **2018**, *20*, 4916–4925. [CrossRef]

117. Carter, K.P.; Kalaj, M.; Cahill, C.L. Harnessing uranyl oxo atoms via halogen bonding interactions in molecular uranyl materials featuring 2,5-di-iodobenzoic acid and N-donor capping ligands. *Inorg. Chem. Front.* **2017**, *4*, 65–78. [CrossRef]

118. Kalaj, M.; Carter, K.P.; Cahill, C.L. Isolating Equatorial and Oxo Based Influences on Uranyl Vibrational Spectroscopy in a Family of Hybrid Materials Featuring Halogen Bonding Interactions with Uranyl Oxo Atoms. *Eur. J. Inorg. Chem.* **2017**, 4702–4713. [CrossRef]

Communication

Hydroquinone-Based Anion Receptors for Redox-Switchable Chloride Binding

Daniel A. McNaughton, Xiaochen Fu, William Lewis, Deanna M. D'Alessandro and Philip A. Gale *

School of Chemistry (F11), The University of Sydney, Sydney, NSW 2006, Australia
* Correspondence: philip.gale@sydney.edu.au

Received: 21 June 2019; Accepted: 8 July 2019; Published: 11 July 2019

Abstract: A series of chloride receptors has been synthesized containing an amide hydrogen bonding site and a hydroquinone motif. It was anticipated that oxidation of the hydroquinone unit to quinone would greatly the diminish chloride binding affinity of these receptors. A conformational switch is promoted in the quinone form through the formation of an intramolecular hydrogen bond between the amide and the quinone carbonyl, which blocks the amide binding site. The reversibility of this oxidation process highlighted the potential of these systems for use as redox-switchable receptors. ^{1}H-NMR binding studies confirmed stronger binding capabilities of the hydroquinone form compared to the quinone; however, X-ray crystal structures of the free hydroquinone receptors revealed the presence of an analogous inhibiting intramolecular hydrogen bond in this state of the receptor. Binding studies also revealed interesting and contrasting trends in chloride affinity when comparing the two switch states, which is dictated by a secondary interaction in the binding mode between the amide carbonyl and the hydroquinone/quinone couple. Additionally, the electrochemical properties of the systems have been explored using cyclic voltammetry and it was observed that the reduction potential of the system was directly related to the expected strength of the internal hydrogen bond.

Keywords: anion binding; chloride receptor; switchable system; hydroquinone; redox switch

1. Introduction

Considerable effort has been devoted to the development of sophisticated and functional molecular machine architectures [1–4]. An integral part of this endeavour has focused on constructing simple molecular switches that mimic elegant examples found in nature [5–7] and expanding the scope of their functional remit to new applications [8–12]. This includes the field of anion recognition chemistry, with systems being applied in the detection and extraction of environmentally damaging and biologically important ions [13–16]. Switchable systems associated with anion recognition have focused mainly on anion-mediated events, where a coordination event induces a detectable change in the molecule which can be sensed by a reporter group [17–20]. However, the field is beginning to expand to include systems where the binding itself can be controlled by external stimuli.

We have previously reported examples of pH-dependent anion receptors which promote chloride efflux under acidic conditions [21–23]. Switchable transporters may allow the movement of anions in healthy cells to be regulated, akin to the mechanism found in many transport proteins [7,24], or find use in targeting cancer cells for efflux induced apoptosis [25,26]. Additionally, a small number of compounds has been developed in which anion binding can be controlled by a photophysical input [27,28]. Photoisomerization can be used to alter the shape of the binding cleft, resulting in a difference in binding affinity between the two states [29,30]. The application of switches governed by electrochemical stimuli provides another method through which anions can be bound and released, yet these systems remain underexplored.

The hydroquinone/quinone redox couple is found extensively throughout biology [31,32], and its well-defined electrochemical properties highlighted its potential candidacy for use in a switchable receptor system [33–35]. With this in mind, we designed a simple yet novel receptor scaffold consisting of a hydroquinone motif with an appended benzamide group. It was envisaged that the convergent hydrogen bonds of the amide and one of the hydroquinone OH group could create a chloride recognition site inspired by similar benzenediol receptors [36]. Subsequent oxidation to the corresponding quinone would induce a rotation around the quinone-benzamide bond due to the favorable formation of an intramolecular hydrogen bond, which is demonstrated in Figure 1. The oxidation event removes the OH groups and causes a conformational change in the orientation of the amide NH and would therefore greatly diminish the chloride affinity of the oxidised form of the receptor.

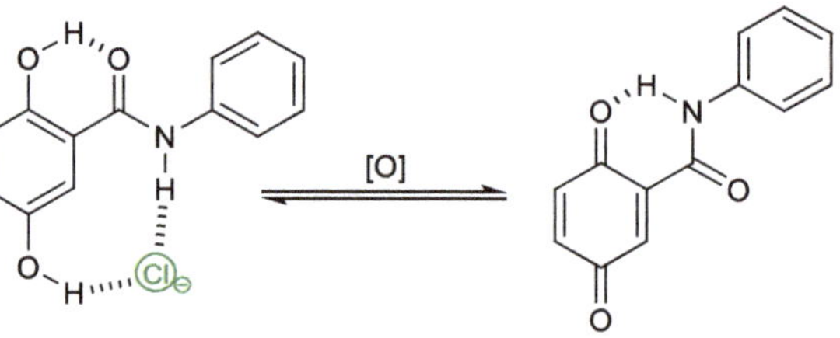

Figure 1. Proposed binding mode of the hydroquinone (left) and proposed intramolecular hydrogen bonded quinone species (right).

A series of hydroquinone-benzamide receptors appended with a variety of electron withdrawing and donating groups was prepared. Subsequently, these molecules were oxidised to afford their quinone forms. Chloride binding affinity has been determined, and the results compared to evaluate the effect of the redox-activated conformational switch on chloride recognition. Additionally, cyclic voltammetry has been used to study the electrochemical properties of the receptors.

2. Results and Discussions

2.1. Synthesis

Hydroquinones **1–4** were prepared following a literature procedure for the parent compound [37], which can be followed in Figure 2. Initially, 2,5-dimethoxybenzoic acid was reacted with oxalyl chloride in dry toluene to afford the acid chloride. Immediate reaction with the respective aniline derivative under basic conditions resulted in the preparation of dimethoxybenzamide compounds **9–12** in yields of between 38–69%. Subsequently, the compounds were reacted with boron tribromide at 0 °C to yield the hydroquinone series. A number of methods were attempted to successfully oxidise the hydroquinones to their quinone forms. Initially, reactions of the hydroquinone compounds with $FeCl_3$ and cerium ammonium nitrate respectively did not lead to the formation of product. This may be due to the instability of the final product in methanol, which caused conversion back to the hydroquinone species, and separation problems as a consequence of additional products. Incomplete conversion to the quinone product in solution led to the formation of a purple compound, identified as a sandwich quinhydrone-like complex which contains one unit each of hydroquinone and quinone [38].

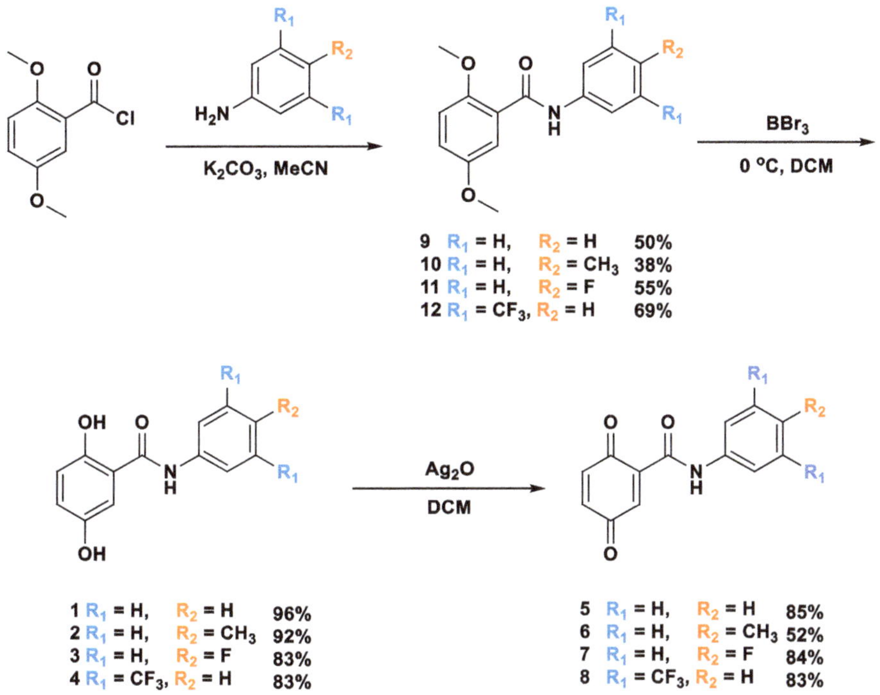

9 R_1 = H, R_2 = H 50%
10 R_1 = H, R_2 = CH$_3$ 38%
11 R_1 = H, R_2 = F 55%
12 R_1 = CF$_3$, R_2 = H 69%

1 R_1 = H, R_2 = H 96%
2 R_1 = H, R_2 = CH$_3$ 92%
3 R_1 = H, R_2 = F 83%
4 R_1 = CF$_3$, R_2 = H 83%

5 R_1 = H, R_2 = H 85%
6 R_1 = H, R_2 = CH$_3$ 52%
7 R_1 = H, R_2 = F 84%
8 R_1 = CF$_3$, R_2 = H 83%

Figure 2. Synthetic procedure for the hydroquinone series and subsequent oxidation to the analogous quinone species.

Conversion was successfully achieved by adding silver oxide to a suspension of the hydroquinone derivatives in dichloromethane [39]. Only the quinone compound enters the solution, meaning the formation of quinhydrone can be avoided. The quinone series was successfully isolated with yields ranging from 52–83%. All new compounds were characterized by ^{1}H-NMR and ^{13}C {^{1}H}-NMR. The dimethoxybenzamide and hydroquinone compounds were characterized using Electrospray Ionization (ESI) mass spectrometry, and the quinones using Atmospheric-Pressure Chemical Ionization (APCI) mass spectrometry. Full details and characterization are provided in the Supplementary Materials.

Crystals of Compound **3** were grown by slow evaporation of an acetonitrile solution of the receptor and the structure was elucidated by single crystal X-ray diffraction (CCDC 1919631). The structure (Figure 3a) shows the receptor adopting the expected conformation in the solid state with an intramolecular hydrogen bond between hydroquinone oxygen O17 and amide oxygen O10 (2.52(4) Å). Crystals of Compound **4** were grown in a similar fashion from a saturated acetonitrile solution and the structure was elucidated by single crystal X-ray diffraction (CCDC 1919630). In this case, the compound crystallized as the acetonitrile solvate with the acetonitrile bound via a hydrogen bond to the hydroquinone oxygen (O2 ⋯ N2 2.818(9) Å) (Figure 3b). There was an additional intramolecular hydrogen bond between amide N1 and hydroquinone O2 (N1 ⋯ O2 2.645(8) Å) stabilizing the solvate in an alternate conformation to that adopted by Compound **3** in the solid state.

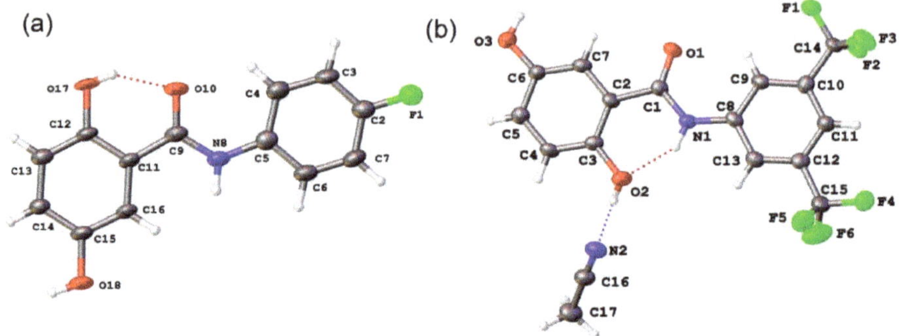

Figure 3. X-ray crystal structures of free receptors **3** (**a**) and **4** (**b**).

2.2. Anion Binding Studies

Binding affinities with chloride were determined for Compounds **1–8** using ^{1}H-NMR spectroscopy titration techniques, with the results displayed in Table 1. Studies for the quinone receptors **5–8** were conducted in pure acetonitrile-d_3 and for the hydroquinones **1–4** in acetonitrile-d_3/1% DMSO-d_6 to assist with solubility. The amide hydrogen of the quinone species **5–8** was followed and fitted to a 1:1 binding model using Bindfit [40]. For titrations with the hydroquinone species, the change in chemical shift of both the hydroxyl and amide hydrogens were followed where possible. The resultant binding for these compounds proved complex and the data could not be fitted adequately to a 1:1 model. However, fitting to a 2:1 binding model provided a greater than 10 times increase in the quality of fit (cov$_{\mathrm{fit}}$, see supporting information), which is evidence in support of the formation of a 2:1 complex in the presence of small amounts of chloride. It is likely that a 1:1 complex is favored as the concentration of chloride in solution is increased following the equilibrium:

$$2\mathrm{H} \underset{+G}{\rightleftharpoons} \mathrm{H_2G} \underset{+G}{\rightleftharpoons} 2[\mathrm{HG}] \tag{1}$$

The interaction parameter, α, was calculated for the series **1–4** and was found to have a value of $\alpha > 14$ in all cases. Values of $\alpha > 1$ describe positive cooperative binding [41], and this can be taken as further evidence for the initial favorable formation of a 2:1 complex at low chloride equivalents.

Compound **5** was also titrated with TBACl in acetonitrile-d_3/1% DMSO-d_6, the solvent mixture used in the hydroquinone receptor experiments. The more complex 2:1 binding exhibited by receptor **1** means direct comparison is difficult, however both K_{21} and K_{11} for **1** are larger than the association constant for equivalent quinone Compound **5** with chloride. Interestingly, the hydroquinone series do not follow the expected trend where increasing the electron-withdrawing power of the motif appended to the amide increases the binding affinities, because a more polarized N-H bond should result in stronger hydrogen bond formation.

Receptor **4** possesses the most strongly electron-withdrawing substituents and the X-ray crystal structure of the free receptor revealed the presence of a surprising alternate conformation, which can be seen in Figure 3b. In this case, a stronger amide hydrogen bond results in a premature switch in the molecule due to the formation of an intramolecular hydrogen bond between the amide proton and the hydroxyl oxygen. This competing interaction blocks the availability of the binding site, adversely affecting chloride affinity with increasing magnitude for receptors with the highest degree of electron-withdrawing substitution. The chloride binding event reverts the conformation back to the anticipated binding mode, which is evidenced by the downfield shift of both hydroxyl protons during the titration experiments. One hydroxyl group is involved in convergent chloride binding with the amide, while the other shifts due to the formation of another intramolecular hydrogen bond with the amide carbonyl upon altering conformation. The second hydroxyl peak shift would not be expected to be as significant if due only to inductive effects of chloride binding. This secondary hydrogen bond can also be considered to play a part in the reduction in chloride affinity as stronger electron-withdrawing groups are appended. Removing electron density from the amide carbonyl will deplete the strength of this stabilizing secondary hydrogen bond and hence diminish the favorability of the binding mode.

Table 1. Overview of the 2:1 association constants for the complexation of hydroquinone receptors **1–4** and Cl^- (as TBA salt) in CD_3CN/1% DMSO-d_6, their interaction parameters (α), and the 1:1 association constants for the complexation of quinone Compounds **5–9** with Cl^- in pure CD_3CN. The 1:1 association constant for 5 and Cl^- in CD_3CN/1% DMSO-d_6 is also reported.

Hydroquinone Receptor	K_{21} [a]	K_{11} [a]	α [b]	Quinone Receptor	K_a [a]
1 [c]	679.5	112.96	24.06	**5** [c]	12.57
				5 [d]	11.2
2 [c]	665.6	191.36	13.91	**6** [d]	11.39
3 [c]	517.76	30.76	67.33	**7** [d]	19.64
4 [c]	336.49	54.44	24.72	**8** [d]	67.72

[a] All errors < 12%. [b] The interaction parameter (α) is calculated by multiplying K_{21} by 4 and dividing by K_{11}. [c] Titrations performed in CD_3CN/1% DMSO-d_6 at 298K. [d] Titrations performed in pure CD_3CN at 298K.

In comparison, the quinones **5–8** have a stronger chloride affinity with increasingly electron-withdrawing appendages. Repulsion between the quinone carbonyls and the approaching chloride anion inhibits binding, and repulsion is reduced when electron density is pulled away from the quinone system. The anticipated intramolecular hydrogen bond between the amide and hydroquinone carbonyl is still expected to interfere with binding, however in this case the chloride binding event is not enhanced by the formation of a new intramolecular hydrogen bond with the amide carbonyl. Instead, it is hindered by additional repulsion between the amide and quinone carbonyls, which is diminished in the presence of more potent electron-withdrawing groups.

The converse trends in binding strengths between the reduced and oxidised forms of the receptors highlight the importance of the relationship between the hydroquinone/quinone couple and the amide carbonyl in dictating chloride affinity. ^{1}H-NMR titrations were also performed for the dimethoxybenzamide Compounds **9–12** with TBACl and fitted to a 1:1 binding model. The association constants were all found to be greater than 7 M^{-1}, highlighting the importance of the hydroquinone hydroxyl proton in creating a convergent anion binding site. Full titration data and fitting for these species can be found in the Supplementary Materials. The likely presence of an amide intramolecular hydrogen bond in the lowest energy conformers of both the hydroquinone and quinone forms of the receptors suggest that this singular contribution cannot result in the differences in binding affinity reported. A combination of the relative strengths of this intramolecular bond, the convergent nature of the hydroquinone binding cleft and the influence of the amide carbonyl over the stability of the binding mode all must be factored in when considering the stronger binding exhibited by the hydroquinone species.

2.3. Electrochemical Studies

Cyclic voltammetry (CV) experiments were employed to assess the electrochemical properties and reversibility of the quinone/hydroquinone couples. The compounds were dissolved in a 0.1 M $TBAPF_6$/CH_3CN solution and the potential was swept at a rate of 100 mVs^{-1} between a range of 800 to -1000 mV. Ferrocene was also added to the solution and used to reference the reduction potentials acquired for each couple, which are displayed in Table 2.

Table 2. Reduction potentials for the quinone species **4–8** obtained using cyclic voltammetry.

Compound	$E_{1/2}$ (mV) [a]
5	−544
6	−552
7	−539
8	−449

[a] Reduction potentials obtained at 100 mVs^{-1} over the range 560 to −1000 mV. Potentials were referenced vs a Fc/Fc$^+$ couple.

The results indicate a direct correlation between the electron-withdrawing power of the functional group and how readily the quinone motif is reduced. The effect of an intramolecular amide hydrogen bond on the quinone reduction potential has been previously reported [34,42], and the trend observed for quinones **5–8** suggests that a similar interaction is in effect. The internal hydrogen bond pulls electron density away from the quinone group, allowing electrons to be accepted more easily by the system. A stronger hydrogen bond will withdraw a greater amount of electron density as evidenced by the reduction potentials of quinones **5** and **8**. The presence of strongly withdrawing bis-CF$_3$ groups in Compound **8** results in an almost 100 mV difference in reduction potential.

The reversibility of the quinone/hydroquinone couple was evaluated by conducting a series of cyclic voltammetry experiments with varying scan rates. The compounds were dissolved in the same solvent mixture used in the previous experiments, and voltammograms were collected between a range of 800 to −1000 mV for a range of scan rates from 20–300 mVs^{-1}, of which an example can be viewed in Figure 4. The Randles–Sevcik equation states that for an electrochemically reversible process, a plot of I_p (peak current) against $v^{1/2}$ (root of scan rate) should return a linear relationship which passes through the origin [43]. Data were plotted for each of the quinone systems (see Supporting Information) and a linear correlation was observed for quinones **5–7**. Compound **8** however did not follow a linear trend and ΔE_p (the separation of the cathodic and anodic peaks) grew larger with increasing scan rate. This indicates that this redox couple is quasireversible, and that the stability of the hydrogen bond may be affecting the kinetics of the electron transfer process.

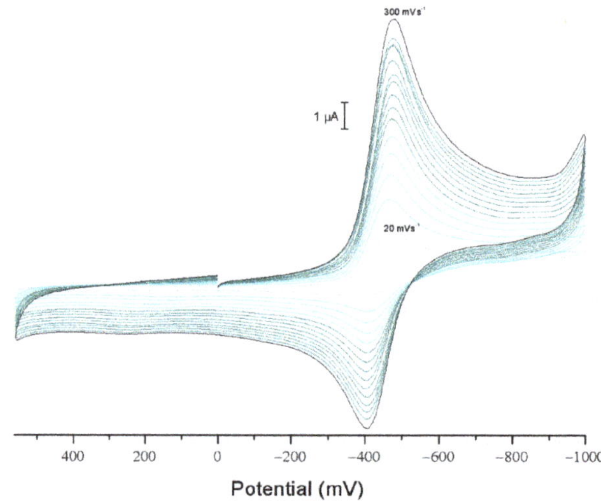

Figure 4. Overlay of cyclic voltammograms of Compound **7** recorded with increasing scan rate from 20 mVs^{-1} to 300 mVs^{-1}. The current response increases with scan rate.

3. Conclusions

We have demonstrated that the hydroquinone/quinone redox couple can be employed in the creation of a redox-switchable chloride receptor. It was discovered that the intramolecular bond initially expected to only be present in the quinone form of the molecule was also present in the hydroquinone receptor, and it is likely this competing interaction led to a reduction in binding capability. Converging trends in binding affinity due to effects dictated by the amide carbonyl resulted in the receptor couples for Compounds **1** and **2** having the greatest difference in binding ability between the two forms of the switch. CV studies highlighted the effect on reduction potential of increasing electron-withdrawing groups, and the results for Compound **8** suggested a loss in reversibility. This can be taken as further evidence of the inhibitory effect of benzamide-appended electron-withdrawing groups on the effectiveness of this class of receptor. Overall, this work has verified the ability of hydroquinone oxidation as a method of reducing anion binding affinity. We are currently exploring the properties of other receptors containing quinoid systems.

Supplementary Materials: The following are available online at http://www.mdpi.com/2624-8549/1/1/7/s1.

Author Contributions: Conceptualization P.A.G. and D.A.M.; synthesis, anion binding studies and electrochemical studies D.A.M.; electrochemical studies, X.F. and D.M.D.; crystallography, W.L.; writing—original draft preparation, D.A.M. and P.A.G.; supervision, P.A.G.

Funding: This research was supported by the Australian Research Council (DP180100612) and the University of Sydney.

Conflicts of Interest: The authors declare no conflicts of interest. The funders had no role in the design of the study; in the collection, analyses, or interpretation of data; in the writing of the manuscript, or in the decision to publish the results.

References

1. Balzani, V.; Credi, A.; Raymo, F.M.; Stoddart, J.F. Artificial molecular machines. *Angew. Chem. Int. Ed.* **2000**, *39*, 3348–3391. [CrossRef]

2. Roke, D.; Wezenberg, S.J.; Feringa, B.L. Molecular rotary motors: Unidirectional motion around double bonds. *Proc. Natl. Acad. Sci. USA* **2018**, *115*, 9423–9431. [CrossRef] [PubMed]

3. Altieri, A.; Gatti, F.G.; Kay, E.R.; Leigh, D.A.; Martel, D.; Paolucci, F.; Slawin, A.M.Z.; Wong, J.K.Y. Electrochemically switchable hydrogen-bonded molecular shuttles. *J. Am. Chem. Soc.* **2003**, *125*, 8644–8654. [CrossRef] [PubMed]

4. Bruns, C.J.; Stoddart, J.F. Molecular machines muscle up. *Nat. Nanotechnol.* **2012**, *8*, 9. [CrossRef] [PubMed]

5. Minamino, T.; Imada, K.; Namba, K. Molecular motors of the bacterial flagella. *Curr. Opin. Struct. Biol.* **2008**, *18*, 693–701. [CrossRef]

6. Spudich, J.L. Variations on a molecular switch: Transport and sensory signalling by archaeal rhodopsins. *Mol. Microbiol.* **1998**, *28*, 1051–1058. [CrossRef] [PubMed]

7. Feng, L.; Campbell, E.B.; Hsiung, Y.; MacKinnon, R. Structure of a Eukaryotic CLC transporter defines an intermediate state in the transport cycle. *Science* **2010**, *330*, 635–641. [CrossRef] [PubMed]

8. Feringa, B.L.; Browne, W.R. *Molecular Switches*; John Wiley & Sons: Hoboken, NJ, USA, 2001.

9. Boelke, J.; Hecht, S. Designing molecular photoswitches for soft materials applications. *Adv. Opt. Mater.* **2019**, 1900404. [CrossRef]

10. Kassem, S.; Lee, A.T.L.; Leigh, D.A.; Markevicius, A.; Solà, J. Pick-up, transport and release of a molecular cargo using a small-molecule robotic arm. *Nat. Chem.* **2015**, *8*, 138. [CrossRef] [PubMed]

11. Kistemaker, J.C.M.; Štacko, P.; Roke, D.; Wolters, A.T.; Heideman, G.H.; Chang, M.C.; van der Meulen, P.; Visser, J.; Otten, E.; Feringa, B.L. Third-generation light-driven symmetric molecular motors. *J. Am. Chem. Soc.* **2017**, *139*, 9650–9661. [CrossRef] [PubMed]

12. Schmittel, M.; De, S.; Pramanik, S. Reversible ON/OFF nanoswitch for organocatalysis: Mimicking the locking and unlocking operation of CaMKII. *Angew. Chem. Int. Ed.* **2012**, *51*, 3832–3836. [CrossRef] [PubMed]

13. Katayev, E.A.; Kolesnikov, G.V.; Sessler, J.L. Molecular recognition of pertechnetate and perrhenate. *Chem. Soc. Rev.* **2009**, *38*, 1572–1586. [CrossRef] [PubMed]

14. Moyer, B.A.; Bonnesen, P.V.; Custelcean, R.; Delmau, L.H.; Hay, B.P. Strategies for using host-guest chemistry in the extractive separations of ionic guests. *Kem. Ind.* **2005**, *54*, 65–87. [CrossRef]

15. Moyer, B.A.; Custelcean, R.; Hay, B.P.; Sessler, J.L.; Bowman-James, K.; Day, V.W.; Kang, S.O. A case for molecular recognition in nuclear separations: Sulfate separation from nuclear wastes. *Inorg. Chem.* **2013**, *52*, 3473–3490. [CrossRef] [PubMed]

16. Busschaert, N.; Caltagirone, C.; Van Rossom, W.; Gale, P.A. Applications of supramolecular anion recognition. *Chem. Rev.* **2015**, *115*, 8038–8155. [CrossRef] [PubMed]

17. García-Garrido, S.E.; Caltagirone, C.; Light, M.E.; Gale, P.A. Acridinone-based anion receptors and sensors. *Chem. Commun.* **2007**, *14*, 1450–1452. [CrossRef] [PubMed]

18. Barendt, T.A.; Rašović, I.; Lebedeva, M.A.; Farrow, G.A.; Auty, A.; Chekulaev, D.; Sazanovich, I.V.; Weinstein, J.A.; Porfyrakis, K.; Beer, P.D. Anion-mediated photophysical behavior in a C60 fullerene [3] rotaxane shuttle. *J. Am. Chem. Soc.* **2018**, *140*, 1924–1936. [CrossRef]

19. Suk, J.; Naidu, V.R.; Liu, X.; Lah, M.S.; Jeong, K.S. A foldamer-based chiroptical molecular switch that displays complete inversion of the helical sense upon anion binding. *J. Am. Chem. Soc.* **2011**, *133*, 13938–13941. [CrossRef]

20. Jones, I.M.; Hamilton, A.D. Anion-dependent switching: Dynamically controlling the conformation of hydrogen-bonded Diphenylacetylenes. *Angew. Chem. Int. Ed.* **2011**, *50*, 4597–4600. [CrossRef]

21. Howe, E.N.W.; Busschaert, N.; Wu, X.; Berry, S.N.; Ho, J.; Light, M.E.; Czech, D.D.; Klein, H.A.; Kitchen, J.A.; Gale, P.A. pH-regulated nonelectrogenic anion transport by Phenylthiosemicarbazones. *J. Am. Chem. Soc.* **2016**, *138*, 8301–8308. [CrossRef]

22. Busschaert, N.; Elmes, R.B.P.; Czech, D.D.; Wu, X.; Kirby, I.L.; Peck, E.M.; Hendzel, K.D.; Shaw, S.K.; Chan, B.; Smith, B.D.; et al. Thiosquaramides: pH switchable anion transporters. *Chem. Sci.* **2014**, *5*, 3617–3626. [CrossRef] [PubMed]

23. Santacroce, P.V.; Davis, J.T.; Light, M.E.; Gale, P.A.; Iglesias-Sánchez, J.C.; Prados, P.; Quesada, R. Conformational control of transmembrane Cl-transport. *J. Am. Chem. Soc.* **2007**, *129*, 1886–1887. [CrossRef] [PubMed]

24. Jasti, J.; Furukawa, H.; Gonzales, E.B.; Gouaux, E. Structure of acid-sensing ion channel 1 at 1.9 Å resolution and low pH. *Nature* **2007**, *449*, 316. [CrossRef] [PubMed]

25. Sessler, J.L.; Eller, L.R.; Cho, W.S.; Nicolaou, S.; Aguilar, A.; Lee, J.T.; Lynch, V.M.; Magda, D.J. Synthesis, Anion-binding properties, and in vitro anticancer activity of prodigiosin analogues. *Angew. Chem. Int. Ed.* **2005**, *44*, 5989–5992. [CrossRef] [PubMed]

26. Ko, S.K.; Kim, S.K.; Share, A.; Lynch, V.M.; Park, J.; Namkung, W.; Van Rossom, W.; Busschaert, N.; Gale, P.A.; Sessler, J.L.; et al. Synthetic ion transporters can induce apoptosis by facilitating chloride anion transport into cells. *Nat. Chem.* **2014**, *6*, 885–892. [CrossRef]

27. Lee, S.; Flood, A.H. Photoresponsive receptors for binding and releasing anions. *J. Phys. Org. Chem.* **2013**, *26*, 79–86. [CrossRef]

28. Vlatković, M.; Feringa, B.L.; Wezenberg, S.J. Dynamic inversion of stereoselective phosphate binding to a bisurea receptor controlled by light and heat. *Angew. Chem. Int. Ed.* **2016**, *55*, 1001–1004. [CrossRef]

29. Li, Z.; Zhang, C.; Ren, Y.; Yin, J.; Liu, S.H. Amide-and urea-functionalized dithienylethene: Synthesis, photochromism, and binding with halide anions. *Org. Lett.* **2011**, *13*, 6022–6025. [CrossRef]

30. Shimasaki, T.; Kato, S.; Ideta, K.; Goto, K.; Shinmyozu, T. Synthesis and structural and photoswitchable properties of novel chiral host molecules: Axis chiral 2, 2'-dihydroxy-1, 1'-binaphthyl-appended stiff-stilbene1. *J. Org. Chem.* **2007**, *72*, 1073–1087. [CrossRef]

31. Nohl, H.; Jordan, W.; Youngman, R.J. Quinones in biology: Functions in electron transfer and oxygen activation. *Adv. Free Radic. Biol. Med.* **1986**, *2*, 211–279. [CrossRef]

32. Monks, T.J.; Hanzlik, R.P.; Cohen, G.M.; Ross, D.; Graham, D.G. Quinone chemistry and toxicity. *Toxicol. Appl. Pharmacol.* **1992**, *112*, 2–16. [CrossRef]

33. Rüssel, C.; Janicke, W. Heterogeneous electron exchange of quinones in aprotic solvents: Part III. The second reduction step of p-benzoquinone and its dependence on the supporting electrolyte. *J. Electroanal. Chem. Interfacial Electrochem.* **1986**, *199*, 139–151. [CrossRef]

34. Eggins, B.R.; Chambers, J.Q. Proton effects in the electrochemistry of the quinone hydroquinone system in aprotic solvents. *J. Electrochem. Soc.* **1970**, *117*, 186–192. [CrossRef]

35. Walczak, M.M.; Dryer, D.A.; Jacobson, D.D.; Foss, M.G.; Flynn, N.T. pH dependent redox couple: An illustration of the Nernst equation. *J. Chem. Educ.* **1997**, *74*, 1195. [CrossRef]

36. Winstanley, K.J.; Sayer, A.M.; Smith, D.K. Anion binding by catechols—an NMR, optical and electrochemical study. *Org. Biomol. Chem.* **2006**, *4*, 1760–1767. [CrossRef] [PubMed]

37. Wachter, V. Chemical Synthesis of Small Molecule Libraries Around the p-Benzoquinone Scaffold. Ph.D. Thesis, Technical University of Braunschweig, Braunschweig, Germany, 2007.

38. Michaelis, L.; Granick, S. Molecular compounds of the quinhydrone type in solution. *J. Am. Chem. Soc.* **1944**, *66*, 1023–1030. [CrossRef]

39. Valderrama, J.A.; Zamorano, C.; González, M.F.; Prina, E.; Fournet, A. Studies on quinones. Part 39: Synthesis and leishmanicidal activity of acylchloroquinones and hydroquinones. *Bioorg. Med. Chem.* **2005**, *13*, 4153–4159. [CrossRef] [PubMed]

40. Bindfit. Available online: http://app.supramolecular.org/bindfit/ (accessed on 15 January 2019).

41. Thordarson, P. Determining association constants from titration experiments in supramolecular chemistry. *Chem. Soc. Rev.* **2011**, *40*, 1305–1323. [CrossRef]

42. Feldman, K.S.; Hester, D.K.; Golbeck, J.H. A relationship between amide hydrogen bond strength and quinone reduction potential: Implications for photosystem I and bacterial reaction center quinone function. *Bioorg. Med. Chem. Lett.* **2007**, *17*, 4891–4894. [CrossRef]

43. Faulkner, L.R.; Bard, A.J. *Electrochemical Methods: Fundamentals and Applications*; John Wiley and Sons: Hoboken, NJ, USA, 2002.

 *chemistry*

Article

A Study of the Interaction between Cucurbit[7]uril and Alkyl Substituted 4-Pyrrolidinopyridinium Salts

Weitao Xu [1], Xinyi Zhu [1], Bing Bian [2], Xin Xiao [1,*], Zhu Tao [1] and Carl Redshaw [3,*]

[1]   Key Laboratory of Macrocyclic and Supramolecular Chemistry of Guizhou Province, Guizhou University, Guiyang 550025, China; xxz_xwt@163.com (W.X.); zxy18355451912@163.com (X.Z.); gzutao@263.net (Z.T.)
[2]   College of Chemistry and Environmental Engineering, Shandong University of Science and Technology, Qingdao 266590, China; bianbing@sdust.edu.cn
[3]   Department of Chemistry & Biochemistry, University of Hull, Hull HU6 7RX, UK
*   Correspondence: gyhxxiaoxin@163.com (X.X.); c.redshaw@hull.ac.uk (C.R.);
     Tel.: +86-155-1908-9928 (X.X.); +44-148-246-5219 (C.R.)

Received: 28 February 2020; Accepted: 8 April 2020; Published: 14 April 2020

Abstract: The interaction between cucurbit[7]uril (Q[7]) and a series of 4-pyrrolidinopyridinium salts bearing aliphatic substituents at the pyridinium nitrogen, namely 4-$(C_4H_8N)C_5H_5NRBr$, where R = H (C0), Et (C2), *n*-butyl (C4), *n*-hexyl (C6), has been studied in aqueous solution by ^{1}H NMR spectroscopy, electronic absorption spectroscopy, and mass spectrometry.

Keywords: host–guest interaction; cucurbit[7]uril; 4-pyrrolidinopyridinium

---

1. Introduction

The behavior of a host toward a guest is typically controlled by non-covalent interactions, and this can impact greatly on the properties exhibited by the guest [1–3]. Given this, supramolecular approaches have been used extensively to construct functional materials, and these have seen applications in a number of areas, for example, in molecular electronics, drug-delivery, optical sensors, and for molecular machines [4–8]. In recent years, new host–guest systems have been reported that employ hosts comprising calix[*n*]arenes, crown ethers, pillararenes, cyclodextrins, or cucurbit[*n*]urils [9–14]. Indeed, during the last two decades, the host–guest chemistry of the cucurbit[*n*]uril (n = 5–8, 10) family has started to flourish [15,16], and this is now impacting on exciting new applications in the fields of materials, biomedicine, sensors, and catalysis [17–20].

In 1983, Mock et al., were the first to study the complexation of alkylammonium and alkyldiammonium ions with Q[6] in aqueous formic acid and to determine their binding affinities [21]. However, there have been few applications of Q[6] in host–guest chemistry due to its small cavity diameter and poor aqueous solubility. In contrast, Q[7] not only has a cavity size amenable to the encapsulation of sizable guest molecules, but also has a much greater aqueous solubility relative to other members of the Q[*n*] family. For example, Q[7] exhibits a water solubility of 30 mM compared to 0.01 mM for Q[8] or Q[10] [18]. This greater solubility, coupled with the larger cavity volume, has resulted in a variety of specific applications of Q[7] systems in aqueous solution [22].

In our previous research, for a series of guests with the same 'central motif', but with different alkyl chain substituents, we found that the length of the alkyl chain determined the mode of interaction with the Q[*n*] [23,24]. Therefore, for the same type of Q[*n*], we can establish different host–guest interaction modes simply by adjusting the length of the alkyl chain, thereby obtaining host–guest materials with different properties. In the case of Q[8], for a series of 4-pyrrolidinopyridinium guests bearing aliphatic substituents at the pyridinium nitrogen, studies in aqueous solution revealed that the alkyl chain at the pyridinium nitrogen can either reside in the Q[8] cavity along with the rest of the guest (as observed for R = *n*-hexyl, *n*-octyl, *n*-dodecyl), can be found outside the Q[8] with the rest of the guest inside (as seen for R = Et), or that the two species can exist in equilibrium for which either the chain or the rest of the guest is encapsulated by the Q[8]. In the solid-state, the structures are somewhat different (in the case of Q[8]@g2, two Q[8] molecules are filled with a centro-symmetric pair of guest molecules) with the cyclic amine encapsulated, and the molecule enters at a rather shallow angle. Interestingly for Q[8]@g3, the two Q[8] molecules behave in different ways. In particular, for one Q[8], the cyclic amine of the guest enters the ring at a rather shallow angle, but for the other Q[8], it is the alkyl chain of the guest that enters the ring with the four carbon atoms of the alkyl chain almost perpendicular to the cavity opening and almost completely encapsulated by the Q[8]). It is well known that the portal diameter and cavity volume of Q[7](7.3 Å, 279 Å^3) are less than Q[8](8.8 Å, 479Å^3) [6,25]. Herein, we examined the interaction of the same family of guests with Q[7] (see Scheme 1) and compared the behavior with that observed for Q[8]. Given that 4-pyrrolidinopyridines have seen widespread use as catalysts in acyl transfer reactions, [26–28] information about their bonding actions can provide insight into their behavior and may inform such catalytic research.

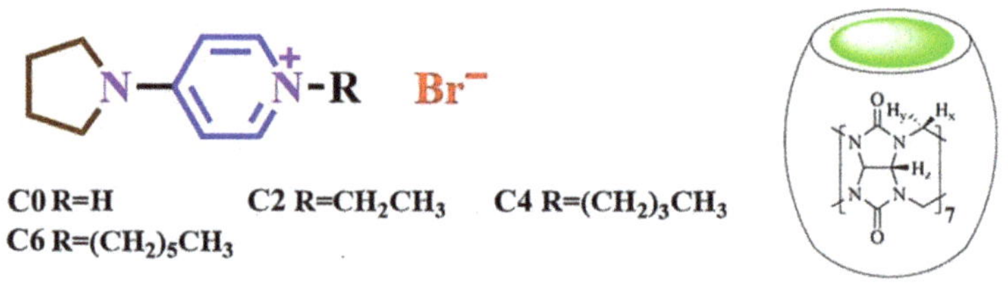

C0 R=H   **C2 R=CH$_2$CH$_3$**   **C4 R=(CH$_2$)$_3$CH$_3$**
C6 R=(CH$_2$)$_5$CH$_3$

Scheme 1. The guests and Q[7] used in this study.

2. Materials and Methods

To analyze the host–guest complexation between Q[7] and C0/C2/C4/C6, 2.0–2.5 × 10^{-3} mmol solutions of Q[7] in 0.5–0.7 mL D$_2$O with Q[7]:C0/C2/C4/C6 ratios ranging between 0 and 2 were prepared. All ^{1}H NMR spectra including those for the titration experiments were recorded at 298.15 K on a JEOL JNM-ECZ400S 400 MHz NMR spectrometer (JEOL, Akishima, Japan) in D$_2$O. D$_2$O was used as a field-frequency lock, and the observed chemical shifts were reported in parts per million (ppm).

All UV–Visible spectra were recorded from samples in 1 cm quartz cells on an Agilent 8453 spectrophotometer, equipped with a thermostat bath (Hewlett Packard, CA, USA). The host and guests were dissolved in distilled water. UV–Visible spectra were obtained at 25 °C at a concentration of 2.00 × 10^{-5} mol·L^{-1} Ci (i = 0, 2, 4, 6) and different Q[7] concentrations for the Q[7]@Ci (i = 0, 2, 4, 6) system. The MALDI-TOF mass spectra were recorded on a Bruker BIFLEX III mass spectrometer with α-cyano-4-hydrox-ycinnamic acid as the matrix.

3. Results and Discussion

3.1. NMR Spectroscopy

The binding interactions between each of the pyrrolidinopyridinium guests and Q[7] can be conveniently monitored using ^{1}H NMR spectroscopic data recorded in neutral D$_2$O solution.

In the case of C0, Figure 1 shows the changes observed in the ^{1}H NMR spectrum of C0 as progressively larger amounts of Q[7] were added to the D$_2$O solution at 25 °C. Spectrum A was obtained in the absence of Q[7], B with Q[7] at 0.326 (B), C with 0.487 equivalents with progressive increasing amounts up to spectrum I with 1.939 equivalents of Q[7] in D$_2$O (D) at 20 °C. The peaks associated with protons a–d all experienced upfield shifts as increasing amounts of Q[7] was added (for specific shift changes, see Table A1, ESI). This is consistent with the complete encapsulation of C0 as depicted in the image shown top right, Figure 1. For the COSY NMR spectrum of this system with 1.939 equivalents of C0, see Figure A1, ESI.

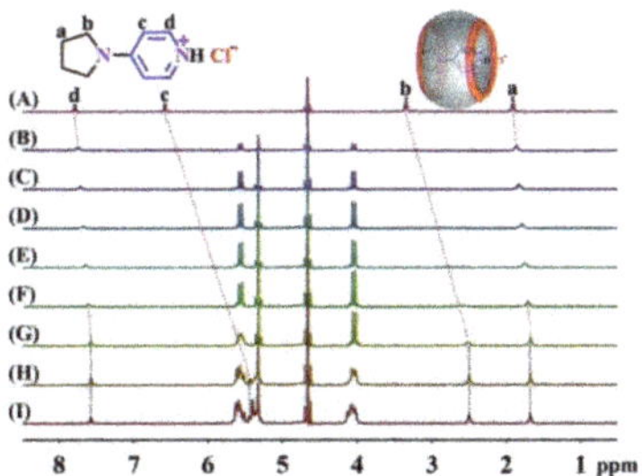

Figure 1. Interaction of C0 and Q[7] (25 °C): ^{1}H NMR spectra (400 MHz, D$_2$O) of C0 (*ca.* 0.5 mM) in the absence (**A**), 0.191 equiv. (**B**), 0.326 equiv. Q[7] (**C**), 0.487 equiv. (**D**), 0.663 equiv. Q[7] (**E**), 0.838 equiv. (**F**), 0.930 equiv. (**G**), 1.570 equiv. (**H**), and 1.939 equiv. of Q[7] (**I**).

In the case of C2, similar analysis (Figure 2) of the system involving Q[7] and C2 revealed that the peaks associated with the ethyl chain, particularly the methyl group (f), did not undergo any significant changes. However, the remaining peaks associated with the protons of the pyridine and pyrrole rings did undergo an upfield shift (for specific shift changes, see Table A2). This situation, whereby the pyridine and pyrrole rings are accommodated within the cavity of Q[7] is reminiscent of that observed for Q[8] [23] and the same guest. There may be slight differences with regard to how much the alkyl chain protrudes out of the cavity for Q[7] *versus* Q[8], but the ^{1}H NMR spectra suggest the difference is small (the change of chemical shifts between these two systems is not obvious).

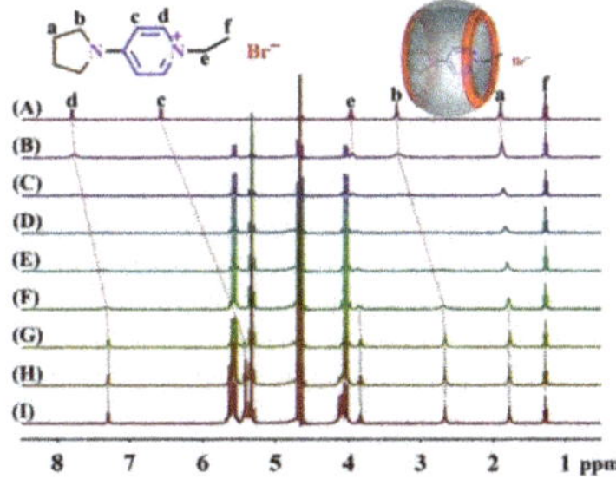

Figure 2. Interaction of C2 and Q[7] (25 °C): ^{1}H NMR spectra (400 MHz, D$_2$O) of C2 (*ca.* 0.5 mM) in the absence (**A**), 0.133 equiv. (**B**), 0.273 equiv. Q[7] (**C**), 0.453 equiv. (**D**), 0.595 equiv. Q[7] (**E**), 0.755 equiv. (**F**), 0.926 equiv. (**G**), 1.415 equiv. (**H**), and 1.733 equiv. of Q[7] (**I**).

In the case of C4, the ^{1}H NMR titration spectra of C4/Q[7] in D$_2$O are presented in Figure 3. There was a clear upfield shift of the signals of all protons in the pyridine ring, whilst the pyrrole ring and alkyl chain protons were also shifted upfield, though to a lesser degree (for specific shift changes, see Table A3, ESI). This indicates that the pyridine ring, pyrrole ring, and the alkyl chain are all accommodated within the cavity of Q[7], and that the Q[7] is capable of shuttling on the guest C4 in a state of dynamic equilibrium. For the COSY NMR spectrum of this system with 1.939 equivalents of C4, see Figure A2, ESI. This shuttling situation is reminiscent of that observed for Q[8] and the same guest.

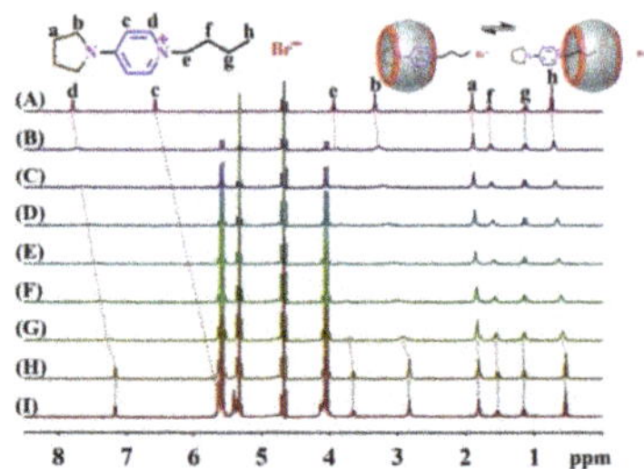

Figure 3. Interaction of C4 and Q[7] (25 °C): ^{1}H NMR spectra (400 MHz, D_2O) of C4 (*ca.* 0.5 mM) in the absence (**A**), 0.116 equiv. (**B**), 0.173 equiv. Q[7] (**C**), 0.331 equiv. (**D**), 0.474 equiv. Q[7] (**E**), 0.641 equiv. (**F**), 0.750 equiv. (**G**), 1.164 equiv. (**H**), and 1.653 equiv. of Q[7] (**I**).

In the case of C6, Figure 4 depicts the ^{1}H NMR titration spectra of the C6/Q[7] in D_2O. All the protons of the guest experienced upfield shifts to varying degrees (for specific shift changes, see Table A4, ESI). There was little change on increasing the Q[7] concentration beyond the addition of 0.997 equivalents. The situation is consistent with the pyridine ring, pyrrole ring, and the alkyl chain all being accommodated within the cavity of Q[7], and with the Q[7] shuttling on the guest C6 in a state of dynamic equilibrium. This contrasts with the situation observed for the same guest and Q[8], in which the pyridine ring, the alkyl chain, and the N part of the pyrrole were accommodated within the cavity of Q[8], and the another part of pyrrole was at its portal; the alkyl chain was buried in the cavity of Q[8] in a twisted form. The reason for the different inclusion modes involving C6 is likely to be related to the smaller cavity size of Q7 versus Q8, which prevents the alkyl chain from bending in the cavity of Q7.

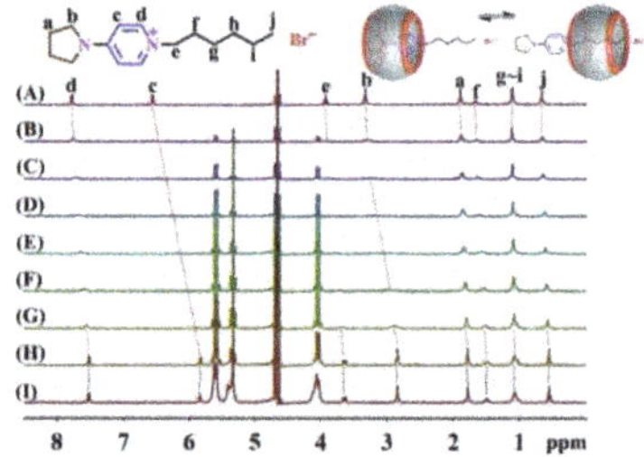

Figure 4. Interaction of C6 and Q[7] (25 °C): ^{1}H NMR spectra (400 MHz, D_2O) of C6 (*ca.* 0.5 mM) in the absence (**A**), 0.110 equiv. (**B**), 0.223 equiv. Q[7] (**C**), 0.363 equiv. (**D**), 0.461 equiv. Q[7] (**E**), 0.646 equiv. (**F**), 0.739 equiv. (**G**), 0.997 equiv. (**H**), and 1.494 equiv. of Q[7] (**I**).

3.2. Ultraviolet (UV) Spectroscopy

To further understand the binding of these 4-pyrrolidinopyridinium salts to Q[7], we also investigated the systems by UV–Vis spectroscopy. The UV spectra were obtained using aqueous solutions containing a fixed concentration of guests C0–C6 and variable concentrations of Q[7] (Figures 5 and A3, Figures A4 and A5). All systems of action showed similar phenomena, and here, only the interactions between Q[7] and guest C2 are described as an example. On gradually increasing the Q[7] concentration in the C2 solution, the absorption band of the guest exhibited a progressively higher absorbance due to the formation of the host–guest complex Q[7]@C2. The absorbance vs. ratio of n(Q[7])/n(C2) data can be fitted to a 1:1 binding model. The pyrrolidinopyridinium part of the guest was encapsulated into the cavity of the Q[7] host, whilst the alkyl moiety remained outside. This generated a 1:1 host–guest inclusion complex. The encapsulation by Q[7] of this guest is presumably due to the favorable ion-dipole interactions between the positively charged guest and the portal oxygen atoms of Q[7] in addition to hydrophobic effects. Moreover, the association constant

(K) is calculated from the UV-vis spectroscopy data according to the modified Benesi–Hildebrand (B–H) equation. For these systems, the binding constants have determined as Ka(Q[7]@C0) = 8.0×10^6, Ka(Q[7]@C2) = 5.7×10^9, Ka(Q[7]@C4) = 3.6×10^7 and Ka(Q[7]@C6) = 2.5×10^9.

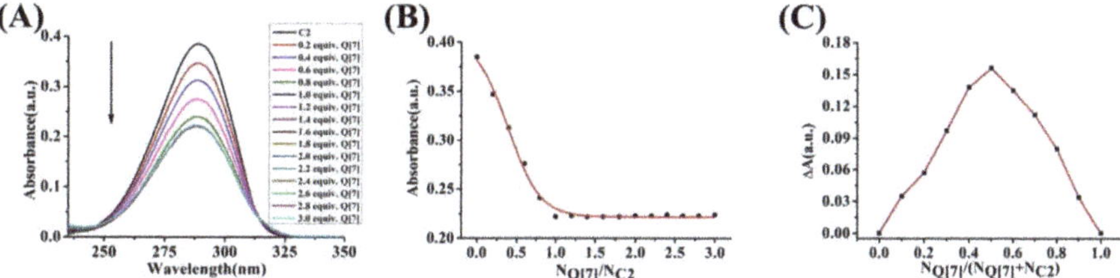

Figure 5. (Color online) (**A**) Electronic absorption of C2 (2×10^{-5} mol L^{-1}) upon addition of increasing amounts (0, 0.2, 0.4⋯⋯2.6, 2.8, 3.0 equiv.) of Q[7]; (**B**) the concentrations and absorbance vs. $N_{Q[7]}/N_{C2}$ plots; (**C**) the corresponding ΔA–$N_{Q[7]}/(N_{Q[7]} + N_{C2})$ curves.

3.3. Mass Spectrometry

The nature of the inclusion complexes between Q[7] and the 4-pyrrolidinopyridinium guests was also established by the use of MALTI-TOF mass spectra, as shown in Figure 6. Intense signals were found at 1311.080, 1339.624, 1367.817, 1395.699, 1423.788, and 1479.762, which corresponded to [(Q[7]@C0)–Cl$^-$]$^+$ (calculated as 1312.196), [(Q[7]@C2)–Br$^-$]$^+$ (calculated as 1340.250), [(Q[7]@C4)–Br$^-$]$^+$ (calculated as 1368.304) and [(Q[7]@C6)–Br$^-$]$^+$ (calculated as 1396.358), respectively, thereby providing support for the formation of 1:1 host–guest inclusion complexes.

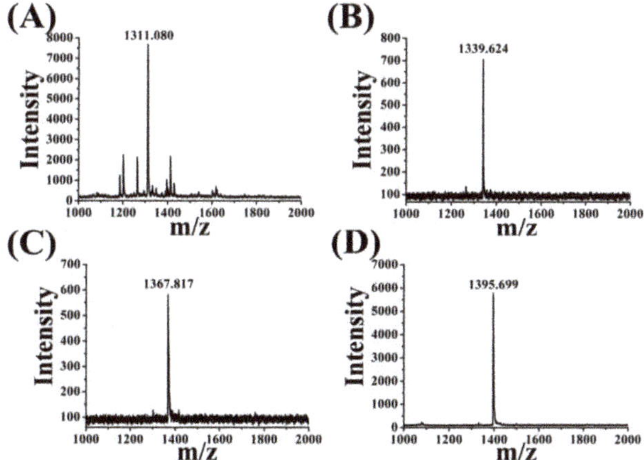

Figure 6. MALDI-TOF mass spectrometry of Q[7]@C0 (**A**), Q[7]@C2 (**B**), Q[7]@C4 (**C**), and Q[7]@C6 (**D**).

4. Conclusions

In summary, we investigated the binding interactions of Q[7] with a series of 4-pyrrolidinopyridinium guests (namely C0, C2, C4, C6) bearing aliphatic substituents at the pyridinium nitrogen by using ^{1}H NMR and UV spectroscopy and mass spectrometry. The results herein revealed that for C0, the entire C0 molecule is encapsulated in the cavity of Q7. For C2, the pyridine and pyrrole rings are accommodated within the cavity of Q[7] and this is reminiscent of the situation observed for Q[8] and the same guest, but the difference between Q[7] and Q[8] is whether the alkyl chain is completely at its portal. For C4, this shuttling situation is reminiscent of that observed for Q[8] and the same guest. For C6, the shuttling situation herein is in contrast with that observed previously for Q[8], where the alkyl chain was twisted and buried in the Q[8] cavity; only part of pyrrole ring protruded from the portal. Based on this work and the previous results for the same guests with Q[8], we found that, for larger guests in particular, the size of the cavity of the cucuribit[*n*]uril dictates the interaction between the cucuribit[*n*]uril and guest. This is illustrated herein for C6, where there is insufficient space in the Q[7] system to allow for the bending of the alkyl chain previously observed for the Q[8] system. Among the Q[*n*]-based rotaxane/pseudorotaxane systems, most of the axle molecules have been built from ammonium, pyridinium ions, and viologen derivatives [17–20]. In comparison with previously reported cucurbit[7]uril-based host–guest systems (such as pyridinium ions, viologen derivatives), the interaction mode is similar [6,25]. This provides more insight into the possible applications of systems such as switchable mechanically pseudorotaxane molecules, for instance, in the area of supramolecular materials science.

Author Contributions: The main work of W.X. and X.Z. was to complete the experiment; B.B. participated in the optimization of the MS experiment; Manuscript writing by X.X., Z.T., and C.R. All authors have read and agree to the published version of the manuscript.

Funding: The Natural Science Foundation of China (21861011), the Major Program for Creative Research Groups of Guizhou Provincial Education Department (2017-028), the Innovation Program for High-level Talents of Guizhou Province (No. 2016-5657), and the Science and Technology Fund of Guizhou Province (No. 2018-5781) are gratefully acknowledged for their financial support. CR thanks the EPSRC (Engineering and Physical Sciences Research Council) for a travel grant (EP/L012804/1).

Conflicts of Interest: The authors declare no conflicts of interest.

Appendix A

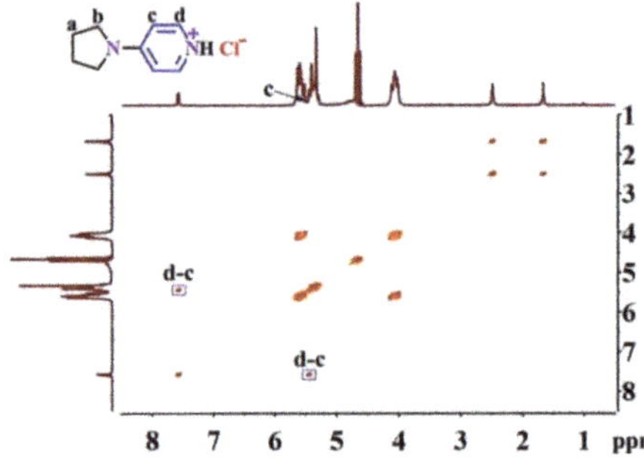

Figure A1. The COSY spectrum of C0 with 1.939 equivalent of Q[7] in D$_2$O (400 MHz).

Table A1. Data from the ^{1}H NMR spectra for the interaction of C0 and different proportions of Q[7].

	Hd	Hc	Hb	Ha
C0	7.80	6.58	3.35	1.91
0.191 Q[7]	7.75	\	\	1.86
0.326 Q[7]	7.72	\	\	1.83
0.487 Q[7]	7.68	\	\	1.79
0.663 Q[7]	7.64	\	\	1.75
0.838 Q[7]	7.61	\	2.60	1.70
0.930 Q[7]	7.58	\	2.51	1.67
1.570 Q[7]	7.58	5.45	2.50	1.67
1.939 Q[7]	7.58	5.45	2.50	1.67

Table A2. Data from the ^{1}H NMR spectra for the interaction of C2 and different proportions of Q[7].

	Hd	Hc	He	Hb	Ha	Hf
C2	7.81	6.57	3.97	3.34	1.90	1.28
0.133 Q[7]	7.78	\	3.95	3.31	1.88	1.28
0.273 Q[7]	\	\	3.93	\	1.86	1.28
0.453 Q[7]	\	\	3.90	\	1.83	1.28
0.595 Q[7]	\	\	3.87	\	1.81	1.28
0.755 Q[7]	7.32	\	3.85	2.69	1.79	1.28
0.926 Q[7]	7.30	5.43	3.84	2.66	1.77	1.28
1.415 Q[7]	7.30	\	3.84	2.66	1.77	1.28
1.733 Q[7]	7.30	\	3.84	2.66	1.77	1.28

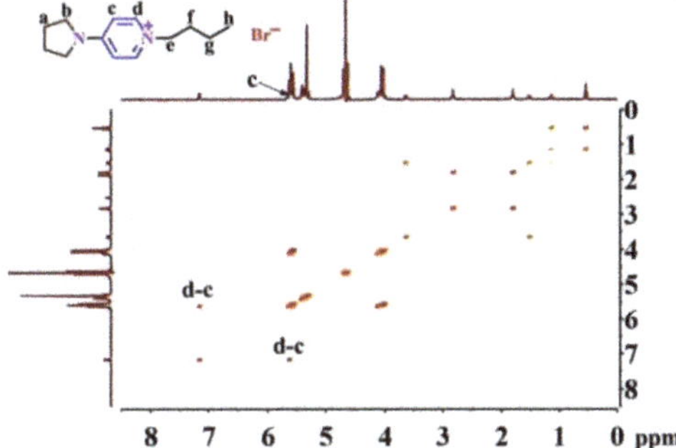

Figure A2. The COSY spectrum of C4 with 1.653 equivalent of Q[7] in D_2O (400 MHz).

Table A3. Data from the ^{1}H NMR spectra for the interaction of C4 and different proportions of Q[7].

	Hd	Hc	He	Hb	Ha	Hf	Hg	Hh
C4	7.79	6.57	3.94	3.33	1.90	1.65	1.13	0.75
0.116 Q[7]	7.73	\	3.90	3.28	1.89	1.63	1.13	0.72
0.173 Q[7]	7.65	\	3.86	3.22	1.87	1.62	1.14	0.69
0.331 Q[7]	7.56	\	3.83	3.14	1.86	1.60	1.14	0.67
0.474 Q[7]	7.46	\	3.78	3.06	1.85	1.58	1.14	0.64
0.641 Q[7]	7.36	\	3.74	2.98	1.83	1.57	1.14	0.60
0.750 Q[7]	7.27	\	3.71	2.91	1.82	1.55	1.15	0.58
1.164 Q[7]	7.16	\	3.64	2.82	1.80	1.53	1.15	0.54
1.653 Q[7]	7.16	5.45	3.64	2.82	1.80	1.53	1.15	0.54

Table A4. Data from the ^{1}H NMR spectra for the interaction of C6 and different proportions of Q[7].

	Hd	Hc	He	Hb	Ha	Hf	Hg-i	Hj
C6	7.79	6.56	3.93	3.33	1.90	1.67	1.12	0.68
0.110 Q[7]	7.76	6.54	3.91	3.30	1.89	1.65	1.11	0.66
0.223 Q[7]	7.72	\	3.87	3.26	1.87	1.63	1.11	0.65
0.363 Q[7]	7.68	\	3.82	3.19	1.85	1.60	1.10	0.63
0.461 Q[7]	7.64	\	3.77	3.04	1.84	1.57	1.09	0.62
0.646 Q[7]	7.59	\	3.72	2.95	1.82	1.55	1.08	0.60
0.739 Q[7]	7.56	5.86	3.68	2.89	1.80	1.52	1.08	0.58
0.997 Q[7]	7.52	5.81	3.64	2.84	1.78	1.49	1.07	0.56
1.494 Q[7]	7.52	5.81	3.64	2.84	1.78	1.49	1.07	0.56

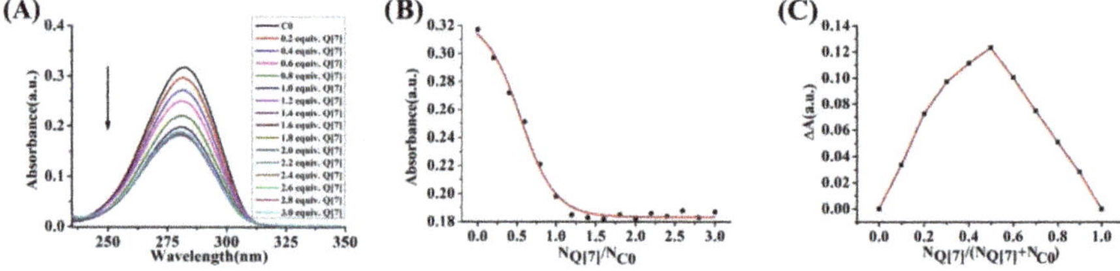

Figure A3. (Color online) (**A**) Electronic absorption of C0 (2×10^{-5} mol L^{-1}) upon addition of increasing amounts (0, 0.2, 0.4······2.6, 2.8, 3.0 equiv.) of Q[7]; (**B**) the concentrations and absorbance *vs.* $N_{Q[7]}/N_{C0}$ plots; (**C**) the corresponding ΔA–$N_{Q[7]}/(N_{Q[7]} + N_{C0})$ curves.

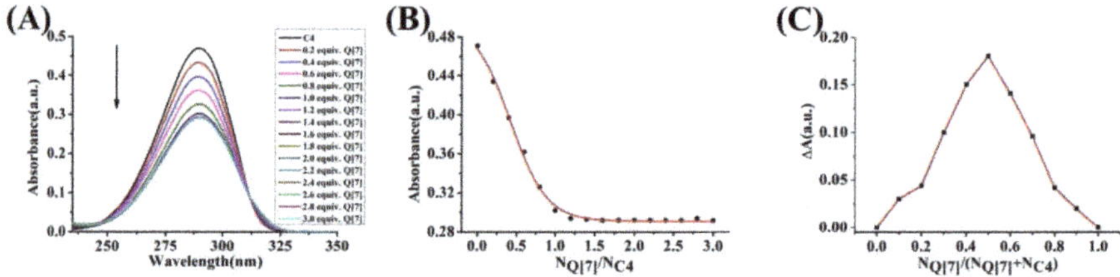

Figure A4. (Color online) (**A**) Electronic absorption of C4 (2×10^{-5} mol L^{-1}) upon addition of increasing amounts (0, 0.2, 0.4······2.6, 2.8, 3.0 equiv.) of Q[7]; (**B**) the concentrations and absorbance *vs.* $N_{Q[7]}/N_{C4}$ plots; (**C**) the corresponding ΔA–$N_{Q[7]}/(N_{Q[7]} + N_{C4})$ curves.

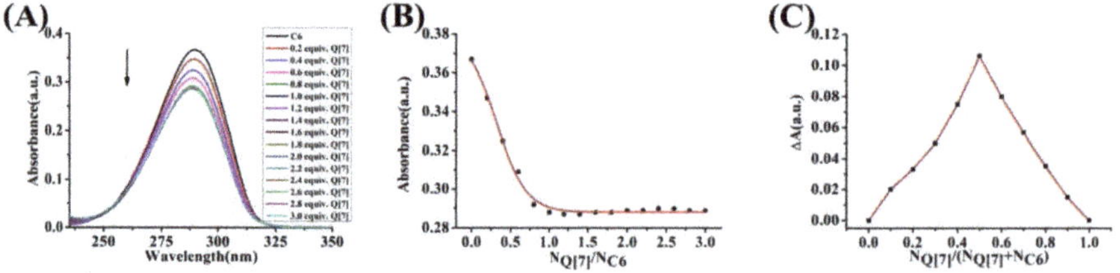

Figure A5. (Color online) (**A**) Electronic absorption of C6 (2×10^{-5} mol L^{-1}) upon addition of increasing amounts (0, 0.2, 0.4······2.6, 2.8, 3.0 equiv.) of Q[7]; (**B**) the concentrations and absorbance *vs.* $N_{Q[7]}/N_{C6}$ plots; (**C**) the corresponding ΔA–$N_{Q[7]}/(N_{Q[7]} + N_{C6})$ curves.

Guests C0, C2, C4, C6 were synthesized by previously reported methods. [23].

Synthesis of guest C0: 4-pyrrolidinopyridine (296 mg, 0.002 mol) and HCl (10 mL) were stirred under an inert nitrogen atmosphere and heated to 80 °C and refluxed for 12 h. The resulting solution was filtered and then the white precipitate was washed with diethyl ether and dried in vacuum to give C0 (331 mg, 90%). ^{1}H NMR (D$_2$O, 400 MHz) δ: 7.79 (d, J = 7.5 Hz, 2H), 6.58 (d, J = 7.5 Hz, 2H), 3.35 (t, J = 6.8 Hz, 4H), 1.94–1.89 (m, 4H). ^{13}C NMR (101 MHz) δ: 154.65, 137.82, 107.26, 48.09, 24.65. Anal. Calcd. for C$_9$H$_{13}$N$_2$Cl: C, 58.54; H, 7.10; N, 15.17; found C, 57.92; H, 7.18; N, 15.93. The ^{1}H NMR (top) (400 MHz) and ^{13}C NMR (below) (100 MHz) spectra of C0 in D$_2$O are presented in Figure A6.

Synthesis of guest C2: 4-pyrrolidinopyridine (296 mg, 0.002 mol) and bromoethane (1.308 g, 0.012 mol) were dissolved in acetonitrile (40 mL). The solution was stirred under an inert nitrogen atmosphere and heated to 80 °C and refluxed for 12 h. The resulting solution was filtered and then the yellow precipitate was washed with diethyl ether and then dried in vacuum to give C2 (437 mg, 85%). ^{1}H NMR (D$_2$O, 400 MHz) δ: 7.78 (d, J = 7.6 Hz, 2H), 6.54 (d, J = 7.5 Hz, 2H), 3.94 (q, J = 7.3 Hz, 2H), 3.30 (m, J = 8.0 Hz, 4H), 1.90–1.84 (m, 4H), 1.25 (t, J = 7.3 Hz, 3H). ^{13}C NMR (101 MHz) δ 153.50, 140.95, 108.23, 52.98, 48.15, 24.82, 15.42. Anal. Calcd. for C$_{11}$H$_{17}$N$_2$Br: C, 51.37; H, 6.66; N, 10.89; found C, 51.29; H, 6.71; N, 10.92. The ^{1}H NMR (top) (400 MHz) and ^{13}C NMR (below) (100 MHz) spectra of C2 in D$_2$O are presented in Figure A7.

Synthesis of guest C4: The same method as for C2 was employed, but using 4-pyrrolidinopyridine (296 mg, 0.002 mol) and bromobutane (1.644 g, 0.012 mol) to give C4 (496 mg, 87%). ^{1}H NMR (D$_2$O, 400 MHz) δ: 7.79 (d, J = 7.6 Hz, 2H), 6.57 (d, J = 7.5 Hz, 2H), 3.94 (t, J = 7.1 Hz, 2H), 3.34 (t, J = 6.7 Hz, 4H), 1.93–1.88 (m, 4H), 1.66 (m, J = 14.8 Hz, 2H), 1.14 (m, J = 14.8 Hz, 2H), 0.75 (t, J = 7.4 Hz, 3H). ^{13}C NMR (D$_2$O, 101 MHz) δ: 153.52, 141.23, 107.82, 57.46, 48.24, 32.12, 24.78, 18.73, 12.69. Anal. Calcd. for C$_{13}$H$_{21}$N$_2$Br: C, 54.74; H, 7.42; N, 9.82; found C, 54.82; H, 7.47; N, 9.75. The ^{1}H NMR (top) (400 MHz) and ^{13}C NMR (below) (100 MHz) spectra of C4 in D$_2$O are as presented in Figure A8.

Synthesis of guest C6: The same method as for C2 was employed, but using 4-pyrrolidinopyridine (296 mg, 0.002 mol) and bromohexane (1.981 g, 0.012 mol) to give C6 (551 mg, 88%). ^{1}H NMR (D$_2$O, 400 MHz) δ: 7.79 (d, J = 7.0 Hz, 2H), 6.57 (d, J = 7.1 Hz, 2H), 3.94 (q, J = 7.0 Hz, 2H), 3.35 (d, J = 6.1 Hz, 4H), 1.91 (m, 4H), 1.68 (m, J = 6.6 Hz, 2H), 1.12 (m, 6H), 0.69 (t, J = 6.4 Hz, 3H). ^{13}C NMR (400 MHz) δ: 153.52, 141.23, 108.22, 57.42, 48.24, 30.42, 29.94, 24.90, 24.77, 21.80, 13.21. Anal. Calcd. for C$_{15}$H$_{25}$N$_2$Br: C, 57.51; H, 8.04; N, 8.94; found C, 57.48; H, 8.11; N, 8.99. The ^{1}H NMR (top) (400 MHz) and 13C NMR (below) (100 MHz) spectra of C6 in D$_2$O are as presented in Figure A9.

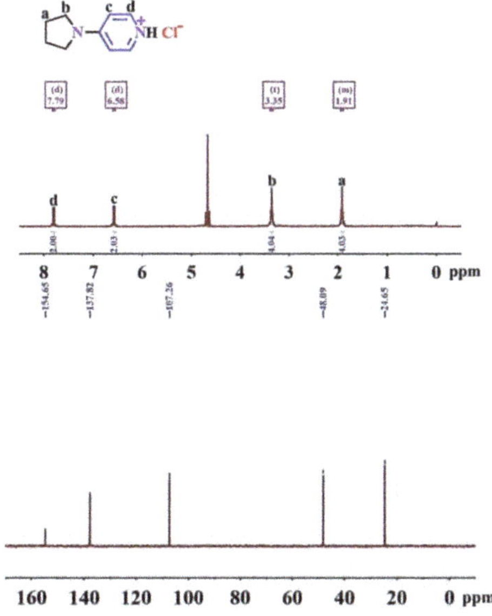

Figure A6. ^{1}H NMR (**top**) (400 MHz) and ^{13}C NMR (**below**) (100 MHz) spectra of C0 in D$_2$O.

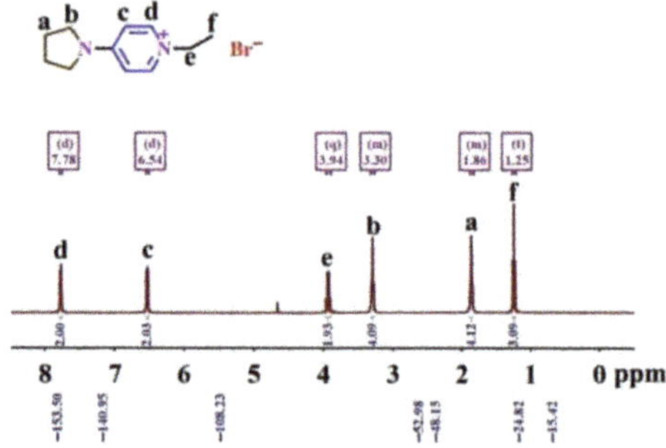

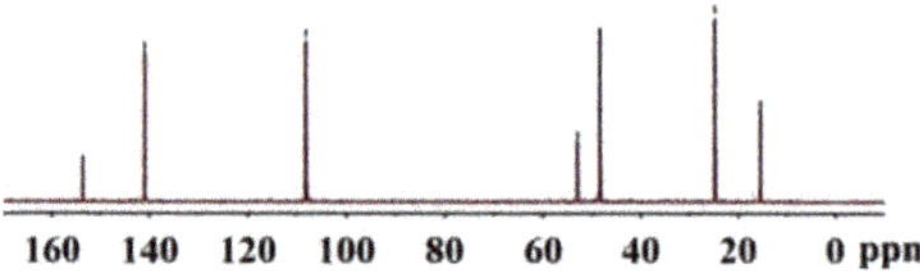

Figure A7. ^{1}H NMR (**top**) (400 MHz) and ^{13}C NMR (**below**) (100 MHz) spectra of C2 in D_2O.

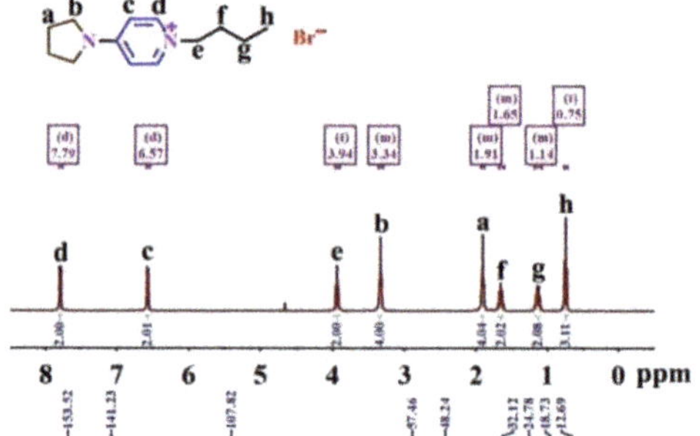

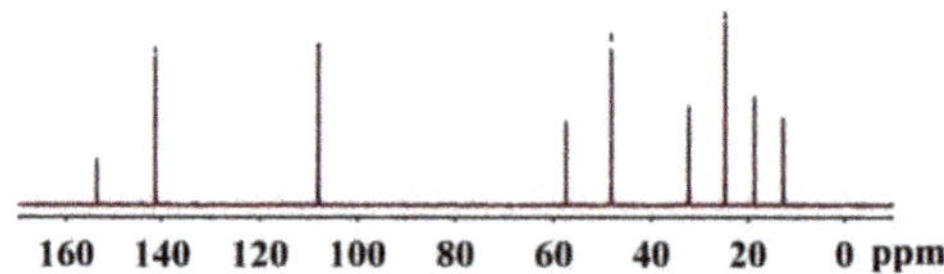

Figure A8. ^{1}H NMR (**top**) (400 MHz) and ^{13}C NMR (**below**) (100 MHz) spectra of C4 in D_2O.

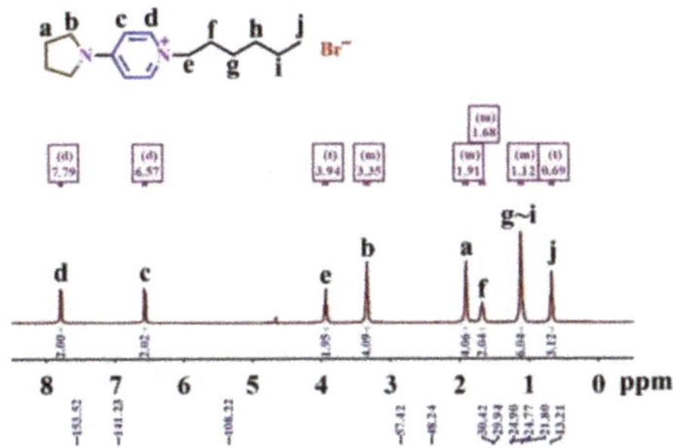

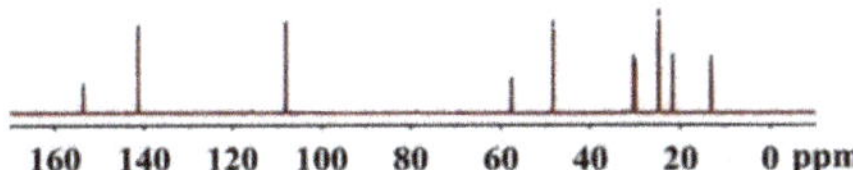

Figure A9. ^{1}H NMR (**top**) (400 MHz) and ^{13}C NMR (**below**) (100 MHz) spectra of C6 in D_2O.

References

1. Zhang, W.; Luo, Y.; Zhou, Y.; Liu, M.; Xu, W.; Bian, B.; Tao, Z.; Xin, X. A highly selective fluorescent chemosensor probe for the detection of Fe^{3+} and Ag^+ based on supramolecular assembly of cucubit[10]uril with a pyrene derivative. *Dyes Pigment.* **2020**, *176*, 108235–108242. [CrossRef]
2. Lucenti, E.; Forni, A.; Botta, C.; Carlucci, L.; Giannini, C.; Marinotto, D.; Previtali, A.; Righetto, S.; Cariati, E. H-aggregates granting crystallization-induced emissive behavior and ultralong phosphorescence from a pure organic molecule. *J. Phys. Chem. Lett.* **2017**, *8*, 1894–1898. [CrossRef]
3. Bian, L.; Shi, H.; Wang, X.; Ling, K.; Ma, H.; Li, M.; Cheng, Z.; Ma, C.; Cai, S.; Wu, Q.; et al. Simultaneously enhancing efficiency and lifetime of ultralong organic phosphorescence materials by molecular self-assembly. *J. Am. Chem. Soc.* **2018**, *140*, 10734–10739. [CrossRef] [PubMed]
4. You, L.; Zha, D.; Anslyn, E.V. Recent advances in supramolecular analytical chemistry using optical sensing. *Chem. Rev.* **2015**, *115*, 7840–7892. [CrossRef] [PubMed]
5. Dale, E.J.; Vermeulen, N.A.; Juricek, M.; Barnes, J.C.; Young, R.M.; Wasielewski, M.R.; Stoddart, J.F. Supramolecular Ex plorations: Ex hibiting the Ex tent of Ex tended cationic cyclophanes. *Acc. Chem. Res.* **2016**, *49*, 262–273. [CrossRef] [PubMed]
6. Barrow, S.J.; Kasera, S.; Rowland, M.J.; del Barrio, J.; Scherman, O.A. Cucurbituril-based molecular recognition. *Chem. Rev.* **2015**, *115*, 12320–12406. [CrossRef]
7. Yu, G.; Jie, K.; Huang, F. Supramolecular amphiphiles based on host–guest molecular recognition motifs. *Chem. Rev.* **2015**, *115*, 7240–7303. [CrossRef]
8. Wu, M.X.; Yang, Y.W. Metal–organic framework (MOF)-based drug/cargo delivery and cancer therapy. *Adv. Mater.* **2017**, *29*, 1606134. [CrossRef]
9. Diederich, F. Complexation of neutral molecules by cyclophane hosts. *Angew. Chem. Int. Ed. Engl.* **1988**, *27*, 362–386. [CrossRef]
10. Xue, M.; Yang, Y.; Chi, X.; Zhang, Z.; Huang, F. Pillararenes, a new class of macrocycles for supramolecular chemistry. *Acc. Chem. Res.* **2012**, *45*, 1294–1308. [CrossRef]
11. Rebek, J. Molecular behavior in small spaces. *Acc. Chem. Res.* **2009**, *42*, 1660–1668. [CrossRef] [PubMed]
12. Zarra, S.; Wood, D.M.; Roberts, D.A.; Nitschke, J.R. Molecular containers in complex chemical systems. *Chem. Soc. Rev.* **2015**, *44*, 419–432. [CrossRef] [PubMed]
13. Szente, L.; Szejtli, J. Highly soluble cyclodextrin derivatives: Chemistry, properties, and trends in development. *Adv. Drug Delivery Rev.* **1999**, *36*, 17–28. [CrossRef]

14. Pedersen, C.J. The discovery of crown ethers (Noble Lecture). *Angew. Chem. Int. Ed. Engl.* **1988**, *27*, 1021–1027. [CrossRef]
15. Masson, E.; Ling, X.; Roymon, J.; Kyeremeh-Mensah, L.; Lu, X. Cucurbituril chemistry: A tale of supramolecular success. *RSC Adv.* **2012**, *2*, 1213–1247. [CrossRef]
16. Kim, K. Mechanically interlocked molecules incorporating cucurbituril and their supramolecular assemblies. *Chem. Soc. Rev.* **2002**, *31*, 96–107. [CrossRef]
17. Assaf, K.I.; Nau, W.M. Cucurbiturils: From synthesis to high-affinity binding and catalysis. *Chem. Soc. Rev.* **2015**, *44*, 394–418. [CrossRef]
18. Lagona, J.; Mukhopadhyay, P.; Chakrabarti, S.; Isaacs, L. The cucurbit[*n*]uril family. *Angew. Chem. Int. Ed.* **2005**, *44*, 4844–4870. [CrossRef]
19. Gomez-Casado, A.; Jonkheijm, P.; Huskens, J. Recognition properties of cucurbit[7]uril self-assembled monolayers studied with force spectroscopy. *Langmuir* **2011**, *27*, 11508–11513. [CrossRef]
20. Bosmans, R.P.; Briels, J.M.; Milroy, L.G.; de Greef, T.F.; Merkx, M.; Brunsveld, L. Supramolecular Control over Split-Luciferase Complementation. *Angew. Chem. Int. Ed.* **2016**, *55*, 8899–8903. [CrossRef]
21. Mock, W.L.; Shih, N.Y. Host-guest binding capacity of cucurbituril. *J. Org. Chem.* **1983**, *48*, 3618. [CrossRef]
22. Walker, S.; Oun, R.; McInnes, F.J.; Wheate, N.J. The potential of cucurbit[*n*]urils in drug delivery. *Isr. J. Chem.* **2011**, *51*, 616–624. [CrossRef]
23. Xu, W.; Kan, J.; Yang, B.; Prior, T.J.; Bian, B.; Xiao, X.; Tao, Z.; Redshaw, C. A Study of the Interaction Between Cucurbit[8]uril and Alkyl-Substituted 4-Pyrrolidinopyridinium Salts. *Chem. Asian J.* **2019**, *14*, 235–242. [CrossRef]
24. Xu, W.; Liu, M.; Escanõ, M.C.; Redshaw, C.; Bian, B.; Fan, Y.; Tao, Z.; Xiao, X. Alkyl substituted 4-pyrrolidinopyridinium salts encapsulated in the cavity of cucurbit[10]uril. *New J. Chem.* **2019**, *43*, 7028–7034. [CrossRef]
25. Ji, L.; Yang, L.; Yu, Z.Y.; Tan, C.S.; Parker, R.M.; Abell, C.; Scherman, O.A. Cucurbit[*n*]uril-based microcapsules self-assembled within microfluidic droplets: A versatile approach for supramolecular architectures and materials. *Accounts Chem. Res.* **2017**, *50*, 208–217.
26. Yamanaka, M.; Yoshida, U.; Sato, M.; Shigeta, T.; Yoshida, K.; Furuta, T.; Kawabata, T.J. Origin of high E-selectivity in 4-pyrrolidinopyridine-catalyzed tetrasubstituted α, α'-alkenediol: A computational and experimental study. *Org. Chem.* **2015**, *80*, 3075–3082. [CrossRef] [PubMed]
27. Sammaki, T.; Hurley, T.B. 2-formyl-4-pyrrolidinopyridine (FPP): A new catalyst for the hydroxyl-directed methanolysis of esters. *J. Am. Chem. Soc.* **1996**, *118*, 8967–8968. [CrossRef]
28. Nguyen, H.V.; Butler, D.C.; Richards, C.J. A metallocene-pyrrolidinopyridine nucleophilic catalyst for asymmetric synthesis. *Org. Lett.* **2006**, *8*, 769–772. [CrossRef]

Article

Selective Proton-Mediated Transport by Electrogenic K$^+$-Binding Macrocycles

Yu-Hao Li [1], Shao-Ping Zheng [1], Dawei Wang [1] and Mihail Barboiu [1,2,*]

[1]   Lehn Institute of Functional Materials, School of Chemistry, Sun Yat-Sen University, Guangzhou 510275, China; yuhaoli.mat@gmail.com (Y.-H.L.); zhshaop@126.com (S.-P.Z.); wdawei@mail.sysu.edu.cn (D.W.)

[2]   Adaptive Supramolecular Nanosystems Group, Institut Européen des Membranes, University of Montpellier ENSCM-UM-CNRS UMR-5635, Place E. Bataillon, CC 047, F-34095 Montpellier, France

*   Correspondence: mihail-dumitru.barboiu@umontpellier.fr

Received: 13 December 2019; Accepted: 16 January 2020; Published: 20 January 2020

Abstract: Synthetic K$^+$-binding macrocycles have potential as therapeutic agents for diseases associated with KcsA K$^+$ channel dysfunction. We recently discovered that artificial self-assembled *n*-alkyl-benzoureido-15-crown-5-ether form selective ion-channels for K$^+$ cations, which are highly preferred to Na$^+$ cations. Here, we describe an impressive selective activation of the K$^+$ transport via electrogenic macrocycles, stimulated by the addition of the carbonyl cyanide-4-(trifluoromethoxy) phenylhydrazone (FCCP) proton carrier. The transport performances show that both the position of branching or the size of appended alkyl arms favor high transport activity and selectivity S$_{K^+/Na^+}$ up to 48.8, one of the best values reported up to now. Our study demonstrates that high K$^+$/Na$^+$ selectivity obtained with natural KcsA K$^+$ channels is achievable using simpler artificial macrocycles displaying constitutional functions.

Keywords: ion-channels; crown-ethers; bilayer membranes; self-assembly; supramolecular chemistry

---

1. Introduction

Facilitated transmembrane transport by molecular carriers that selectively recognize and transport ionic species is an important and complex physiological process [1]. Nature has evolved millions of years to generate highly selective cation carriers, for which the transport mechanisms are controlled via a membrane potential (i.e., electrogenic valinomycin) or pH gradient (i.e., neutral monensin) [2]. Most of the previously reported artificial systems cannot achieve this type of stable potential, due to their poor selectivity. Macrocyclic crown-ethers have already proven to be efficient cation channels [2,3]. Whether these macrocycles are covalently connected [3–7] or self-assembled via H-bonding [8–19], they form ion-channels, performing effective transport of ions across lipid bilayers. The structural variation between closely related synthetic carriers or channels could lead to a huge difference in activities [8–14].

Artificial ion-channels presenting high K$^+$/Na$^+$ selectivity are rare. We unexpectedly discovered that benzo-15-crown-5-ethers are showing far superior selectivity for K$^+$ cation when compared with benzo-18-crown-6 congeners [15–19]. So far, research efforts have mainly focused on the electrogenic polarization across the membrane, related to a net transfer of charge via neutral alkyl-benzoureido-15-crown-5-ether channels [15]. Moreover, we also demonstrated that cholesteric or squalene moieties appended to cation binding benzo-15-crown-5-ether show the most efficient transport among similar structures to a certain extent [16]. Under the same conditions, squalene-benzoamido-15-crown-5-ether has the highest selectivity for K$^+$ over Na$^+$, S$_{K+/Na+}$ = 58.3, [17] while hexylbenzo-ureido-15-crown-5-ether has lower S$_{K+/Na+}$ = 17 [15].

We know from previous studies [20–24] that the transport activity has an optimal relationship with lipophilicity of carriers, while the disposition of the alkyl groups on the carrier backbone proved to be important too. The self-assembly behaviors of the functional transporting system are directly determining the structural dynamics that control the self-organized superstructures along cation recognition and transport pathways. The transport mechanism is determined by the optimal coordination rather than classical dimensional compatibility between crown-ether and cation, and systematic changes of the structure lead to adaptive selection in cation-transport activity [25].

As far as we know the alkyl skeleton isomerism was rarely considered to be a determinant factor to influence the transport activity. The structural variability of the alkyl tails of the macrocyclic superstructures at the interface within the membrane have been not specifically studied in our previous work. Herein, we continue our exploration and we serendipitously found that simple small variations on the structure of the linear or branched octyl tails in octyl-benzoureido-15-crown-5-ethers **1, r2, s2**, and racemic **3** (Figure 1, Scheme S1) are strongly influencing the transport activities of monovalent cations. Together with this result, we confirmed another unexpected phenomenon; the carrier-induced electrogenic influx of K^+ cations is considerably boosted when coupled with proton transporter like carbonyl cyanide-4-(trifluoromethoxy) phenylhydrazone (FCCP) [25].

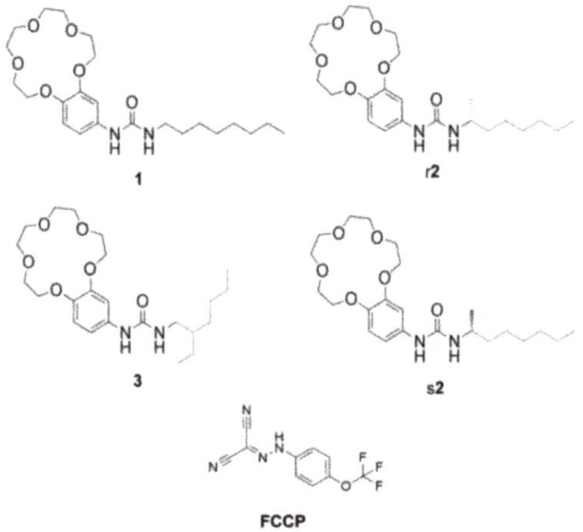

Figure 1. Crown-ether compounds **1, r2, s2, 3** and the H^+ transporter carbonyl cyanide-4-(trifluoromethoxy)phenylhydrazone (FCCP) reported in this article.

Simultaneous addition of **1, r2, s2, 3** and FCCP leads to the induction of K^+/H^+ antiport via the formation of an electrical potentials across the membrane. This coupling of K^+ and H^+ fluxes can proceed without a potential via non-electrogenic K^+/H^+ exchanges across the membrane. We observe an unprecedented increase in the selectivity [15,16] from $S_{K+/Na+}$ = 3.6 to 48.8 in the best case.

2. Materials and Methods

2.1. Materials

Octyl isocyanate, (*R*)-(−)-2-octyl isocyanate, (*S*)-(+)-2-octyl isocyanate, 2-ethylhexyl isocyanate, and FCCP were purchased from Sigma Aldrich, Paris, France. 8-hydroxypyrene-1,3,6-trisulfonic acid trisodium salt (HPTS) was purchased from Fluka, Paris, France. 4′-Aminobenzo-15-crown-5 was purchased from TCI Europe, Zwijndrecht, Belgium. L-α-Phosphatidylcholine EYPC was purchased from Avanti Polar Lipids, Alabaster, AL, USA and used as received.

2.2. Methods

^{1}H NMR spectra were recorded on an ARX 300 MHz Bruker Spectrometer (Bruker Daltonics, Billerica, MA, USA). Chemical shifts are reported as δ values (ppm) with corresponding deuterated solvent peak as an internal standard. Mass spectrometric analysis was performed in the positive ion mode using a quadrupole mass spectrometer (Micromass, Platform II spectrometer). Fluorescence spectra were recorded in Perkin Elmer LS-55 spectrometer (Perkin Elmer Inc., Waltham, MA, USA).

2.3. General Synthetic Procedure

All compounds have been synthesized as follows: 0.2 mmol 4'-aminobenzo-15-crown-5 are mixed with the corresponding 1.1 equiv. isocyanate in 10 mL of chloroform. The solution was stirred under reflux overnight. Solution was concentrated under rotary evaporator. Hexane was introduced to precipitate the product, after filtration, the filter cake was dried in the air and yields as final product, yields are around 65%.

1-(Benzo-15-crown-5-15-yl)-3-octylurea (**1**), ^{1}H NMR (DMSO-d_6, 300 MHz, 298 K) δ 8.19 (s, 1H), 7.13 (d, J = 2.1 Hz, 1H), 6.77 (dt, J = 8.6, 5.4 Hz, 2H), 6.00 (t, J = 5.6 Hz, 1H), 3.98 (dd, J = 8.6, 4.1 Hz, 4H), 3.74 (dd, J = 10.7, 6.0 Hz, 4H), 3.60 (s, 8H), 3.04 (dd, J = 12.6, 6.6 Hz, 2H), 1.40 (s, 2H), 1.26 (s, 10H), 0.86 (t, J = 6.6 Hz, 3H). ^{13}C NMR (75 MHz, CDCl$_3$, 298 K) δ 156.89, 149.56, 145.71, 132.33, 114.99, 114.49, 108.70, 70.89, 70.79, 70.42, 70.26, 69.61, 69.38, 68.55, 40.43, 31.81, 30.12, 29.31, 29.25, 26.93, 22.64, 14.09. ESI-MS Calcd. for C$_{23}$H$_{39}$N$_2$O$_6$ (M + H$^+$) 439.28, Found 439.38; Calcd. for C$_{23}$H$_{38}$N$_2$O$_6$Na (M + Na$^+$) 461.55, Found 461.35. IR (cm^{-1}): 3297, 2924, 2855, 1627, 1559, 1516, 1457, 1417, 1269, 1233, 1135, 985, 937, 847, 628. HR-MS (ESI+) calcd. for C$_{23}$H$_{39}$N$_2$O$_6$ 439.2808 found *m/z* 439.2814.

(R)-(Benzo-15-crown-5-15-yl)-3-(octan-2-yl)urea (**r2**), ^{1}H NMR (300 MHz, 298 K, DMSO-d_6) δ 8.08 (s, 1H), 7.14 (d, J = 2.3 Hz, 1H), 6.81 (d, J = 8.7 Hz, 1H), 6.73 (dd, J = 8.6, 2.3 Hz, 1H), 5.83 (d, J = 8.1 Hz, 1H), 3.98 (dd, J = 8.9, 4.5 Hz, 4H), 3.75 (dd, J = 10.8, 6.1 Hz, 4H), 3.60 (s, 9H), 1.31 (d, J = 29.3 Hz, 10H), 1.05 (d, J = 6.5 Hz, 3H), 0.86 (t, J = 6.6 Hz, 3H). ^{13}C NMR (75 MHz, 298 K, CDCl$_3$) δ 156.19, 149.77, 145.87, 132.69, 115.25, 114.69, 108.97, 71.09, 71.00, 70.62, 70.45, 69.79, 69.54, 68.77, 46.29, 37.46, 31.94, 29.36, 26.22, 22.74, 21.60, 14.22. ESI-MS Calcd. for C$_{23}$H$_{39}$N$_2$O$_6$ (M + H$^+$) 439.3, Found 439.3; Calcd. for C$_{23}$H$_{38}$N$_2$O$_6$Na (M + Na$^+$) 461.3, Found 461.3; Calcd for C$_{23}$H$_{38}$N$_2$O$_6$K (M + K$^+$) 477.2, Found 477.3. HR-MS (ESI+) calcd. for C$_{23}$H$_{39}$N$_2$O$_6$ 439.2808 found *m/z* 439.2813.

(S)-(Benzo-15-crown-5-15-yl)-3-(octan-2-yl)urea (**s2**), ^{1}H NMR (300 MHz, 298 K, DMSO-d_6) δ 8.09 (s, 1H), 7.14 (s, 1H), 6.81 (d, J = 8.7 Hz, 1H), 6.73 (d, J = 8.5 Hz, 1H), 5.84 (d, J = 8.1 Hz, 1H), 3.97 (s, 4H), 3.75 (s, 4H), 3.60 (s, 9H), 1.30 (d, J = 29.2 Hz, 10H), 1.05 (d, J = 6.5 Hz, 3H), 0.86 (t, J = 6.3 Hz, 3H). ^{13}C NMR (75 MHz, 298 K, CDCl$_3$) δ 156.09, 149.73, 145.63, 133.07, 115.34, 114.42, 108.74, 71.04, 70.94, 70.59, 70.41, 69.86, 69.78, 69.52, 68.74, 46.19, 37.49, 31.95, 29.38, 26.23, 22.74, 21.62, 14.22. ESI-MS Calcd. for C$_{23}$H$_{39}$N$_2$O$_6$ (M + H$^+$) 439.3, Found 439.4; Calcd. for C$_{23}$H$_{38}$N$_2$O$_6$K (M + K$^+$) 477.2, Found 477.3. HR-MS (ESI+) calcd. for C$_{23}$H$_{39}$N$_2$O$_6$ 439.2808 found *m/z* 439.2810.

1-(2-Ethylhexyl)-3-(benzo-15-crown-5-15-yl)urea (**3**), ^{1}H NMR (300 MHz, 298 K, DMSO-d_6) δ 8.18 (s, 1H), 7.13 (d, J = 2.1 Hz, 1H), 6.80 (d, J = 8.6 Hz, 1H), 6.74 (dd, J = 8.6, 2.1 Hz, 1H), 5.98 (t, J = 5.5 Hz, 1H), 4.07 – 3.93 (m, 4H), 3.74 (dd, J = 10.6, 5.9 Hz, 4H), 3.60 (s, 8H), 3.01 (d, J = 3.2 Hz, 2H), 1.40 – 1.13 (m, 9H), 0.86 (q, J = 7.0 Hz, 6H). ^{13}C NMR (75 MHz, 298 K, CDCl$_3$) δ 156.81, 149.72, 145.73, 132.87, 115.18, 114.47, 108.78, 71.06, 70.95, 70.61, 70.43, 69.78, 69.53, 68.72, 43.28, 39.77, 31.07, 29.00, 24.28, 23.18, 14.21, 10.99. ESI-MS Calcd. for C$_{23}$H$_{39}$N$_2$O$_6$ (M + H$^+$) 439.3, Found 439.3; Calcd for C$_{23}$H$_{38}$N$_2$O$_6$Na (M + Na$^+$) 461.3, Found 461.3; Calcd. for C$_{23}$H$_{38}$N$_2$O$_6$K (M + K$^+$) 477.2, Found 477.2. HR-MS (ESI+) calcd. for C$_{23}$H$_{39}$N$_2$O$_6$ 439.2808 found *m/z* 439.2809.

2.4. Lipid Bilayer Transport Experiments

2.4.1. LUV Preparation for HPTS Experiments

The large unilamellar vesicles (LUVs) were formed using egg yolk ʟ-α-phosphatidylcholine (EYPC chloroform solution, 800 μL, 20 mg). To this solution was added 800 μL of methanol MeOH and the solvent was slowly removed by evaporation under vacuum at room temperature and dried overnight under high vacuum. The resulting thin film was hydrated with 400 μL of buffer (10 mM sodium phosphate, pH 6.4, 100 mM NaCl) containing 10 μM HPTS. During hydration, the suspension was submitted to seven freeze-thaw cycles (liquid nitrogen, water at room temperature). The obtained white suspension was extruded 21 times through a 0.1 μm polycarbonate membrane in order to transform the large multilamellar liposome suspension (LMVs) into LUVs with an average diameter of 100 nm. The LUVs suspension was separated from extra-vesicular HPTS dye by using size exclusion chromatography (SEC, stationary phase Sephadex G-50, mobile phase: phosphate buffer with 100 mM NaCl) and diluted with mobile phase to give 2.8 mL of 11mM lipid stock solution (considering all the lipids have been incorporated) [15].

2.4.2. Cation Transport Experiments

The fast filter method was used for data collection. 100 μL of stock vesicle solution was suspended in 1.85 mL of the corresponding buffer (10 mM PBS pH 6.4 containing 100 mM of the analyzed cation) and placed into a quartz fluorimetric cell. The emission of HPTS at 510 nm was monitored at two excitation wavelengths (403 and 460 nm) simultaneously. 0.464 mM lipid was used in each experiment. The final concentration of the compound is 0.06 mM for 12.9 mol% of compound/lipid ratio. This concentration is high enough for crown ethers to assemble as channels in lipid bilayers, assuming the insertion is complete. At 20 s, 29 μL of aqueous NaOH (0.5 M) was added, and macrocyclic compound added at 50 s, detergent was added after 340 s, and the measurement lasted another 50 s. For experiments involving FCCP, the compound was added at 40 s, and the tested macrocyclic compound was added at 60 s. The lysis started at 350 s for 50 s to finish the measurements this time. Experiments without FCCP (from 50 s to 340 s) are compared to experiments with FCCP (from 60 s to 350 s). The extent of transport was monitored using the ratio of the emission intensities of HPTS at 460 and 403 nm (I_{460} and I_{403}). We calculated the first-order initial rate constant from the slopes of the plot of ln ($[H^+]_{in} - [H^+]_{out}$) versus time, where $[H^+]_{in}$ and $[H^+]_{out}$ are the intravesicular and the extravesicular proton concentrations, respectively. The $[H^+]_{out}$ was assumed to remain constant during the experiment (pH 7.4), while $[H^+]_{in}$ values were calculated for each point from HPTS emission intensity using the equation pH = $1.1684 \times \log(I_{460}/I_{403}) + 6.9807$. The initial and final pH are pH at $t = 0$ s and at the end of the experiment, respectively. In the picture, I refer to the intensity at 460 nm, and I_0 refers to intensity at 403 nm. To calculate EC_{50} and Hill coefficient n, we used the fractional activity Y. Y was calculated for each curve using the normalized value of I_{460}/I_{403} (just before lysis of the vesicles). We expressed Y as a function of time, and we performed fittings using a 2-parameter equation, which is, Hill equation: $Y = Y_{blank} + (Y_{max} - Y_{blank})/(1 + (EC_{50}/[C])^n)$, from which Y_{blank} is the fractional activity from the experiment of blank and Y_{max} is the largest fractional activity from the same series of experiments.

3. Results

Three octyl-benzoureido-15-crown-5-ethers have been prepared for the studies described here. The octyl-isocyanate (*R*)-(−)-2-octyl isocyanate, (*S*)-(+)-2-octyl isocyanate, 2-ethylhexyl isocyanate were treated with the corresponding 4-aminobenzo-15-crown-5-ether ($CHCl_3$, 65 °C, overnight) to afford after precipitation from hexane **1**, **r2**, **s2** and **3**, respectively, as powders (Scheme S1). ^1H-, ^{13}C-NMR, ESI-MS, and HR-MS spectroscopic data are in accord with proposed structures (see Supplementary Materials for details, Figures S1–S8).

The generation of functional transporting superstructures of **1**, **r2**, **s2** and **3** is based on three encoded features: (1) they contain macrocyclic cation-binding moieties [3–8]; (2) the supramolecular guiding interaction is the urea head-to-tail H-bond association, reminiscent with the amide H-bonding moiety in protein [9–15]; (3) the alkyl tails control the H-bonding self-assembly and induce variable hydrophobic stabilization at the interface with the bilayer membrane.

^{1}H NMR dilutions experiments on solutions in CDCl$_3$ of **1**, **r2**, **s2**, **3** show a variable downfield shift of both NH protons upon increasing the concentration, which is indicative of self-association through intermolecular H-bonding (Figure 2, Figures S9–S11). To quantify the degree of the aggregation, the NH shifts at different concentrations were fitted to the NMR CoEK aggregation model (Nelder–Mead method) using Bindfit [9]. The determined values of the aggregation constants, as well as the cooperativity factors, are given in Table 1. There is encouraging strong cooperative H-bonding effect. The most associated compound **1** contains the linear *n*-octyl sidearm while the formation of aggregates is hindered when crowded supplementary residues block the H-bonding and slightly deactivate (~25% decrease) the self-association in **r2**, **s2** and **3**.

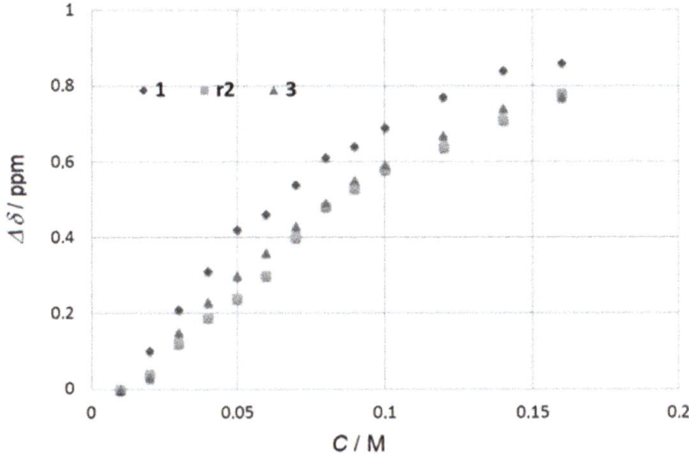

Figure 2. The difference in chemical shifts of the urea-type Ha protons relative to chemical shifts δ at 0.01 M plotted against the concentration of intermolecular H-bonded oligomers **1**, **r2**, and **3**, in CDCl$_3$ at 25 °C.

Table 1. Association Constants Ke for Numerical Fits a Monomer/Dimer/Aggregate model, obtained from NMR dilution experiments in CDCl$_3$ at 25 °C [9,13].

Compound	1	r2	3
K$_e$	19.3	15.3	14.6
ρ [a]	0.23	0.10	0.20
K$_e$ error (%)	2.64	3.89	2.30

[a] The factor ρ = K$_d$/K$_e$ highlight the cooperativity of aggregation with ρ < 1 positive cooperativity, ρ = 1, no cooperativity and ρ > 1 negative cooperativity for K$_e$ aggregation vs. K$_d$ dimerization.

The ion-transport activities were evaluated by the HPTS assay [24,25]. L-α-phosphatidylcholine, EYPC liposomes (Large Unilamellar Vesicles-LUV, 100 nm) were filled with a pH-sensitive dye, HPTS and 100 mM NaCl in a phosphate buffer (10 mM, pH 6.4). The liposomes were then suspended in an external phosphate buffer (10 mM, pH 6.4) containing 100 mM of MCl, M$^+$ = Na$^+$, K$^+$. Then, after addition of **1**, **r2**, **s2**, **3** in the bilayer membrane via injection of 20 µL of compound aliquots from stock 10 mM Dimethylsulfoxide (DMSO) solutions, see the final concentration values (% mol of the compound/mol of lipid) in hill plot analyses in Figures S12–S25), an external pH gradient was created by addition of NaOH. The internal pH change inside the liposome was monitored by the change in the fluorescence of HPTS (Figure 3a,b) [26].

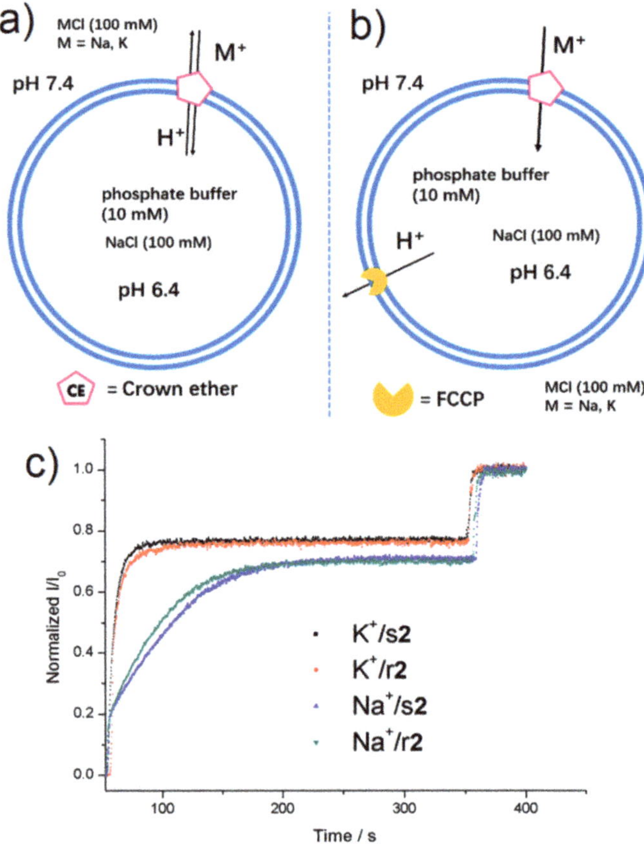

Figure 3. Schematic representation of the translocation mechanism in the presence of macrocycles (**a**) without or (**b**) with FCCP proton transporter. (**c**) Normalized I_{460}/I_{403} (I/I_0) versus time profiles for the transport of K^+ or Na^+ across the bilayer membrane promoted by **r2** or **s2** (32.3 mol%).

Compounds **1**, **r2**, **s2**, and **3** are presenting subtle variations on the transport activities of monovalent K^+ and Na^+ cations, depending the structure of isomeric octyl substituents of ureido-benzo-15-crown-5 macrocycle. We note that enantiomers **s2** and **r2** have rather same activity for translocation of K^+ and Na^+ at the same concentration (Figure 3c). Under the same conditions, compounds **1**, **r2**, **s2** and **3** (12.9 mol%) present one order of magnitude higher initial transport rate for K^+ cations than for Na^+, with kinetic selectivity of $S_{K+/Na+}$ = 3 to 6.6, depending on the compound (Tables S1 and S2 and Figure 4a).

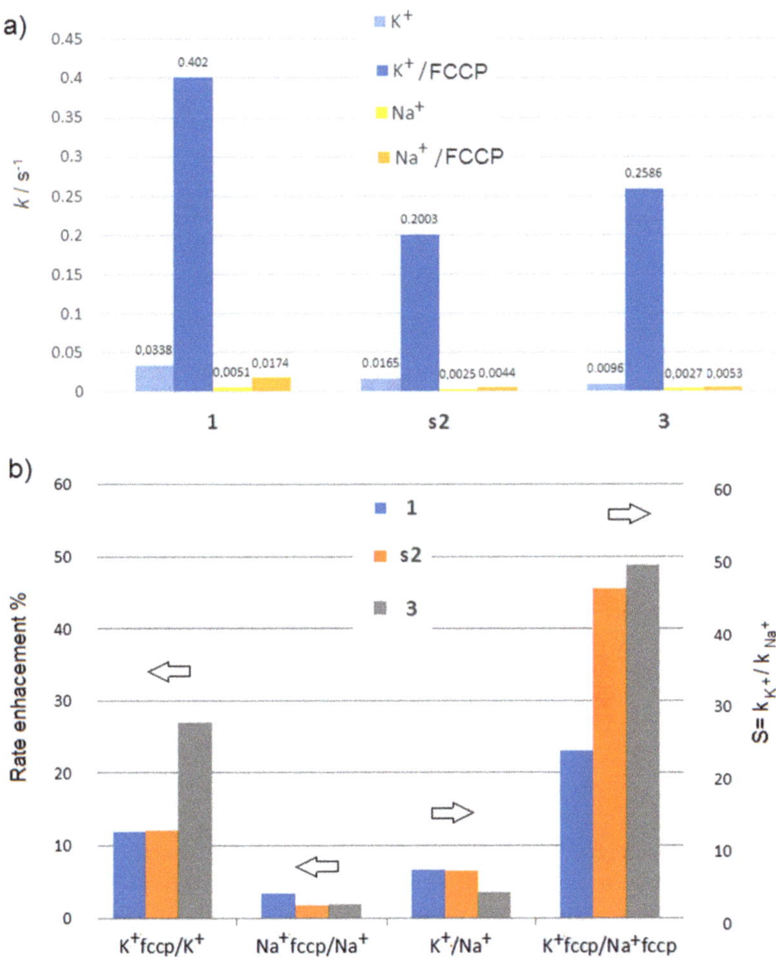

Figure 4. Bar graphs showing (**a**) the pseudo-first-order rate constants k (s^{-1}) for the transport of K$^+$ and Na$^+$ cations and (**b**) rate enhancement and K$^+$/Na$^+$ selectivity, for the macrocyclic carriers **1**, **s2**, **3**, (12.9 mol%) in presence or absence of protogenic FCCP carrier.

Hill analysis [27] (Table 2, Figures S12–S25) revealed all macrocycles present K$^+$ over Na$^+$ selectivity, which is consistent with previous results. [8,15] Compound **1** is the most active, as it has the lowest EC$_{50}$ for both Na$^+$ (10.6 mol%), and K$^+$ (4.1 mol%), much lower than its isomers following the transport activity sequence of **1** > **s2** > **3** within the µM concentration range. Compared with Valinomicin (EC$_{50K+}$ = 5 pM) they are several orders of magnitude for K$^+$ activity. All the Hill coefficients are above 2, indicative of a formation of self-assembled aggregates containing more than two molecules of macrocycles, which transport cooperatively the cations, contrarily to Valinomycin acting as a carrier with a Hill number = 1 [27].

Table 2. Hill analysis results of K^+ and Na^+ transport with crown ethers **1, r2, s2, 3**, in the presence or absence of FCCP. EC_{50} values expressed as mol% (% molar of compound/lipid needed to obtain 50% ion transport activity) and n is the Hill coefficient.

	K^+		K^+/FCCP		Na^+		Na^+/FCCP	
	EC_{50} [a] (mol%)	n [b]	EC_{50} [a] (mol%)	n	EC_{50} (mol%)	n	EC_{50} (mol%)	n
1	4.1	2.4	1.9	2.2	10.9	5.3	7.1	3.8
r2	/	/	3.0	3.2	/	/	10.7	4.1
s2	6.4	3.3	3.0	2.9	14.9	3.4	11.2	4.6
3	8.3	3.6	1.8	2.9	15.2	5.1	10.1	3.8

[a] EC_{50} was determined by the Hill plot, using the fractional activities at 340 s–290 s after the addition of the compound; [b] Hill coefficient.

According to our previous results, the addition of alkyl-benzoureido-15-crown-5-ether channels in the membrane generate a quite stable transmembrane potential, following a $K^+ > H^+$ electrogenic antiport through the lipid bilayer [15]. We also know that the simultaneous addition of macrocycles and FCCP leads to the electrical coupling of fluxes and the transport can proceed without a potential across the membrane via a non-electrogenic exchange [28,29].

As can be seen from Figure 4, the electrogenic component of the macrocyclic compound-mediated cation influx is stimulated by the addition of 0.1 mol% FCCP proton carrier as well, which overcome the proton transport rate-limiting barrier with a special emphasis for the transport of K^+ cations for which the transport rate show a 12 to 27 fold increase. This result is valid for all studied macrocycles **1, r2, s2, 3**, for which the fractional activities Y for K^+ cations are also higher. This is also obvious from the EC_{50} values, which are strongly improving for K^+ but not for Na^+. FCCP improves transport efficiency, but more impressively, we observe an increase of S_{K^+/Na^+} selectivity from 6.6 to 45.5 in the case of **2** and from 3.6 to 48.8 in the case of **3**.

For all compounds, FCCP did improve transport activity towards Na^+, although not as much as K^+, and the rate of transport follows the sequence: $K^+ > Na^+ > H^+$. As far as we know, simple artificial systems presenting so high K^+/Na^+ selectivity are rare. It can be explained by several reasons: (i) The fact that these macrocycles can achieve this type of stable ionic potential, suited to drive protonic gradients, is simply due to their high selectivity for K^+. We know from our previous studies that similar compounds are selective for the transport of K^+ cations, even in the presence of Na^+ cations. [15] (ii) the formation of complexes is highly controlled via the formation of carrier dimers (benzo-15-crown-5-ether$_2$$K^+$) or higher oligomeric channels (benzo-15-crown-5-ether$_n$$K^+$). [13] Log P = 3.63–3.66 values, which are often used for evaluating the lipophilicity of the compound, are quite similar for all studied compounds **1, r2, s2, 3**, thus, their partitions within the membrane might be similar (Table 3). Within this context, the branched alkyl chains might have a negative impact on the formation of the aggregates inside the lipid bilayer. Branched alkyl tails may favor the presence of low dimensional self-assembles species of **r2, s2,** or **3**. They show lower permeability than **1** (Figure 4a), but they lead to an increased S_{K^+/Na^+} selectivity via proton-mediated electrogenic conductance across the membrane (Figure 4b).

Table 3. Calculated logP (cLogP) values were obtained using VCC labs online calculator ALOGPS 2.1 to assess the lipophilicity of the compounds [30].

Compounds	logP
1	3.66
r2	3.64
s2	3.64
3	3.63

4. Conclusions

In conclusion, we demonstrated that macrocycles containing benzoureido-15-crown-5-ether cation-binding head and bearing linear or branched alkyl side-chains with almost identical lipophilicity, present completely different transport activities for translocations of cations across the lipid bilayer. The isomeric octyl-benzoureido-15-crown-5-ethers **1**, **r2**, **s2**, **3** described here, are very intriguing electrogenic macrocycles presenting K^+ over Na^+ kinetic selectivity of $S_{K+/Na+} = 3$ to 6.6. Specifically, we have demonstrated that simple structural variation from linear to branched octyl chains are strongly influencing the transport activities of K^+ cations when compared with Na^+ cations. The electrogenic macrocyclic compound-mediated K^+ influx is highly increased up to 27 fold by the addition of FCCP proton carrier with a special emphasis for the transport of K^+ cations, for which the S_{K^+/Na^+} selectivity, is increasing from 6.6 to 45.5 in the case of **s2** and from 3.6 to 48.8 in the case of **3**. This is a significant step forward toward the development of electrogenic macrocycles with high K^+/Na^+ selectivity.

Supplementary Materials: The following are available online at http://www.mdpi.com/2624-8549/2/1/3/s1, Scheme S1: Synthesis of compounds **1**, **r**, **s2**, **3**, Figure S1: ^{1}H NMR spectrum of **1** (300 MHz, 298K, DMSO-d6), Figure S2: ^{13}C NMR spectrum of **1** (75 MHz, 298K DMSO-d6), Figure S3 ^{1}H NMR spectrum of **r2** (300 MHz, 298 K, DMSO-d6), Figure S4 ^{13}C NMR spectrum of **r2** (75 MHz, 298 K, CDCl3), Figure S5 ^{1}H NMR spectrum of **s2** (300 MHz, 298 K, DMSO-d_6), Figure S6 ^{13}C NMR spectrum of **s2** (75 MHz, 298 K, CDCl$_3$), Figure S7 ^{1}H NMR spectrum of **3** (300 MHz, 298 K, DMSO-d_6), Figure S8 ^{13}C NMR spectrum of **3** (75 MHz, 298 K, CDCl$_3$), Figure S9 ^{1}H NMR dilution experiments in CDCl$_3$ of **1** (0.01 to 0.16 M, from top to bottom), Figure S10 ^{1}H NMR dilution experiments in CDCl$_3$ of **r2** (0.01 to 0.16 M, from top to bottom), Figure S11 ^{1}H NMR dilution experiments in CDCl$_3$ of **3** (0.01 to 0.16 M, from top to bottom), Figure S12. Normalized I_{460}/I_{403} for transporting K^+ across the bilayer membrane facilitated by different amount of **1** and hill plot analysis of K^+/H^+ antiport, Figure S13. Normalized I_{460}/I_{403} for transporting Na^+ across the bilayer membrane facilitated by different amount of **1** and hill plot analysis of Na^+/H^+ antiport, Figure S14 Normalized I_{460}/I_{403} for transporting K^+ across the bilayer membrane facilitated by different amount of **1** coupled with FCCP (0.1 mol%) and hill plot analysis of K^+/H^+ antiport, Figure S15 Normalized I_{460}/I_{403} for transporting Na^+ across the bilayer membrane facilitated by different amount of **1** coupled with FCCP (0.1 mol%) and hill plot analysis of Na^+/H^+ antiport, Figure S16 Normalized I_{460}/I_{403} for transporting K^+ across the bilayer membrane facilitated by different amount of **s2** and Hill plot analysis of K^+/H^+ antiport, Figure S17 Normalized I_{460}/I_{403} for transporting Na^+ across the bilayer membrane facilitated by different amount of **s2** and hill plot analysis of Na^+/H^+ antiport, Figure S18 Normalized I_{460}/I_{403} for transporting K^+ across the bilayer membrane facilitated by different amount of **s2** coupled with FCCP (0.1 mol%) and hill plot analysis of K^+/H^+ antiport, Figure S19 Normalized I_{460}/I_{403} for transporting Na^+ across the bilayer membrane facilitated by different amount of **s2** coupled with FCCP (0.1 mol%) and hill plot analysis of Na^+/H^+ antiport, Figure S20 Normalized I_{460}/I_{403} for transporting K^+ across the bilayer membrane facilitated by different amount of **r2** coupled with FCCP (0.1 mol%) and hill plot analysis of K^+/H^+ antiport, Figure S21 Normalized I_{460}/I_{403} for transporting Na^+ across the bilayer membrane facilitated by different amount of **r2** coupled with FCCP (0.1 mol%) and hill plot analysis of Na^+/H^+ antiport, Figure S22 Normalized I_{460}/I_{403} for transporting K^+ across the bilayer membrane facilitated by different amount of **3** and hill plot analysis of K^+/H^+ antiport, Figure S23 Normalized I_{460}/I_{403} for transporting Na^+ across the bilayer membrane facilitated by different amount of **3** and hill plot analysis of Na^+/H^+ antiport, Figure S24 Normalized I_{460}/I_{403} for transporting K^+ across the bilayer membrane facilitated by different amount of **3** coupled with FCCP (0.1 mol%) and hill plot analysis of K^+/H^+ antiport, Figure S25 Normalized I_{460}/I_{403} for transporting Na^+ across the bilayer membrane facilitated by different amount of **3** coupled with FCCP (0.1 mol%) and hill plot analysis of Na^+/H^+ antiport, Figure S26. HR-MS spectra of compounds **1**, **r2**, **s2**, **3**, Table S1 Pseudo first-order rate constants k (s^{-1}) for the transport of K^+/H^+ through LUVs at different concentrations of the compounds to lipid without or with proton transporter FCCP, Table S2 Pseudo first-order rate constants k (s^{-1}) for the transport of Na^+/H^+ through LUVs at different concentrations of the compounds to lipid without or with proton transporter FCCP. The initial rate for the blank has already been subtracted from all the rates.

Author Contributions: Conceptualization—M.B. Synthesis, Cation Transport—Y.-H.L. and S.-P.Z.; Investigation—D.W.; Writing—Original Draft Preparation—Y.-H.L. and M.B.; Supervision—M.B.; Funding Acquisition—M.B. All authors have read and agreed to the published version of the manuscript.

Funding: This work was also conducted within ANR-18-CE06-0004-02, WATERCHANNELS, and 1000 Talent Plan, WQ20144400255 of SAFEA, China. Y.-H.L. wishes to thank the Innovation and Talent Introduction Base of Photoelectronic and Functional Molecular Solids Material (Grant No. 90002-18011002) from Sun Yat-Sen University, and also the China Scholarship Council (No. 201606380054) for the financial support. NSFC (National Natural Science Foundation of China, 21720102007), China. S.-P.Z. wishes to thank the China Scholarship Council for financial support.

Conflicts of Interest: The authors declare no conflict of interest. The funders had no role in the design of the study; in the collection, analyses, or interpretation of data; in the writing of the manuscript, or in the decision to publish the results.

References

1. Hille, B. *Ion Channels of Excitable Membranes*, 3rd ed.; Sinauer Associates: Sunderland, MA, USA, 2001.
2. Sakai, N.; Gerard, D.; Matile, S. Electrostatics of Cell Membrane Recognition: Structure and Activity of Neutral and Cationic Rigid Push-Pull Rods in Isoelectric, Anionic, and Polarized Lipid Bilayer Membranes. *J. Am. Chem. Soc.* **2001**, *123*, 2517–2524. [CrossRef]
3. Barboiu, M. Supramolecular Polymeric Macrocyclic Receptors—Hybrid Carrier vs. Channel Transporters in Bulk Liquid Membranes. *J. Incl. Phenom. Macrocycl. Chem.* **2004**, *49*, 133–137. [CrossRef]
4. Otis, F.; Racine-Berthiaume, C.; Voyer, N. How Far Can a Sodium Ion Travel within a Lipid Bilayer? *J. Am. Chem. Soc.* **2011**, *133*, 6481–6483. [CrossRef]
5. Otis, F.; Auger, M.; Voyer, N. Exploiting Peptide Nanostructures to Construct Functional Artificial Ion Channels. *Acc. Chem. Res.* **2013**, *46*, 2934–2943. [CrossRef]
6. Weber, M.E.; Schlesinger, P.H.; Gokel, G.W. Dynamic Assessment of Bilayer Thickness by Varying Phospholipid and Hydraphile Synthetic Channel Chain Lengths. *J. Am. Chem. Soc.* **2005**, *127*, 636–642. [CrossRef] [PubMed]
7. Gokel, G.W.; Negin, S. Synthetic membrane active amphiphiles. *Adv. Drug Deliv. Rev.* **2012**, *64*, 784–796. [CrossRef] [PubMed]
8. Ren, C.; Shen, J.; Zeng, H. Combinatorial Evolution of Fast-Conducting Highly Selective K^+-Channels via Modularly Tunable Directional Assembly of Crown Ethers. *J. Am. Chem. Soc.* **2017**, *139*, 12338–12341. [CrossRef] [PubMed]
9. Webb, J.E.A.; Crossley, M.J.; Turner, P.; Thordarson, P. Pyromellitamide Aggregates and Their Response to Anion Stimuli. *J. Am. Chem. Soc.* **2007**, *129*, 7155–7162. [CrossRef] [PubMed]
10. Zheng, S.P.; Li, Y.H.; Jiang, J.J.; van der Lee, A.; Dumitrescu, D.; Barboiu, M. Self-Assembled Columnar Triazole-Quartets—An example of synergetic H-bonding/Anion-π Channels. *Angew. Chem. Int. Ed.* **2019**, *58*, 12037–12042. [CrossRef] [PubMed]
11. Barboiu, M.; Cerneaux, S.; Van der Lee, A.; Vaughan, G. Ion-driven ATP-pump by Self-Organized Hybrid Membrane Materials. *J. Am. Chem. Soc.* **2004**, *126*, 3545–3550. [CrossRef] [PubMed]
12. Mihai, S.; Cazacu, A.; Arnal-Herault, C.; Nasr, G.; Meffre, A.; van der Lee, A.; Barboiu, M. Supramolecular self-organization in constitutional hybrid materials. *New J. Chem.* **2009**, *33*, 2335–2343. [CrossRef]
13. Cazacu, A.; Tong, C.; van der Lee, A.; Fyles, T.M.; Barboiu, M. Columnar Self-Assembled Ureidocrown-ethers—An Example of Ion-channel Organization in Lipid Bilayers. *J. Am. Chem. Soc.* **2006**, *128*, 9541–9548. [CrossRef]
14. Cazacu, A.; Legrand, Y.M.; Pasc, A.; Nasr, G.; Van der Lee, A.; Mahon, E.; Barboiu, M. Dynamic hybrid materials for constitutional selective membranes. *Proc. Natl. Acad. Sci. USA* **2009**, *106*, 8117–8122. [CrossRef] [PubMed]
15. Gilles, A.; Barboiu, M. Highly Selective Artificial K^+ Channels: An Example of Selectivity- Induced Transmembrane Potential. *J. Am. Chem. Soc.* **2016**, *138*, 426–432. [CrossRef] [PubMed]
16. Sun, Z.; Barboiu, M.; Legrand, Y.M.; Petit, E.; Rotaru, A. Selective Artificial Cholesteryl Crown Ether K^+-Channels. *Angew. Chem. Int. Ed.* **2015**, *54*, 14473–14477. [CrossRef] [PubMed]
17. Sun, Z.; Gilles, A.; Kocsis, I.; Legrand, Y.M.; Petit, E.; Barboiu, M. Squalene Crown-Ether Self-assembled Conjugates—An example of highly selective artificial K^+-channels. *Chem. Eur. J.* **2016**, *22*, 2158–2164. [CrossRef]

18. Feng, W.X.; Sun, Z.; Zhang, Y.; Legrand, Y.M.; Petit, E.; Su, C.Y.; Barboiu, M. Bis-15-Crown-5-Ether-Pillar[5]arene K$^+$-Responsive Channels. *Org. Lett.* **2017**, *19*, 1438–1441. [CrossRef]
19. Schneider, S.; Licsandru, E.D.; Kocsis, I.; Gilles, A.; Dumitru, F.; Moulin, E.; Tan, J.J.; Lehn, J.M.; Giuseppone, N.; Barboiu, M. Columnar Self-Assemblies of Triarylamines as Scaffolds for Artificial Biomimetic Channels for Ion and for Water Transport. *J. Am. Chem. Soc.* **2017**, *139*, 3721–3727. [CrossRef]
20. Spooner, M.J.; Gale, P.A. Anion transport across varying lipid membranes—The effect of lipophilicity. *Chem. Commun.* **2015**, *51*, 4883–4886. [CrossRef]
21. Knight, N.J.; Hernando, E.; Haynes, C.J.E.; Busschaert, N.; Clarke, H.J.; Takimoto, K.; Garcia-Valverde, M.; Frey, J.G.; Quesada, R.; Gale, P.A. QSAR analysis of substituent effects on tambjamine anion transporters. *Chem. Sci.* **2016**, *7*, 1600–1608. [CrossRef]
22. Saggiomo, V.; Otto, S.; Marques, I.; Felix, V.; Torroba, T.; Quesada, R. The role of lipophilicity in transmembrane anion transport. *Chem. Commun.* **2012**, *48*, 5274–5276. [CrossRef] [PubMed]
23. Li, Z.; Deng, L.Q.; Chen, J.X.; Zhou, C.Q.; Chen, W.H. Does lipophilicity affect the effectiveness of a transmembrane anion transporter? Insight from squaramido-functionalized bis(choloyl) conjugates. *Org. Biomol. Chem.* **2015**, *13*, 11761–11769. [CrossRef] [PubMed]
24. Berezin, S.K.; Davis, J.T. Catechols as Membrane Anion Transporters. *J. Am. Chem. Soc.* **2009**, *131*, 2458–2459. [CrossRef] [PubMed]
25. Li, Y.H.; Zheng, S.P.; Legrand, Y.M.; Gilles, A.; Van der Lee, A.; Barboiu, M. Structure-driven selection of adaptive transmembrane Na$^+$ carriers or K$^+$ channels. *Angew. Chem. Int. Ed.* **2018**, *57*, 10520–10524. [CrossRef] [PubMed]
26. Matile, S.; Sakai, N. *Analytical Methods in Supramolecular Chemistry*; Schalley, C.A., Ed.; Wiley-VCH: Weinheim, Germany, 2007; pp. 381–418.
27. Bhosale, S.; Matile, S. A Simple Method to Identify Supramolecules in Action: Hill Coefficients for Exergonic Self-Assembly. *Chirality* **2006**, *18*, 849–856. [CrossRef] [PubMed]
28. Valkenier, H.; Haynes, C.J.E.; Herniman, J.; Gale, P.A.; Davis, A.P. Lipophilic balance—A new design principle for transmembrane anion carriers. *Chem. Sci.* **2014**, *5*, 1128–1134. [CrossRef]
29. Wu, X.; Judd, L.W.; Howe, E.N.W.; Withecombe, A.M.; Soto-Cerrato, V.; Li, H.; Busschaert, N.; Valkenier, H.; Pérez-Tomás, R.; Sheppard, D.N.; et al. Nonprotonophoric Electrogenic Cl$^-$ Transport Mediated by Valinomycin-like Carriers. *Chem* **2016**, *1*, 127–146. [CrossRef]
30. Varnek, A.; Gaudin, C.; Marcou, G.; Baskin, I.; Pandey, A.K.; Tetko, I.V. Inductive Transfer of Knowledge: Application of Multi-Task Learning and Feature Net Approaches to Model Tissue-Air Partition Coefficients. *J. Chem. Inf. Model.* **2009**, *49*, 133–144. [CrossRef]

 *chemistry*

Communication

A Brucite-Like Mixed-Valent Cluster Capped by [Mn^{III}p-^{t}Bu-calix[4]arene]$^-$ Moieties

Marco Coletta [1], Maria A. Palacios [1,†], Euan K. Brechin [1,*] and Scott J. Dalgarno [2,*]

[1] EaStCHEM School of Chemistry, University of Edinburgh, Joseph Black Building, David Brewster Road, Edinburgh EH9 3FJ, UK; marco.coletta@ed.ac.uk (M.C.); mpalacios@ugr.es (M.A.P.)

[2] Institute of Chemical Sciences, School of Engineering and Physical Sciences, Heriot-Watt University, Riccarton, Edinburgh EH14 4AS, UK

* Correspondence: E.Brechin@ed.ac.uk (E.K.B.); S.J.Dalgarno@hw.ac.uk (S.J.D.); Tel.: +44-131-650-7545 (E.K.B.); +44-131-451-8025 (S.J.D.)

† Current address: Departamento de Química Inorgánica, Facultad de Ciencias, Universidad de Granada, 18071 Granada, Spain.

Received: 30 March 2020; Accepted: 8 April 2020; Published: 9 April 2020

Abstract: p-^{t}Bu-calix[4]arene (H$_4$TBC[4]) has proven to be an incredibly versatile ligand for the synthesis of 3d- and 3d/4f- clusters, in particular those containing mixed-valent Mn ions. These are of interest to the magnetochemist for the diversity of magnetic behaviours that can be shown, along with a huge variety of nuclearities and topologies accessible, which allow one to outline magneto-structural correlations and a quantitative understanding of their properties. This contribution reports the synthesis, analysis and magnetic properties of a Brucite-like Mn-oxo/hydroxo octanuclear fragment encapsulated within/capped by four [Mn^{III}-TBC[4]]$^-$ moieties. A diol coligand in the reaction mixture plays a seemingly important role in determining the outcome, though it is not incorporated in the final structure.

Keywords: calixarenes; coordination clusters; manganese; molecular magnetism

---

1. Introduction

Manganese continues to play a prominent role in the chemistry of 3d transition metals, owing to its significance across a breadth of research areas, including bioinorganic [1] and biomedicinal chemistry [2], catalysis [3], nanomaterials [4], spectroscopy [5], and molecular magnetism [6]. In the latter category, the ability of the Mn ion to exist in a variety of stable oxidation states (II-IV) allows for the construction of polymetallic cluster compounds exhibiting a variety of interesting magnetic behaviours, including the stabilization of large spin ground states [7], the slow relaxation of magnetization [8], spin frustration [9], vibrational coherences [10], and enhanced magnetocaloric effects [11]. A key component in understanding the physical properties of all Mn-based molecular magnets is the construction of large families of related compounds so that structure-magnetism relationships can be quantitatively rationalised, and this requires the careful design and exploitation of specific organic bridging ligands.

We have been exploring the coordination chemistry of calix[n]arenes (C[n]s) with Mn (as well as other metals), as these molecules hold the potential to isolate coordination clusters in the solid state in various different ways, for example, by exploiting the wedge shape of p-^{t}Bu-calix[4]arene (H$_4$TBC[4]). As can be seen from the acetonitrile (CH$_3$CN) solvate of H$_4$TBC[4] (Figure 1a) [12], the shape of the building block exerts strong influence over assembly and typically results in antiparallel bilayer formation in the solid state. The hydrophobic cavities are offset in this case and are occupied by acetonitrile of crystallization, though the same phenomenon is also observed for other solvates such as dmf [13]. With respect to cluster formation, p-^{t}Bu-calix[4]arene (H$_4$TBC[4]) has proven to be a particularly versatile platform for the synthesis of a wide range of different topologies, with nuclearities reaching up to fourteen [13–16]. A recurring structural theme we have noticed in this work is that the [Mn[III]TBC[4]][−] moieties act as bridges to metal ions within the cluster through their phenolate groups (Figure 1b), but also as polyhedral capping units (Figure 1c) [17]. The latter suggests that these can be used to encapsulate small metal-oxo-hyroxo fragments growing in 2- or 3-dimensions. Thia, sulfonyl and sulfinyl calix[4]arenes have also been employed in this way, though they give access to markedly different topologies due to the presence of donor atoms at the bridge positions [18–23]. In all cases, new species isolated with methylene- or heteroatom-bridged calix[4]arenes would be of particular interest to magnetochemists researching topological spin frustration [24]. Herein, we discuss the synthesis, structure and magnetic behaviour of a mixed-valent [Mn[III]8Mn[II]4] species built with TBC[4], the core metallic skeleton of which is related to the hydroxide-based mineral Brucite.

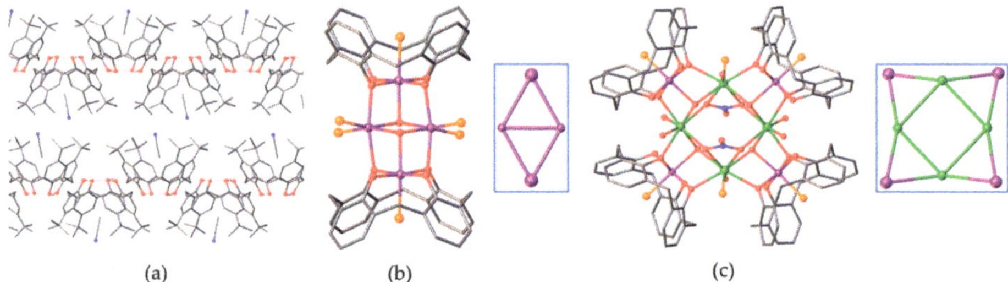

(a)   (b)   (c)

Figure 1. (**a**) Section of the extended structure found in the CH$_3$CN solvate of H$_4$TBC[4], showing the antiparallel bi-layer assembly and offset head-to-head arrangement of the host cavities [12]. (**b**) Partial single crystal X-ray structure of a mixed-valence TBC[4]-supported manganese cluster containing a butterfly-like [Mn[III]$_2$Mn[II]$_2$(OH)$_2$] core [13]. Mn[III] ions occupy the tetraphenolato pocket of the TBC[4] and act as capping units (shown as larger spheres in the inset diagram). (**c**) Partial single crystal X-ray structure of a C[4]-supported 3*d*-4*f* cluster topology that can be isolated with a range of lanthanides [14]. Analogous capping behaviour to that found in (**b**) is also observed here and is represented by the larger spheres in the inset diagram. Colour code: Mn[III]—purple; Ln[III]—green; O—red; C—grey; N—blue; ligated solvent—orange. ^{t}Bu groups, hydrogen atoms, and solvent of crystallization omitted for clarity.

2. Results and Discussion

2.1. Synthesis and Structural Studies

Reaction of H$_4$TBC[4] with MnCl$_2$·6H$_2$O, H$_2$bd (1,4-butanediol), [NH$_4$]ClO$_4$, and NEt$_3$ in a mixture of CH$_3$CN and dmf, followed by vapour diffusion of diethyl ether into the mother liquor, afforded single crystals suitable for X-ray diffraction of formula [Mn[III]$_8$Mn[II]$_4$(μ_4-O)$_2$(μ_3-OH)$_6$(μ-OH)$_4$(μ-Cl)$_2$(TBC[4])$_4$(dmf)$_8$(H$_2$O)$_4$(CH$_3$CN)$_2$]·2dmf·6CH$_3$CN (**1**, Figure 2, please also see Appendix A). The crystals were found to be in a triclinic cell, and structure solution was carried out in the space group *P*-1, with the asymmetric unit (ASU) comprising half of the cluster (Mn1–Mn6, Figure 3). Before moving to the structural description of **1**, it is interesting to note that inspection of Figure 2 shows that 1,4-butanediol is not incorporated in the prevailing structure. Our original intention was to form a heteroleptic TBC[4]-diol cage, and although 1,4-butanediol is not incorporated, its presence in the reaction mixture is required for **1** to form. Indeed, analogous experiments in the absence of diol result in the formation of the well-known TBC[4]-supported [Mn[III]$_2$Mn[II]$_2$] cluster topology shown in Figure 1B; this particular cluster can be isolated quantitatively in under 1 h, further indicating that the diol plays a crucial role in the formation of **1**.

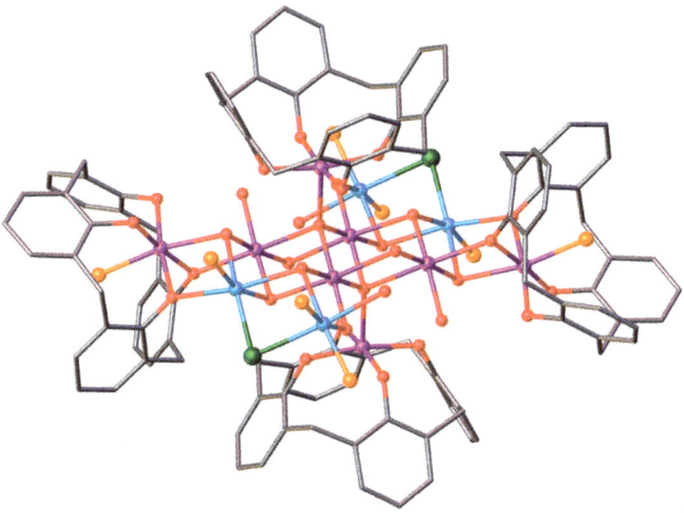

Figure 2. Partial single crystal X-ray structure of **1**. Colour code: Mn^{III}—purple; Mn^{II}—pale blue; O—red; C—grey; Cl—green; ligated dmf—orange sphere. ^{t}Bu groups, hydrogen atoms, solvent of crystallization and atoms other than oxygen of ligated dmf are omitted for clarity.

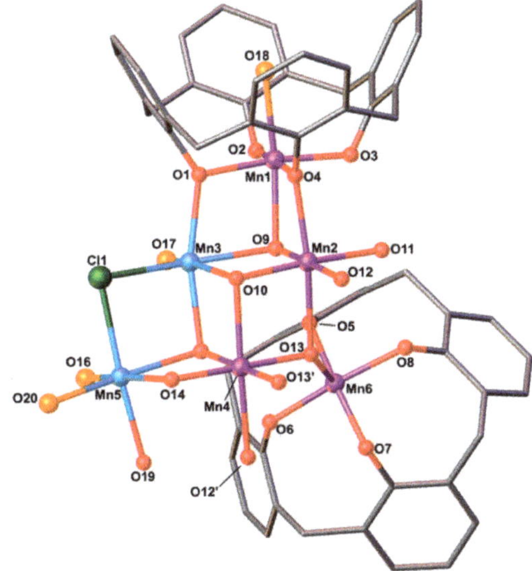

Figure 3. Partial single crystal X-ray structure of **1** showing the asymmetric unit with selected labels shown. Colour code: Mn^{III}—purple; Mn^{II}—pale blue; O—red; C—grey; Cl—green; dmf—orange. ^{t}Bu groups, hydrogen atoms, and solvent of crystallization omitted for clarity. Two s.e. atoms (O12′ and O13′) are included in order to show the octahedral coordination sphere around Mn4.

The central metal–oxygen core of **1** describes a near-planar, mixed-valent [Mn$^{III}_6$Mn$^{II}_4$(μ_4-O)$_2$(μ_3-OH)$_6$(μ-OH)$_4$(Cl)$_2$] sheet-like structure (Figures 2 and 4). The Mn^{III} ions in **1** (Mn1, Mn2, Mn4, and symmetry equivalent, s.e.) are easily distinguished through the presence of coparallel Jahn-Teller (JT) axes, oriented approximately 40° from the plane of the metal ions. The two Cl ions (Cl1 and s.e.) bridge between neighbouring Mn^{II} ions (Mn3-Cl-Mn5 and s.e.,

86.21°). Mn1 (and s.e.) is bound in a TBC[4] lower-rim tetraphenolato pocket (Mn1-O1-4, with bond distances in the range of 1.882(2)–1.973(2) Å), with its distorted octahedral geometry completed by a ligated dmf molecule residing within the TBC[4] cavity (Mn1–O18, 2.271(3) Å), and a μ_3-OH (Mn1–O9, 2.216(2) Å) that also bridges Mn2 and Mn3 (Mn2–O9, 1.931(2) Å and Mn3–O9, 2.209(3) Å). The remaining Mn^{III} ion (Mn6 and s.e.) is also bound within a TBC[4] lower-rim tetraphenolato pocket (Mn6-O5-8, with bond distances in the range of 1.930(2)–1.948(2) Å) sitting above/below the [Mn$_{10}$] plane (Figure 2), bonded to Mn2 and Mn4 (and s.e.) through a μ_4-O^{2-} ion (Mn6–O13, 2.144(2) Å). Each of these O-atoms is H-bonded to a bridging hydroxide in the [Mn$_{10}$] plane (O⋯O, 2.643–2.870 Å).

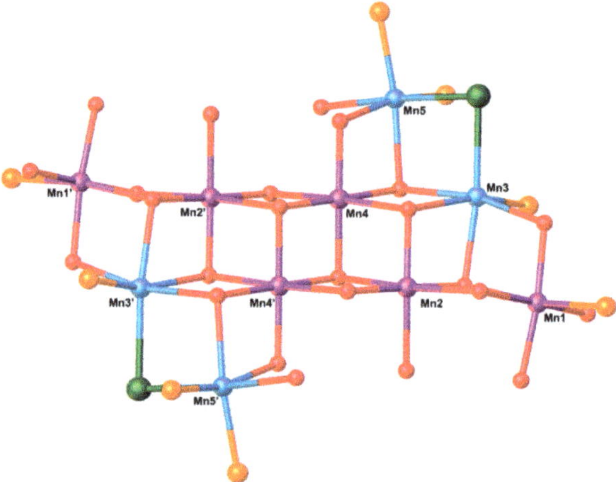

Figure 4. The near-planar, mixed-valent [Mn^{III}$_6$Mn^{II}$_4$(μ_4-O)$_2$(μ_3-OH)$_6$(μ-OH)$_4$(Cl)$_2$] sheet-like structure at the core of complex **1**. Colour code: Mn^{III}—purple; Mn^{II}—pale blue; O – red; Cl—green; dmf—orange. Carbon, nitrogen and hydrogen atoms, as well as solvent of crystallization omitted for clarity.

Interestingly, Mn6 (and s.e.) is formally five-coordinate and in square-pyramidal geometry (Figure 3), with the CH$_3$CN molecule occupying the TBC[4] cavity and interacting with the Ph rings through CH⋯π interactions. We have observed such behaviour in other TBC[4]-supported Mn cages [25], and this is also reminiscent of the host-guest chemistry found in the CH$_3$CN solvate of TBC[4] (Figure 1a) [12]. The preference of the TBC[4] ligands to host Mn^{III} ions over Mn^{II} ions in **1** is entirely consistent with our previously published empirical binding rules for this ligand [17].

One of the goals behind our use of C[*n*]s in cluster synthesis is to isolate or 'dilute' these species in the solid state. This has been achieved efficiently in the formation of **1**, with TBC[4]s protecting the cluster core (Figure 5). The closest intermolecular interactions found between s.e. of **1** in the extended structure occur between Cl ions at a Cl⋯Cl distance of ~3.3 Å, and between dmf molecules at an N⋯O distance of ~3.6 Å, directing chains of [Mn$_{12}$] cages along the *a*-axis of the cell. Inspection of Figure 5 and comparison with Figure 1A shows similar assembly behaviours despite the fact that the overall shape of **1** is markedly different to that of the parent H$_4$TBC[4]. The 'coating' of **1** with [Mn^{III}TBC[4]]$^-$ capping moieties results in head-to-head packing dominated by the calixarenes. This is another common structural trend that is emerging as this work progresses, another example being that trigonal planar enneanuclear [Cu^{II}$_9$] clusters adapt to display similar packing behaviour in sheets, even though the clusters are unable to form the typical bi-layers. Interestingly, the metal–oxygen core present in **1** (Figure 4) is reminiscent of a portion of the Brucite ([Mg(OH)$_2$]) lattice, and is similar to that observed in [Mn$_{6-12}$] "rods" built with the tripodal alcohol ligands H$_3$thme (1,1,1-tris(hydroxymethyl)ethane) and H$_3$tmp (1,1,1-tris(hydroxymethyl)propane) [26], and to the [Mn^{III}$_6$Mn^{II}$_4$] "planar discs" built with the ligands 2-amino-2-methyl-1,3-propanediol (ampH$_2$) and 2-amino-2-ethyl-1,3-propanediol (aepH$_2$) [27].

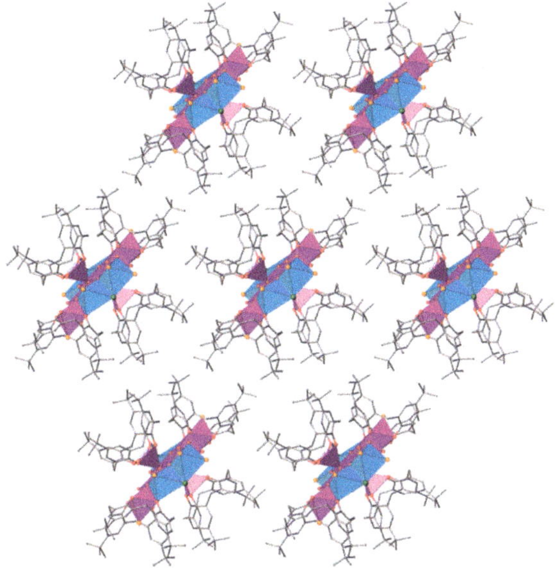

Figure 5. Extended structure of **1** looking down the *a*-axis. Packing of neighbouring clusters is dominated by TBC[4]s, which mimic head-to-head dimer assembly akin to that shown for the CH_3CN solvate of $H_4TBC[4]$ in Figure 1a [12]. Colour code: Mn^{III}—purple polyhedra; Mn^{II}—pale blue polyhedra; O—red; C—grey; hydrogen atoms, solvent of crystallization and ligated solvent omitted for clarity.

2.2. Magnetic Behaviour

Magnetic susceptibility (χ_M) data were measured on a powdered, restrained crystalline sample of **1** under a 0.1 T dc field in the T = 5–300 K temperature range. The temperature dependence of the $\chi_M T$ product is shown in Figure 6. At room temperature, the $\chi_M T$ value of 40.82 cm^3 K mol^{-1} is close to the Curie constant expected for eight Mn^{III} and four Mn^{II} ions, assuming a *g*-value of 2.00 (41.5 cm^3 K mol^{-1}). The $\chi_M T$ value remains approximately constant as the temperature is decreased until ~150 K, where it starts to decrease, reaching a minimum value of 13.27 cm^3 K mol^{-1} at 5 K. This behaviour is indicative of the presence of predominantly weak antiferromagnetic interactions between the constituent metal ions. A fit of the $1/\chi_M$ versus T data to the Curie–Weiss law afforded the Weiss constant, θ = −10 K. Magnetization measurements, performed in fields between 0.5–7 T (Figure 7), are in agreement with this observation, showing M increasing in a near linear like fashion with H, indicative of the field-induced population of low-lying excited states with larger magnetic moments. Previous magneto-structural correlations in alkoxide-bridged Mn^{III} dimers in which the JT axes are coparallel, as seen in the central [Mn$_{10}$] planar core (Figure 4), predict borderline and weak ferro- or antiferromagnetic exchange interactions, as observed [28]. No signals were observed in ac susceptibility measurements for data collected in the T = 1.8–10 K temperature range in frequencies up to 1500 Hz.

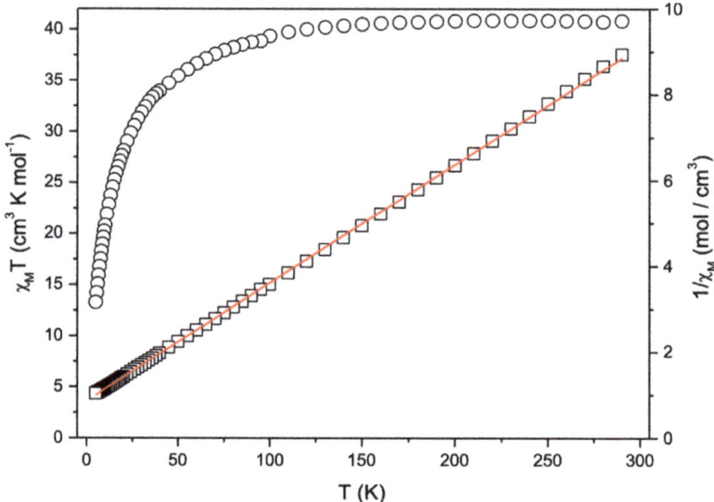

Figure 6. Plot of $\chi_M T$ vs T (o) and $1/\chi_M$ vs T ($\square$) in the range T = 5–300 K in an applied field of 0.1 T. The red line is a fit of the Curie–Weiss law. See text for details.

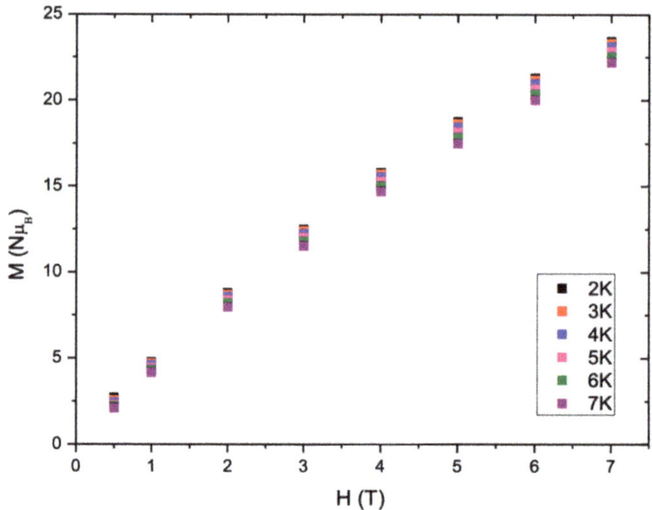

Figure 7. Field dependence of the magnetization (M) measured in the range T = 2–7 K and H = 0.5–7.0 T.

3. Conclusions

The reaction between $MnCl_2 \cdot 6H_2O$, $H_4TBC[4]$, H_2bd, $[NH_4][ClO_4]$ and NEt_3 in CH_3CN / dmf affords the dodecametallic, mixed-valent cluster $[Mn^{III}_8Mn^{II}_4(\mu_4\text{-}O)_2(\mu_3\text{-}OH)_6(\mu\text{-}OH)_4(\mu\text{-}Cl)_2(TBC[4])_4(dmf)_8(H_2O)_4(CH_3CN)_2]$. The structure describes a central, near-planer Mn-oxo-hydroxo moiety, reminiscent of the Brucite lattice, encapsulated by $[Mn^{III}\text{-}TBC[4]]^-$ moieties. Magnetic susceptibility and magnetization measurements reveal dominant and weak antiferromagnetic exchange interactions between the metal centres.

Somewhat surprisingly, 1,4-butanediol is a crucial component required for the formation of **1** despite the fact that it is not incorporated in the resulting cluster. Absence of this key reactant affords the previously reported mixed-valence TBC[4]-supported Mn-butterfly cluster, suggesting that this chelate plays a crucial role in (possibly hindering) cluster formation and producing a significantly higher nuclearity species in the process; the TBC[4]-supported Mn-butterfly cluster (Figure 1b) can be formed rapidly in approximately 15 min, and quantitatively precipitates from solution within an hour. The development reported here suggests that the use of other potential coligands with similar chemical make-up may offer the possibility to access/isolate other large, discrete high-nuclearity species. Results from this study will be reported in due course and will ideally shed additional light on the important role of such non-innocent reactants.

Author Contributions: Conceptualization, S.J.D. and E.K.B.; methodology, M.A.P.; formal analysis, M.A.P., M.C.; writing, M.C., E.K.B. and S.J.D.; visualization, M.C., S.J.D.; funding acquisition, S.J.D., E.K.B. All authors have read and agreed to the published version of the manuscript.

Funding: This research was funded by the EPSRC through grant reference numbers EP/I03255X/1 and EP/I031421/1.

Conflicts of Interest: The authors declare no conflict of interest.

Appendix A

MnCl$_2$·4H$_2$O (99 mg, 0.5 mmol, Sigma Aldrich, UK), H$_4$TBC[4] (97 mg, 0.15 mmol, Sigma Aldrich, UK) and NH$_4$ClO$_4$ (140 mg, 1.2 mmol, Sigma Aldrich, UK) were dissolved in a mixture of dmf (10 mL, Fisher Scientific, UK) and CH$_3$CN (10 mL, Fisher Scientific, UK). 1,4 butanediol (90 mg, 1 mmol, Sigma Aldrich, UK) was then added with stirring. After five minutes, NEt$_3$ (0.42 mL, 3 mmol, Sigma Aldrich, UK) was added dropwise and the reaction mixture stirred for two hours. X-ray quality crystals were obtained after 3 days in ~20% yield following filtration of the mother liquor with subsequent diffusion of diethyl ether (or hexane) into the resulting solution. Elemental analysis: found (calc. %) for C$_{222}$H$_{320}$Cl$_2$Mn$_{12}$N$_{18}$O$_{42}$: C 57.61 (57.43), H 6.94 (6.95), N 5.35 (5.43). Crystal Data (CCDC 1993320): C$_{204}$H$_{288}$Cl$_2$Mn$_{12}$N$_{10}$O$_{40}$ (M =4250.61 g/mol): triclinic, space group P-1 (no. 2), a = 16.144(16) Å, b = 19.437(19) Å, c = 22.33(2) Å, α = 110.119(15)°, β = 104.06(2)°, γ = 90.752(15)°, V = 6347(11) Å^3, Z = 1, T = 100(2) K, μ(MoKα) = 0.656 mm^{-1}, $Dcalc$ = 1.112 g/cm^3, 98657 reflections measured (3.342° $\leq$ 2Θ $\leq$ 52.744°), 25871 unique (R_{int} = 0.0328, R$_{sigma}$ = 0.0396), which were used in all calculations. The final R_1 was 0.0564 (I > 2σ(I)) and wR_2 was 0.1810 (all data). Magnetic properties were determined using a MPMS-XL SQUID magnetometer (Quantum Design Inc., San Diego, CA, USA) for direct current (dc) and alternating current (ac) measurements. The powdered microcrystalline sample was immobilised in eicosane. Experimental dc data were recorded at 0.1 T in the temperature range 5.0–300 K and at 2.0–7.0 K in the field range 0.5–7.0 T. Experimental ac data were collected in the temperature range 1.8–10 K and frequency range 0–1500 Hz using an amplitude of B_{ac} = 3 G. All data were corrected for the sample holder contributions and intrinsic diamagnetic contributions that were determined from Pascal's constants.

References

1. Ferreira, K.N.; Iverson, T.M.; Maghlaoui, K.; Barber, J.; Iwata, S. Architecture of the photosynthetic oxygen-evolving center. *Science* **2004**, *303*, 1831–1838. [CrossRef]
2. Ali, B.; Iqbal, M.A. Coordination complexes of manganese and their biomedical applications. *ChemistrySelect* **2017**, *2*, 1586–1604. [CrossRef]
3. Sampson, M.D.; Nguyen, A.D.; Grice, K.A.; Moore, C.E.; Rheingold, A.L.; Kubiak, C.P. Manganese catalysts with bulky bipyridine ligands for the electrocatalytic reduction of carbon dioxide: Eliminating dimerization and altering catalysis. *J. Am. Chem. Soc.* **2014**, *136*, 5460–5471. [CrossRef] [PubMed]
4. Diehl, S.J.; Lehmann, K.J.; Gaa, J.; McGill, S.; Hoffmann, V.; Georgi, M. MR Imaging of pancreatic lesions. Comparison of manganese-DPDP and gadolinium chelate. *Invest. Radiol.* **1999**, *34*, 589–595. [CrossRef] [PubMed]
5. Davis, K.M.; Palenik, M.C.; Yan, L.; Smith, P.F.; Seidler, G.T.; Dismukes, G.C.; Pushkar, N. X-ray Emission Spectroscopy of Mn Coordination Complexes towards Interpreting the Electronic Structure of the Oxygen Evolving Complex of Photosystem II. *J. Phys. Chem. C* **2016**, *120*, 3326–3333. [CrossRef]

6. Sessoli, R.; Boulon, M.-E.; Caneschi, A.; Mannini, M.; Poggini, L.; Wilhelm, F.; Rogalev, A. Strong magneto-chiral dichroism in a paramagnetic molecular helix observed by hard X-ray. *Nat. Phys.* **2015**, *11*, 69–74. [CrossRef]

7. Shiga, T.; Nojiri, H.; Oshio, H. A Ferromagnetically Coupled Octanuclear Manganese(III) Cluster: A Single-Molecule Magnet with a Spin Ground State of S = 16. *Inorg. Chem.* **2020**. [CrossRef]

8. Sessoli, R.; Gatteschi, D.; Caneschi, A.; Novak, M.A. Magnetic bistability in a metal-ion cluster. *Nature* **1993**, *365*, 141–143. [CrossRef]

9. McCusker, J.K.; Schmitt, E.A.; Hendrickson, D.N. High Spin Inorganic Clusters: Spin Frustration in Polynuclear Manganese and Iron Complexes. In *Magnetic Molecular Materials*; Gatteschi, D., Kahn, O., Miller, J.S., Palacio, F., Eds.; NATO ASI Series (Series E: Applied Sciences); Springer: Dordrecht, The Netherlands, 1991; Volume 198.

10. Liedy, F.; Eng, J.; McNab, R.; Inglis, R.; Penfold, T.J.; Brechin, E.K.; Johansson, J.O. Vibrational coherences in manganese single-molecule magnets after ultrafast photoexcitation. *Nat. Chem.* **2020**. [CrossRef]

11. Evangelisti, M.; Brechin, E.K. Recipes for enhanced molecular cooling. *Dalton Trans.* **2010**, *39*, 4672–4676. [CrossRef]

12. Asfari, F.; Bilyk, A.; Bond, C.; Harrowfield, J.M.; Koutsantonis, G.A.; Langkeek, N.; Mocerino, M.; Skelton, B.W.; Sovolev, A.N.; Strano, S.; et al. Factors influencing solvent adduct formation by calixarenes in the solid state. *Org. Biomol. Chem.* **2004**, *2*, 387–396. [CrossRef] [PubMed]

13. Karotsis, G.; Teat, S.J.; Wernsdorfer, W.; Piligkos, S.; Dalgarno, S.J.; Brechin, E.K. Calix[4]arene-based single-molecule magnest. *Angew. Chem. Int. Ed.* **2009**, *48*, 8285–8288. [CrossRef] [PubMed]

14. Karotsis, G.; Evangelisti, M.; Dalgarno, S.J.; Brechin, E.K. A calix[4]arene 3d/4f magnetic cooler. *Angew. Chem. Int. Ed.* **2009**, *48*, 9928–9931. [CrossRef] [PubMed]

15. Taylor, S.M.; McIntosh, R.D.; Piligkos, S.; Dalgarno, S.J.; Brechin, E.K. Calixarene-supported clusters: Employment of complementary cluster ligands for the construction of a ferromagnetic [Mn5] cage. *Chem. Commun.* **2012**, *48*, 11190–11192. [CrossRef]

16. McLellan, R.; Palacios, M.A.; Sanz, S.; Brechin, E.K.; Dalgarno, S.J. Importance of Steric Influences in the Construction of Multicomponent Hybrid Polymetallic Clusters. *Inorg. Chem.* **2017**, *56*, 10044–10053. [CrossRef]

17. Coletta, M.; Brechin, E.K.; Dalgarno, S.J. Structural trends in calix[4]arene-supported cluster chemistry. In *Calixarenes and Beyond*; Neri, P., Sessler, J.L., Wang, M.-X., Eds.; Springer International Publishing: Cham, Switzerland, 2016; Chapter 25; pp. 671–689.

18. Sone, T.; Ohba, Y.; Moriya, K.; Kumada, H.; Kazuski, I. Synthesis and properties of sulfur-bridged analogs of *p-tert*-Butylcalix[4]arene. *Tetrahedron* **1997**, *53*, 10689–10698. [CrossRef]

19. Mislin, G.; Graf, E.; Hosseini, M.W.; Bilyk, A.; Hall, A.K.; Harrowfield, J.M.; Skelton, B.W.; White, A.H. Thiacalixarenes as cluster keepers: Synthesis and structural analysis of a magnetically coupled tetracopper(II) square. *Chem. Commun.* **1999**, 373–374. [CrossRef]

20. Iki, N.; Kabuto, C.; Fukushima, T.; Kumagai, H.; Takeya, H.; Miyanari, S.; Miyashi, T.; Miyano, S. Synthesis of *p-tert*-Butylthiacalix[4]arene and its Inclusion Property. *Tetrahedron* **2000**, *56*, 1437–1443. [CrossRef]

21. Morohashi, N.; Iki, N.; Sugawara, A.; Miyano, S. Selective oxidation of thiacalix[4]arenes to the sulfinyl and sulfonyl counterparts and their complexation abilities toward metal ions as studied by solvent extraction. *Tetrahedron* **2001**, *57*, 5557–5563. [CrossRef]

22. Bi, Y.F.; Du, S.C.; Liao, W.P. Thiacalixarene-based nanoscale polyhedral coordination cages. *Coord. Chem. Rev.* **2014**, *276*, 61–72. [CrossRef]

23. Su, K.Z.; Jiang, F.L.; Qian, J.J.; Chen, L.; Pang, J.D.; Bawaked, S.M.; Mokhtar, M.; A-Thabaiti, S.A.; Hong, M.C. Stepwise Construction of Extra-Large Heterometallic Calixarene-Based Cages. *Inorg. Chem.* **2015**, *54*, 3183–3188. [CrossRef] [PubMed]

24. Schnack, J. Effects of frustration on magnetic molecules: A survey from Olivier Kahn until today. *Dalton Trans.* **2010**, *39*, 4677–4686. [CrossRef] [PubMed]

25. Coletta, M.; Sanz, S.; McCormick, L.J.; Teat, S.J.; Brechin, E.K.; Dalgarno, S.J. The remarkable influence of *N,O*-ligands in the assembly of a bis-calix[4]arene-supported [$Mn^{IV}_2Mn^{III}_{10}Mn^{II}_8$] cluster. *Dalton Trans.* **2017**, *46*, 16807–16811. [CrossRef] [PubMed]

26. Rajaraman, G.; Murugesu, M.; Sañudo, E.C.; Soler, M.; Wernsdorfer, W.; Helliwell, M.; Muryn, C.; Raftery, J.; Teat, S.J.; Christou, G.; et al. A Family of Manganese Rods: Syntheses, Structures, and Magnetic Properties. *J. Am. Chem. Soc.* **2004**, *126*, 15445–15457. [CrossRef] [PubMed]

27. Manoli, M.; Collins, A.; Parsons, S.; Candini, A.; Evangelisti, M.; Brechin, E.K. Mixed-Valent Mn Supertetrahedra and Planar Discs as Enhanced Magnetic Coolers. *J. Am. Chem. Soc.* **2008**, *130*, 11129–11139. [CrossRef]

28. Berg, N.; Rajeshkumar, T.; Taylor, S.M.; Brechin, E.K.; Rajaraman, G.; Jones, L.F. What Controls the Magnetic Interaction in bis-μ-Alkoxo Mn^{III}Dimers? A Combined Experimental and Theoretical Exploration. *Chem. Eur. J.* **2012**, *18*, 5906–5918. [CrossRef]

 *chemistry*

Article

Catalysis of an Aldol Condensation Using a Coordination Cage

Cristina Mozaceanu, Christopher G. P. Taylor, Jerico R. Piper, Stephen P. Argent and Michael D. Ward *

Department of Chemistry, University of Warwick, Coventry CV4 7AL, UK;
Cristina.Mozaceanu@warwick.ac.uk (C.M.); C.Taylor.10@warwick.ac.uk (C.G.P.T.);
Jerico.Piper@warwick.ac.uk (J.R.P.); stephen.argent@nottingham.ac.uk (S.P.A.)
* Correspondence: m.d.ward@warwick.ac.uk

Received: 7 January 2020; Accepted: 24 January 2020; Published: 25 January 2020

Abstract: The aldol condensation of indane-1,3-dione (**ID**) to give 'bindone' in water is catalysed by an M_8L_{12} cubic coordination cage (**H^w**). The absolute rate of reaction is slow under weakly acidic conditions (pH 3–4), but in the absence of a catalyst it is undetectable. In water, the binding constant of **ID** in the cavity of **H^w** is ca. $2.4\ (\pm1.2) \times 10^3\ M^{-1}$, giving a ΔG for the binding of $-19.3\ (\pm1.2)\ kJ\ mol^{-1}$. The crystal structure of the complex revealed the presence of two molecules of the guest **ID** stacked inside the cavity, giving a packing coefficient of 74% as well as another molecule hydrogen-bonded to the cage's exterior surface. We suggest that the catalysis occurs due to the stabilisation of the enolate anion of **ID** by the 16+ surface of the cage, which also attracts molecules of neutral **ID** to the surface because of its hydrophobicity. The cage, therefore, brings together neutral **ID** and its enolate anion via two different interactions to catalyse the reaction, which—as the control experiments show—occurs at the exterior surface of the cage and not inside the cage cavity.

Keywords: host–guest chemistry; coordination cage; catalysis; crystal structure; supramolecular chemistry

---

1. Introduction

The use of self-assembled molecular containers such as coordination cages as catalysts for reactions that occur in the central cavity has provided some remarkable examples of synthetic hosts providing enzyme-like levels of the rate acceleration of reactions. The range of reactions that has been shown to be catalysed is now extensive [1–10].

Many examples of cage-based catalysis rely on the steric properties of the cavity to provide the catalytic effect. Thus, the early examples of the acceleration of Diels–Alder reactions occurred on the basis that co-location of the two components in the same cavity provided a high local concentration of the two reacting partners [11–14]. Unimolecular pericyclic reactions can be accelerated because the folding of the guest allows it to bind in the cage cavity, resulting in a conformation that is close to the transition state [15–19]. Catalytic effects based on the electronic properties of the cage have also emerged, with photoinduced electron transfer between components of the cage walls and a bound guest, triggering useful reactions [20–23]; and an improved artificial 'Diels–Alderase' has been demonstrated, based on the electronic activation of the dienophile component by hydrogen-bonding interactions between the cage and guest, showing substantial rate enhancements without the need for the diene to be co-located in the cavity [24]. Possibilities for cage-based catalysis have been extended by the encapsulation of small-molecule catalysts, from mononuclear organometallic species to polyoxometallates, inside cage cavities [25–27]: in these cases, the cage itself is not the catalyst, but it modifies the behaviour of the bound catalyst that operates inside a constricted environment quite different from that in the bulk solution.

We have recently demonstrated examples of catalysis using an octanuclear, approximately cubic, M_8L_{12} coordination cage host denoted **H** or **H^w**, depending on external substituents (Figure 1) [10,28–30], which in water binds a wide range of hydrophobic guests in the central cavity, driven principally by the hydrophobic effect [31–33]. We note that a diverse range of octanuclear cages with the capacity to bind guests in the cavity is known [34–37]. The basis of guest binding in our hosts **H/H^w** in solution is well understood to the extent that we have developed a reliable predictive model for quantifying the guest binding free energies [38,39]. In addition to binding guests in the central cavity, the high positive charge [+16, arising from eight Co(II) ions] results in the accumulation of anions around the cage surface, resulting in a high local concentration of anions surrounding the guest, which is the basis for the catalysis [29,30]. We also showed that the binding of anions to the cage surface depends on how readily the anion can be desolvated, with chloride ions displacing hydroxide, and in turn, phenolate anions displacing chloride ions [30], allowing the nature of the anionic reaction partner surrounding a guest to be controlled. Thus, the cage offers the possibility to co-locate (i) a substrate that binds via the hydrophobic effect, with (ii) a high concentration of anions that accumulate around it via ion-pairing, two orthogonal interactions that, in combination, could promote a wide range of catalysed reactions between organic substrates and anions in water. We note that this accumulation of counter-ions around charged cages that can participate in catalytic reactions has also been exploited by Raymond et al. in the opposite sense: they used highly anionic cages to stabilise protonated forms of cavity-bound substrates, even at high pH values [18,40–42].

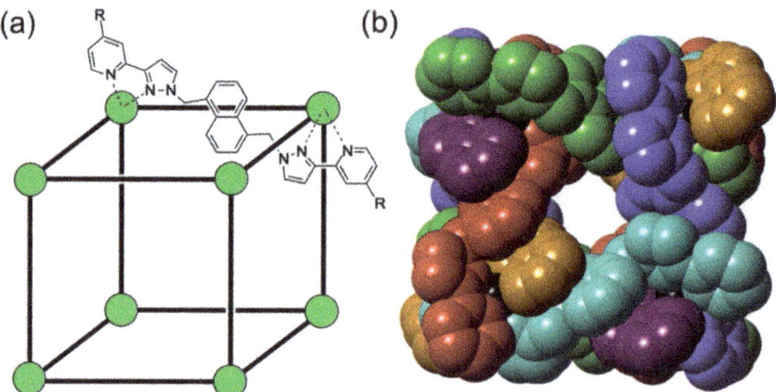

Figure 1. The octanuclear $[Co_8L_{12}](BF_4)_{16}$ cages used in this work (**H**, R = H, [28]; **H^w**, R = CH$_2$OH, [31]). (**a**) A sketch showing the approximate arrangement of metal ions and the structural formula of the bridging ligands, which span every edge of the cubic array of Co(II) ions; (**b**) view of the cationic cage cavity with each ligand coloured separately (from [28]).

We report here that our cage system can catalyse an aldol reaction: the conversion of indane-1,3-dione to bindone (Scheme 1) [43–46]. This was discovered by accident when we were evaluating the binding constants of a range of possible guests using spectroscopic titrations in solution; the addition of indane-1,3-dione (abbreviated hereafter as **ID**) to an aqueous solution of **H^w** resulted in the gradual appearance of a purple colour, which interfered with the titration experiment, but signalled the formation of the condensation product bindone. This did not occur in the absence of the cage under the same conditions. The facile aldol condensation of **ID** to give not just bindone, but also higher oligomers by multiple aldol-type reactions, has been known for over a century [45,46] and was recently re-studied in detail [43,44]. As **ID** has a p*Ka* of close to 7, the reaction can occur under very mild conditions and can even be catalysed by the surface of laboratory glassware, meaning that spectroscopic studies need to be prepared and performed in either plastic or quartz vessels.

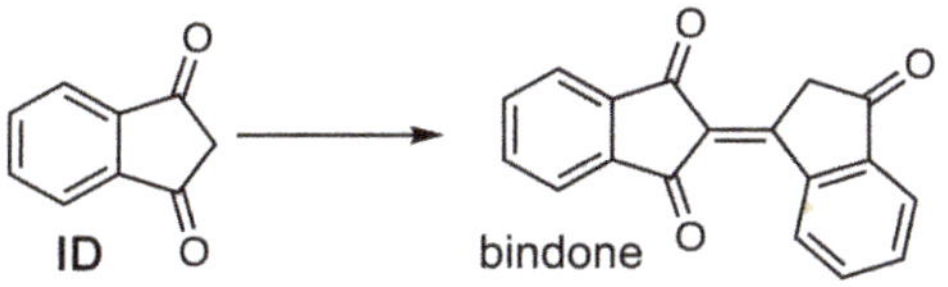

Scheme 1. Aldol condensation of indane-1,3-dione (**ID**) to bindone.

2. Materials and Methods

The Co(II)-based cage $\mathbf{H^w}$ was prepared as its fluoroborate salt through a previously-published method [28]. Indane-1,3-dione was purchased from Sigma-Aldrich; it reacts slowly with atmospheric moisture, so was dried under high vacuum, and stored in a desiccator. The instrumentation used for routine spectroscopic measurements was as follows: ^{1}H-NMR spectroscopy, a Bruker Avance 300 MHz instrument; UV/Vis absorption spectra, an IMPLEN NanoPhotometer C40 cuvette reader, or BMG CLARIOstar plate-reader. Solution pH measurements were performed with a Hamilton Spintrode pH combination electrode calibrated with standards at pH 4.01 and 7.00.

Measurement of the binding constant of **ID** in $\mathbf{H^w}$ was performed as follows. A series of 13 NMR tubes was prepared containing 0.6 mL of a D_2O solution containing 0.2 mM $\mathbf{H^w}$ at pD = 3.8, with the **ID** concentration varying from 0 to 1 mM (i.e., 0 to 5 equivalents) across the series. Spectra were recorded at 298 K and signals where free and bound $\mathbf{H^w}$ could be seen separately were deconvoluted and integrated to allow the calculation of K (see main text).

Catalysis experiments were performed at 298 K using aqueous solutions containing $\mathbf{H^w}$ (0.09 mM) and **ID** (0.9 mM) at pH 3.4 in a 1 cm path-length quartz cuvette. Progress was monitored on a UV/Vis spectrophotometer by growth in the absorbance of the product bindone at 550 nm (see main text).

The X-ray crystallographic data for the $\mathbf{H^w}$/**ID** complex were collected in Experiment Hutch 1 of beamline I-19 at the UK Diamond Light Source synchrotron facility [47]. Full details of the instrumentation, methods used for data collection, and for the solution and refinement of the structure, are as recently published [48]. Crystallographic, data collection, and refinement parameters are collected in Table 1. CCDC deposition number: 1979819.

Table 1. Summary of crystallographic, data collection, and refinement parameters for $\mathbf{H \bullet (ID)_3}$

Empirical formula	$C_{398.15}H_{435.65}B_{16}Co_8F_{64}N_{74.85}O_{39.35}$
Formula weight	8759.54
T/K	100(1)
Crystal system	Monoclinic
Space group	$C2/c$
Crystal size/mm^3	$0.04 \times 0.04 \times 0.04$
a/Å	32.77108(17)
b/Å	30.00687(16)
c/Å	40.3365(2)
β/degrees	96.1279(5)
V/Å^3	39,438.6(3)
Z	4
ρ_{calc}/g cm^{-3}	1.475
μ/mm^{-1}	0.405
Radiation	Synchrotron (λ = 0.6889)
Reflections collected	337,794
Data/restraints/parameters	62,798/5962/2330
Final R indexes [$I \geq 2\sigma(I)$]	$R_1 = 0.0651$, $wR_2 = 0.2094$
Final R indexes (all data)	$R_1 = 0.1117$, $wR_2 = 0.2292$

3. Results

To probe the cage-based catalysis of this reaction, we first examined the pH profile of the uncatalysed 'background' reaction, looking for conditions where this was as slow as possible to allow the effects of any catalysis to be most apparent. Monitoring the formation of bindone by UV/Vis spectroscopy (it has a strong absorption maximum at 510 nm in water) [44] at a range of different pH values showed that the reaction proceeds quickest in the pH range 6–7 where, based on the pKa of the starting material, there will be substantial amounts of both nucleophilic enolate anions and electrophilic neutral **ID** present; the reaction becomes slower at greater extremes of high or low pH. To facilitate the analysis of the reaction by UV/Vis spectroscopy, we therefore performed our subsequent experiments in the range of pH 3–4, reasoning that—as with the Kemp elimination reactions we examined earlier [29,30]—the high positive charge of the cage should stabilise the anionic enolate form of **ID**, even under relatively acidic conditions. Furthermore, at this pH, the intense yellow colour of the enolate anion, which would make the observation of the emerging colour of bindone more challenging, was not present.

We measured the binding constant of **ID** in the cage cavity with a ^{1}H NMR titration, adding several equivalents of **ID** in small aliquots to a D_2O solution of **H^w** at pD 3.8 (when the uncatalysed aldol reaction is extremely slow). The evolution of the ^{1}H NMR spectra is shown in Figure 2. It is clear that **ID** binds in the cage cavity in slow exchange with free guest in solution, as separate signals for free cage **H^w**, and the complex **H^w•ID** could be observed with the former progressively reducing in intensity and the latter increasing during the titration. The occurrence of a single signal for each host proton in the complex **H^w•ID** implies rapid motion of the **ID** guest in the cavity, such that the symmetry of the cage is preserved on the NMR timescale when the guest binds [49].

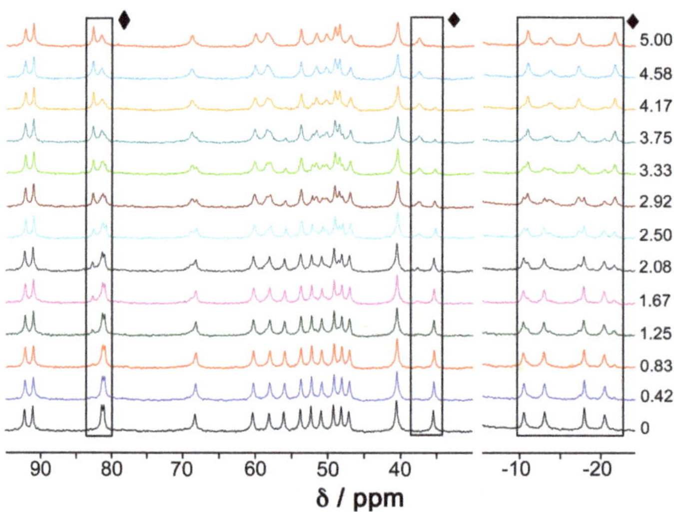

Figure 2. Evolution of paramagnetic ^{1}H NMR spectra during additions of **ID** (0–5 equivalents, indicated for each spectrum on the right) to a solution of **H^w** in D_2O (298 K). Signals associated with empty **H^w** (bottom spectrum) are replaced by new signals for the **H^w•ID** complex as the titration proceeds. Regions where this is particularly clear, and separate signals for the free and bound cage can be deconvoluted and integrated, are shown by the black diamonds.

Estimates of the binding constant K could be obtained by deconvoluting and integrating the separate (but closely spaced) **H^w** and **H^w•ID** signals at several different points across the ^{1}H NMR spectrum for known concentrations of cage and added guest. In this case, the close overlap of signals for the free and bound cage, coupled with uncertainties associated with deconvoluting and integrating broad signals from a paramagnetic compound, resulted in a high uncertainty: the average value of K obtained from several such measurements was $2.4(\pm1.2) \times 10^3$ M^{-1}, where the e.s.d. quoted is double the standard deviation obtained from averaging multiple measurements. We note that our algorithm for estimating binding constants using molecular docking software with a customised scoring function suggested a binding constant of 1200 M^{-1} [38,39], which is in good agreement with our estimate, and many structurally similar guests have binding constants in the 10^3–10^4 M^{-1} range [31,38].

A crystal structure of the cage/guest **H/ID** complex could be obtained using the crystalline sponge method that we have used in previous work by preparing crystals of the free host cage **H** by a solvothermal synthesis, followed by slow cooling [28], and immersing them in a concentrated solution of **ID** in MeOH for several hours, which resulted in guest uptake without loss of crystallinity [48]. Details of the structure are shown in Figures 3–5.

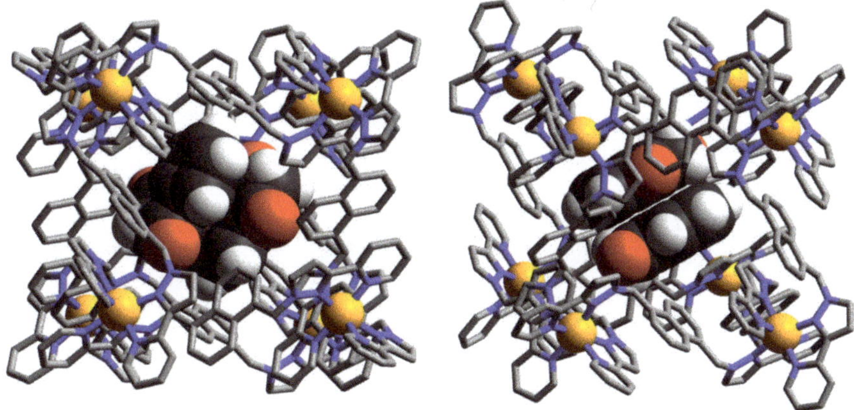

Figure 3. Two views of the crystal structure of the complex of host **H** with **ID**, showing the presence of a stacked pair of **ID** guests (which are shown in space-filling mode) in the cavity (host cage shown in wireframe) lying astride an inversion centre (N atoms, blue; O atoms, red; Co atoms, orange; C atoms, grey).

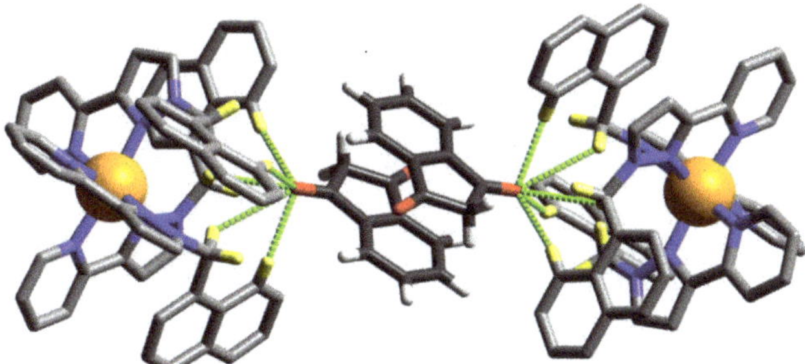

Figure 4. Partial view of the structure of the **H/ID** complex showing how each **ID** guest molecule forms a network of CH···O contacts (shown by green dotted lines) with a convergent set of CH groups associated with the cage interior surface around the two *fac* tris-chelate metal complex vertices of **H**.

Rather than the expected one molecule of guest, which we have often observed in previous work, we found a pair of guests stacked across an inversion centre in the cage cavity (Figure 3). This is not a 50:50 disorder as each guest molecule (the two are crystallographically equivalent) refines with unit site occupancy, and the distance between them is typical of π-stacking (3.48 Å). The combined volume of two **ID** guests (74%) significantly exceeded the value of 55 ± 9% of the host cavity volume (409 Å^3), which Rebek showed a while ago afforded optimal guest binding in solution [50,51]. However, a crystalline sponge experiment was performed under highly forcing and non-equilibrium conditions using a large excess of the guest; we [48] and others [52–54] have observed packing coefficients for guests inside supramolecular host cavities of >80% when favourable interactions such as π-stacking between multiple guests and favourable interactions between guests and the cage interior surface result in a particularly compact guest array. In dilute solution—the conditions under which guest binding is normally evaluated—we can imagine that for this reason, the second binding constant K_2 would be substantially smaller than the first binding constant K_1, in which case, the single-guest binding would dominate the solution speciation behaviour [48].

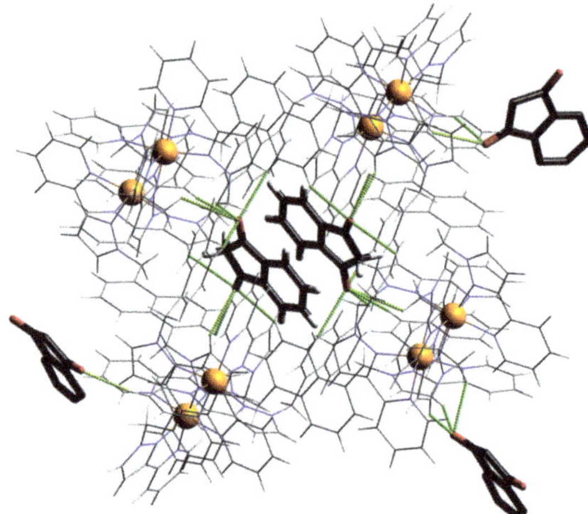

Figure 5. A view of the crystal structure of the complex of host **H** with **ID**, emphasising how the host cage brings together molecules at binding sites inside the cage cavity (*cf.* Figure 3) and around the exterior cage surface. It is not possible to tell whether the external guests are neutral **ID** or are the enolate anions stabilised by the high positive charge of the cage surface.

The (crystallographically equivalent) guests interact with the cage interior surface through multiple CH···O hydrogen bonds between the ketone O atoms, which are weak hydrogen-bond acceptors, and inwardly-directed C–H bonds from the ligand set, whose ability to act as weak H-bond donors is improved by the high positive charge of the assembly (Figure 4) [49,55]. One of the O atoms [O(14G)] lies in an H-bond donor pocket close to a *fac* tris-chelate metal ion, which contains several convergent CH groups (from methylene CH_2 and naphthyl CH protons) whose combined H-bond donor effect is comparable to a phenol group in terms of its overall hydrogen-bond donor strength [49,55]. The penetration of one of the C=O groups of the **ID** guest into this pocket results in non-bonded H···O distances associated with these CH···O interactions in the range of 2.5–3 Å. The other carbonyl O atom [O(15G)] likewise forms CH···O interactions with inwardly-directed naphthyl and pyrazolyl CH protons from the cage surface, with O···H distances of 2.8–2.9 Å.

There is an additional molecule of **ID** in each asymmetric unit (which contains half of a complete cage unit). This was refined with a site occupancy of 0.42; it lies in the space between cage molecules, and interacts with the cage exterior surface (Figure 5) through a similar set of CH···O hydrogen-bonds, as we saw for the interior guests, with O···H distances in the range 2.5–2.7 Å [thus the overall formulation, ignoring solvent molecules, is **H•(ID)**$_{2.84}$]. We have assumed, for the purposes of the crystallographic refinement, that this is a neutral **ID** molecule incorporating a CH_2 fragment between the two ketones. However, the possibility exists that this could be the enolate anion of **ID**, stabilised by the highly cationic cage surface. For a structure of this type (a large supramolecular assembly with weak scattering due to solvent/anion disorder), there is no way to ascertain this crystallographically.

Marginal differences in the C–C and C–O distances between the neutral and enolate forms of **ID** are not meaningful, given the extensive use of geometric restraints in the refinement (see CIF for details, Supplementary Materials), and charge balance considerations are not helpful either due to the solvent/anion disorder that required the use of SQUEEZE [56] to eliminate diffuse electron density from the refinement. We note, however, that the CH···O H-bonding interactions between the carbonyl oxygen atoms of the external **ID** molecule and the CH groups of the cage surface are shorter, on average, than those of the cavity-bound guest, consistent with the 'external' guest being in its anionic form. Whether the 'external' **ID** molecules in the crystal structure are in the neutral form or are actually the enolate anions, this structure provides a nice illustration of how the cage host can simultaneously co-locate a hydrophobic guest (in the cavity) and additional reaction partners around the exterior surface [30].

Catalysis experiments were performed with an aqueous solution of H^W (0.09 mM) and up to 10 equivalents of **ID** at pH 3.4 (Figure 6). Under these conditions, in the absence of the cage, no measurable conversion of **ID** to bindone was seen over prolonged periods (days), presumably due to the absence of any significant amount of the enolate anion of **ID**. Metal fluoroborate salts on their own, likewise had no catalytic effect. However, in the presence of cage H^W, steady growth in the absorbance associated with bindone was seen. This has a maximum at 510 nm, but it was monitored at 550 nm to avoid any possible competition from absorbance associated with the enolate anions, which is significant at 510 nm. The virtually straight-line growth of bindone over this period of time means that the reaction progress cannot meaningfully be fitted to a specific kinetic model; leaving the reaction for longer to let more bindone form results in solutions becoming cloudy as the product has poor water solubility. We can say, however, that after 12 h, around 3% of the **ID** was converted to bindone, increasing to 10% after 36 h, corresponding to approximately one turnover per catalyst molecule over 36 h. Although the absolute rate of the formation of bindone catalysed by the cage is therefore extremely low, compared to the undetectable formation of bindone in the absence of catalyst at this pH, the catalysis of the reaction under these conditions is clear.

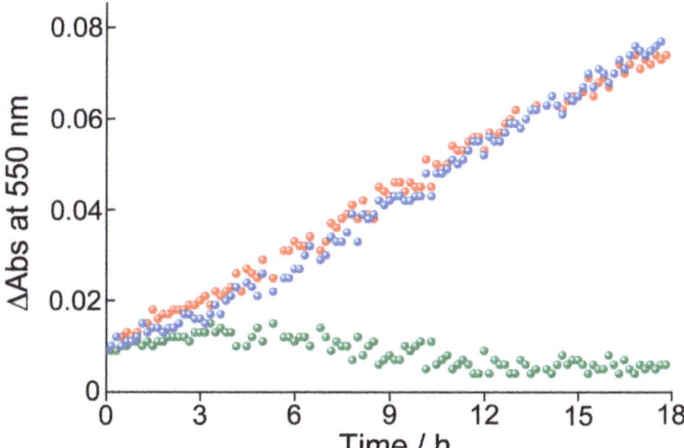

Figure 6. Cage-catalysed conversion of **ID** to bindone (0.9 mM **ID** at start; 298 K, pH 3.4), performed in a UV/Vis cuvette and monitored by measuring the increase in optical density at 550 nm arising from product formation. Green circles represent the background reaction (no significant reaction in the absence of cage H^W). Red circles represent the progress of the reaction in the presence of 0.09 mM H^W (i.e., the catalysed reaction). Blue circles represent catalysis under the same conditions as the red circles, but with 1.8 mM cycloundecanone added to block the cage cavity, showing that blocking the host cavity does not inhibit the catalysis, which must therefore occur at the external surface of the cage (see main text).

Control experiments suggested that the reaction does not actually occur inside the cage cavity, but at the external surface, which is a possibility that has very recently emerged from related studies on catalysis with this cage system [57]. In the confined space of the cavity, any successful reaction would require an ideal configuration of the cavity-bound and surface-bound reacting partners, and when this happens, it can lead to very large rate enhancements [10,29]. However, hydrophobic substrates that do not bind in the cavity, or that cannot react for geometric reasons whilst inside the cavity, can also interact with the cage exterior surface, which is just as hydrophobic as the interior surface and has a greater area: these substrates can thereby be brought into the region around the cationic surface where anions have accumulated because of ion-pairing effects [57]. The key control experiment here is to perform the reaction in the presence of an excess of cycloundecanone. This binds strongly in the cage cavity ($>10^6$ M^{-1}) [32] and therefore prevents the substrate **ID** from binding, however, it has no effect on the rate at which bindone is formed. This clearly indicates that cavity-based binding of **ID** is not necessary for the catalysis to occur, therefore it follows that the catalysed reaction occurs at the external surface of the cage where enolate anions of **ID** accumulate [57].

4. Discussion

Clearly the aldol condensation of two molecules of **ID** to bindone is catalysed by the presence of the cage. A plausible mechanism is that the hydrophobic, but cationic surface of the cage stabilises the enolate anion of **ID**, effectively reducing its pK_a, so that when close to the cage, it can be deprotonated even at pH 3.4 when the expected pK_a is 7. We emphasise that 'soft' anions such as phenolates have already been shown to have a higher affinity for the cage surface than more highly solvated anions such as hydroxide or chloride [9]. It is also significant that Raymond and co-workers observed that the basicity of an amine could be increased by >4 pK_a units when the amine is bound inside a cage host with a charge of -12 [42]: the charge of $+16$ on **H**^{w} implies that a comparable increase in the acidity of **ID** (i.e., stabilisation of the enolate anion) is plausible if the anion is interacting with the cage surface. As the cage can also bind neutral **ID** inside the cavity through the hydrophobic effect (*cf.* the crystal structure), we can see how the cage plays the role of bringing together a neutral substrate and an anion with which it can react through two orthogonal interactions, but this time resulting in a C–C bond forming reaction.

This particular reaction is an old one [45,46] and in itself not of the highest importance. The catalysis we observed is very slow in absolute terms; and looking at the reaction in more detail is made difficult by the poor solubility of bindone in water. The importance of these results, however, is that they open the possibility of using the cage to stabilise enolate anions and accumulate them around substrates attracted to the cage via the hydrophobic effect. Whether the reaction occurs inside the cavity or outside, this opens up the general possibility of using the cationic/hydrophobic surface of this cage and others like it as a general catalyst for aldol-type reactions, which would have substantial synthetic utility. We note that Mukherjee et al. have recently reported the condensation reactions of relatively acidic ketones with hydrophobic aldehydes using a trigonal prismatic [Pd^{2+}]$_6$ coordination cage as a catalyst, which allows the reactions to proceed under much milder conditions than in the absence of a catalyst [58]. This can be attributed in part to the hydrophobicity of the cavity that increased the thermodynamic driving force for the elimination of water, and its expulsion into the bulk solvent, but the role of the positive charge of the cage in stabilising anionic intermediates has also been suggested [58], which is exactly in agreement with what we propose.

Supplementary Materials: The following are available online at http://www.mdpi.com/2624-8549/2/1/4/s1.

Author Contributions: Synthesis and catalysis measurements: C.M., C.G.P.T., and J.R.P.; data analysis, C.M.; crystallography, S.P.A. and C.G.P.T.; project supervision and manuscript preparation: M.D.W. All authors have read and agreed to the published version of the manuscript.

Funding: This research was funded by the EPSRC, grant number EP/K003224/1; and the European Union (H2020-MSCA-ITN grant 'NOAH', project ref. 765297).

Acknowledgments: We thank the Diamond Light Source for the X-ray beamtime (proposal MT19876) and the staff of beamline I-19 for their assistance. Mr. Mark Cooper is thanked for his assistance with some of the early experimental measurements as part of an undergraduate project.

Conflicts of Interest: The authors declare no conflict of interest. The funders had no role in the design of the study; in the collection, analyses, or interpretation of data; in the writing of the manuscript, or in the decision to publish the results.

References

1. Brown, C.J.; Toste, F.D.; Bergman, R.G.; Raymond, K.N. Supramolecular catalysis in metal-ligand cluster hosts. *Chem. Rev.* **2015**, *115*, 3012–3025. [CrossRef]
2. Fang, Y.; Powell, J.A.; Li, E.; Wang, Q.; Perry, Z.; Kirchon, A.; Yang, X.; Xiao, Z.; Zhu, C.; Zhang, L.; et al. Catalytic reactions within the cavity of coordination cages. *Chem. Soc. Rev.* **2019**, *48*, 4707–4730. [CrossRef] [PubMed]
3. Yoshizawa, M.; Klosterman, J.K.; Fujita, M. Functional molecular flasks: New properties and reactions within discrete, self-assembled hosts. *Angew. Chem. Int. Ed.* **2009**, *48*, 3418–3438. [CrossRef] [PubMed]
4. Otte, M. Size-selective molecular flasks. *ACS Catal.* **2016**, *6*, 6491–6510. [CrossRef]
5. Zhang, D.; Ronson, T.K.; Nitschke, J.R. Functional capsules via subcomponent self-assembly. *Acc. Chem. Res.* **2018**, *51*, 2423–2436. [CrossRef] [PubMed]
6. Jing, X.; He, C.; Zhao, L.; Duan, C. Photochemical properties of host-guest supramolecular systems with structurally confined metal-organic capsules. *Acc. Chem. Res.* **2019**, *52*, 100–109. [CrossRef]
7. Hong, C.M.; Bergman, R.G.; Raymond, K.N.; Toste, F.D. Self-Assembled Tetrahedral Hosts as Supramolecular Catalysts. *Acc. Chem. Res.* **2018**, *51*, 2447–2455. [CrossRef]
8. Catti, L.; Zhang, Q.; Tiefenbacher, K. Advantages of catalysis in self-assembled molecular capsules. *Chem. Eur. J.* **2016**, *22*, 9060–9066. [CrossRef]
9. Gao, W.-X.; Zhang, H.-N.; Jin, G.-X. Supramolecular catalysis based on discrete heterometallic coordination-driven metallacycles and metallacages. *Coord. Chem. Rev.* **2019**, *386*, 69–84. [CrossRef]
10. Ward, M.D.; Hunter, C.A.; Williams, N.H. Coordination cages based on bis(pyrazolylpyridine) ligands: Structures, dynamic behavior, guest binding, and catalysis. *Acc. Chem. Res.* **2018**, *51*, 2073–2082. [CrossRef]
11. Yoshizawa, M.; Tamura, M.; Fujita, M. Diels-alder in aqueous molecular hosts: Unusual regioselectivity and efficient catalysis. *Science* **2006**, *312*, 251–254. [CrossRef] [PubMed]
12. Murase, T.; Horiuchi, S.; Fujita, M. Naphthalene Diels-Alder in a self-assembled molecular flask. *J. Am. Chem. Soc.* **2010**, *132*, 2866–2867. [CrossRef]
13. Kang, J.; Rebek, J., Jr. Acceleration of a Diels–Alder reaction by a self-assembled molecular capsule. *Nature* **1997**, *385*, 50–52. [CrossRef] [PubMed]
14. Kang, J.; Hilmersson, G.; Santamaría, J.; Rebek, J., Jr. Diels-Alder reactions through reversible encapsulation. *J. Am. Chem. Soc.* **1998**, *120*, 3650–3656. [CrossRef]
15. Kaphan, D.M.; Toste, F.D.; Bergman, R.G.; Raymond, K.N. Enabling new modes of reactivity via constrictive binding in a supramolecular-assembly-catalyzed aza-Prins cyclization. *J. Am. Chem. Soc.* **2015**, *137*, 9202–9205. [CrossRef] [PubMed]
16. Hart-Cooper, W.M.; Zhao, C.; Triano, R.M.; Yaghoubi, P.; Ozores, H.L.; Burford, K.N.; Toste, F.D.; Bergman, R.G.; Raymond, K.N. The effect of host structure on the selectivity and mechanism of supramolecular catalysis of Prins cyclizations. *Chem. Sci.* **2015**, *6*, 1383–1393. [CrossRef]
17. Hastings, C.J.; Bergman, R.G.; Raymond, K.N. Origins of large rate enhancements in the Nazarov cyclisation catalysed by supramolecular encapsulation. *Chem. Eur. J.* **2014**, *20*, 3966–3973. [CrossRef]
18. Hastings, C.J.; Pluth, M.D.; Bergman, R.G.; Raymond, K.N. Enzymelike catalysis of the Nazarov cyclization by supramolecular encapsulation. *J. Am. Chem. Soc.* **2010**, *132*, 6938–6940. [CrossRef]
19. Fiedler, D.; van Halbeek, H.; Bergman, R.G.; Raymond, K.N. Supramolecular catalysis of unimolecular rearrangements: Substrate scope and mechanistic insights. *J. Am. Chem. Soc.* **2006**, *128*, 10240–10252. [CrossRef]
20. Cullen, W.; Takezawa, H.; Fujita, M. Demethylenation of cyclopropanes via photoinduced guest-to-host electron transfer in an M_6L_4 cage. *Angew. Chem. Int. Ed.* **2019**, *58*, 9171. [CrossRef]

21. Dalton, D.M.; Ellis, S.R.; Nichols, E.M.; Mathies, R.A.; Toste, F.D.; Bergman, R.G.; Raymond, K.N. Supramolecular $[Ga_4L_6]^{12-}$ cage photosensitizes 1,3-rearrangement of encapsulated guest via photoinduced electron transfer. *J. Am. Chem. Soc.* **2015**, *137*, 10128–10131. [CrossRef] [PubMed]

22. Murase, T.; Takezawa, H.; Fujita, M. Photo-driven anti-Markovnikoz alkyne hydration in self-assembled hollow complexes. *Chem. Commun.* **2011**, *47*, 10960–10962. [CrossRef] [PubMed]

23. Murase, T.; Nishijima, Y.; Fujita, M. Unusual photoreaction of triquinacene within self-assembled hosts. *Chem. Asian J.* **2012**, *7*, 826–829. [CrossRef] [PubMed]

24. Martí-Centelles, V.; Lawrence, A.L.; Lusby, P.J. High activity and efficient turnover by a simple, self-assembled "artificial Diels–Alderase". *J. Am. Chem. Soc.* **2018**, *140*, 2862–2868. [CrossRef] [PubMed]

25. Cai, L.-X.; Li, S.-C.; Yan, D.-N.; Zhou, L.-P.; Guo, F.; Sun, Q.-F. Water-soluble redox-active cage hosting polyoxometalates for selective desulfurization catalysis. *J. Am. Chem. Soc.* **2018**, *140*, 4869–4876. [CrossRef] [PubMed]

26. Bender, T.A.; Bergman, R.G.; Raymond, K.N.; Toste, F.D. A supramolecular strategy for selective catalytic hydrogenation independent of remote chain length. *J. Am. Chem. Soc.* **2019**, *141*, 11806–11810. [CrossRef]

27. Jongkind, L.J.; Caumes, X.; Hartendorp, A.P.T.; Reek, J.N.H. Ligand template strategies for catalyst encapsulation. *Acc. Chem. Res.* **2018**, *51*, 2115–2128. [CrossRef]

28. Tidmarsh, I.S.; Faust, T.B.; Adams, H.; Harding, L.P.; Russo, L.; Clegg, W.; Ward, M.D. Octanuclear cubic coordination cages. *J. Am. Chem. Soc.* **2008**, *130*, 15167–15175. [CrossRef]

29. Cullen, W.; Misuraca, M.C.; Hunter, C.A.; Williams, N.H.; Ward, M.D. Highly efficient catalysis of the Kemp elimination in the cavity of a cubic coordination cage. *Nat. Chem.* **2016**, *8*, 231–236. [CrossRef]

30. Cullen, W.; Metherell, A.J.; Wragg, A.B.; Taylor, C.G.P.; Williams, N.H.; Ward, M.D. Catalysis in a cationic coordination cage using a cavity-bound guest and surface-bound anions: Inhibition, activation, and autocatalysis. *J. Am. Chem. Soc.* **2018**, *140*, 2821–2828. [CrossRef]

31. Whitehead, M.; Turega, S.; Stephenson, A.; Hunter, C.A.; Ward, M.D. Quantification of solvent effects on molecular recognition in polyhedral coordination cage hosts. *Chem. Sci.* **2013**, *4*, 2744–2751. [CrossRef]

32. Turega, S.; Cullen, W.; Whitehead, M.; Hunter, C.A.; Ward, M.D. Mapping the internal recognition surface of an octanuclear coordination cage using guest libraries. *J. Am. Chem. Soc.* **2014**, *136*, 8475–8483. [CrossRef] [PubMed]

33. Metherell, A.J.; Cullen, W.; Williams, N.H.; Ward, M.D. Binding of hydrophobic guests in a coordination cage cavity is driven by liberation of 'high-energy' water. *Chem. Eur. J.* **2018**, *24*, 1554–1560. [CrossRef] [PubMed]

34. Yang, X.; Hahn, B.P.; Jones, R.A.; Wong, W.-K.; Stevenson, K.J. Synthesis of an octanuclear Eu(III) cage from Eu_4^{2+}: Chloride anion encapsulation, luminescence, and reversible MeOH adsorption via a porous supramolecular architecture. *Inorg. Chem.* **2007**, *46*, 7050–7054. [CrossRef] [PubMed]

35. Bilyachenko, A.N.; Khrustalev, V.N.; Zubavichus, Y.V.; Vologzhanina, A.V.; Astakhov, G.S.; Gutsul, E.I.; Shubina, E.S.; Levitsky, M.M. High-nuclearity (Cu_8-based) cage silsesquioxanes: Synthesis and structural study. *Cryst Growth Des.* **2018**, *18*, 2452–2457. [CrossRef]

36. Elahi, S.M.; Rajasekharan, M.V. Spherical octanuclear clusters in a Na-Ln-dipic system: Encapsulation of a nitrate ion and incorporation of water nonamers and dodecamers. *CrystEngComm* **2015**, *17*, 7191–7198. [CrossRef]

37. Freudenreich, J.; Dalvit, C.; Süss-Fink, G.; Therrien, B. Encapsulation of photosensitizers in hexa- and octanuclear organometallic cages: Synthesis and characterization of carceplex and host-guest systems in solution. *Organometallics* **2013**, *32*, 3018–3033. [CrossRef]

38. Cullen, W.; Turega, S.; Hunter, C.A.; Ward, M.D. Virtual screening for high affinity guests for synthetic supramolecular receptors. *Chem. Sci.* **2015**, *6*, 2790–2794. [CrossRef]

39. Taylor, C.G.P.; Cullen, W.; Collier, O.M.; Ward, M.D. A quantitative study of the effects of guest flexibility on binding inside a coordination cage host. *Chem. Eur. J.* **2017**, *23*, 206–213. [CrossRef]

40. Pluth, M.D.; Bergman, R.G.; Raymond, K.N. Acid catalysis in basic solution: A supramolecular host promotes orthoformate hydrolysis. *Science* **2007**, *316*, 85–88. [CrossRef]

41. Pluth, M.D.; Bergman, R.G.; Raymond, K.N. The acid hydrolysis mechanism of acetals catalyzed by a supramolecular assembly in basic solution. *J. Org. Chem.* **2009**, *74*, 58–63. [CrossRef] [PubMed]

42. Pluth, M.D.; Bergman, R.G.; Raymond, K.N. Making amines strong bases: Thermodynamic stabilization of protonated guests in a highly-charged supramolecular host. *J. Am. Chem. Soc.* **2007**, *129*, 11459–11467. [CrossRef]

43. Jacob, J.; Sigalov, M.; Becker, J.Y.; Ellern, A.; Khodorkovsky, V. Self-condensation of 1,3-indandione: A reinvestigation. *Eur. J. Org. Chem.* **2000**, 2047–2055. [CrossRef]

44. Sigalov, M.; Krief, P.; Shapiro, L.; Khodorkovsky, V. Inter- and intramolecular C-H···O bonding in the anions of 1,3-indandione derivatives. *Eur. J. Org. Chem.* **2008**, 673–683. [CrossRef]

45. Wislicenus, W. Ueber die vereinigung verschiedener ester durch natrium. *Ber. Dtsch. Chem. Ges.* **1887**, *20*, 589. [CrossRef]

46. Wislicenus, W. Einwirkung von essigester auf phtalsäureester. *Annalen* **1888**, *246*, 347–355.

47. Allan, D.R.; Nowell, H.; Barnett, S.A.; Warren, M.R.; Wilcox, A.; Christensen, J.; Saunders, L.K.; Peach, A.; Hooper, M.T.; Zaja, L.; et al. A novel dial air-bearing fixed-χ diffractometer for small-molecular single-crystal X-ray diffraction on beamline I-19 at Diamond Light Source. *Crystals* **2017**, *7*, 336. [CrossRef]

48. Taylor, C.G.P.; Argent, S.P.; Ludden, M.D.; Piper, J.R.; Mozaceanu, C.; Barnett, S.A.; Ward, M.D. One guest or two? A crystallographic and solution study of guest binding in a cubic coordination cage. *Chem. Eur. J.* **2019**, in press. [CrossRef]

49. Turega, S.; Whitehead, M.; Hall, B.R.; Meijer, A.J.H.M.; Hunter, C.A.; Ward, M.D. Shape-, size- and functional group-selective binding of small organic guests in a paramagnetic coordination cage. *Inorg. Chem.* **2013**, *52*, 1122–1132. [CrossRef]

50. Mecozzi, S.; Rebek, J. The 55% solution: A formula for molecular recognition in the liquid state. *Chem. Eur. J.* **1998**, *4*, 1016–1022. [CrossRef]

51. Rebek, J. Molecular behaviour in small spaces. *Acc. Chem. Res.* **2009**, *42*, 1660–1668. [CrossRef] [PubMed]

52. Puttreddy, R.; Beyeh, N.K.; Kalenius, E.; Ras, R.H.A.; Rissanen, K. 2-Methylresorcinarene: A very high packing coefficient in a mono-anion based dimeric capsule and the X-ray crystal structure of the tetra-anion. *Chem. Commun.* **2016**, *52*, 8115–8118. [CrossRef] [PubMed]

53. Zhang, W.; Yang, D.; Zhao, J.; Hou, L.; Sessler, J.L.; Yang, X.-J.; Wu, B. Controlling the recognition and reactivity of alkyl ammonium guests using an anion coordination-based tetrahedral cage. *J. Am. Chem. Soc.* **2018**, *140*, 5248–5256. [CrossRef] [PubMed]

54. Djemili, R.; Kocher, L.; Durot, S.; Peuronen, A.; Rissanen, K.; Heitz, V. Positive allosteric control of guests encapsulation by metal binding to covalent porphyrin cages. *Chem. Eur. J.* **2019**, *25*, 1481–1487. [CrossRef] [PubMed]

55. Metherell, A.J.; Ward, M.D. Geometric isomerism in coordination cages based on *tris*-chelate vertices: A tool to control both assembly and host/guest chemistry. *Dalton Trans.* **2016**, *45*, 16096–16111. [CrossRef]

56. Spek, A.L. PLATON SQUEEZE: A tool for the calculation of disordered solvent contribution to the calculated structure factors. *Acta Crystallogr. C. Sruct. Chem.* **2015**, *71*, 9–18. [CrossRef]

57. Taylor, C.G.P.; Metherell, A.J.; Argent, S.P.; Ashour, F.M.; Williams, N.H.; Ward, M.D. Coordination cage catalysed hydrolysis of organophosphates: Cavity or surface based? *Chem. Eur. J.* **2019**, in press. [CrossRef]

58. Das, P.; Kumar, A.; Howlader, P.; Mukherjee, P.S. A self-assembled trigonal prismatic molecular vessel for catalytic dehydration reactions in water. *Chem. Eur. J.* **2017**, *23*, 12565–12574. [CrossRef]

 *chemistry*

Article

Complementarity in Cyclotricatechylene Assemblies: Symmetric Cages Linked within 3D Cubic Hydrogen Bonded Networks

Jessica L. Holmes [1], Steven M. Russell [1], Brendan F. Abrahams [1,*], Timothy A. Hudson [1] and Keith F. White [2]

[1] School of Chemistry, University of Melbourne, Parkville VIC 3010, Australia;
    jlholmesphd@gmail.com (J.L.H.); sm.russell@live.com.au (S.M.R.); timothy.hudson@unimelb.edu.au (T.A.H.)
[2] School of Molecular Science, La Trobe University, Wodonga VIC 3690, Australia; K.White2@latrobe.edu.au
* Correspondence: bfa@unimelb.edu.au

Received: 30 April 2020; Accepted: 26 May 2020; Published: 11 June 2020

Abstract: A serendipitous discovery has led to the generation of a family of four compounds in which six components combine to form symmetric metal-cyclotricatechylene (H_6ctc) cages. The four compounds, which have the compositions, $[Cs((CH_3)_2CO)_6][K_4(H_6ctc)_4(H_2O)_8][Cs_4(H_2O)_6](PO_4)_3$, $[Rb((CH_3)_2CO)_6][Rb_2K_2(H_6ctc)_4(H_2O)_6][Rb_4(H_2O)_6](PO_4)_3$, $[Cs((CH_3)_2CO)_6][K_4(H_6ctc)_4(H_2O)_8]$-$[Cs(H_2O)_9](SO_4)_3$ and $[Rb((CH_3)_2CO)_6][Rb_2K_2(H_6ctc)_4(H_2O)_6][Rb(H_2O)_9](SO_4)_3$ possess cubic symmetry that arises from the complementary interactions that govern the assembly of the components. The cage cavities contain water molecules and either one or four large alkali metal ions (either Rb^+ or Cs^+) which interact with the internal aromatic surfaces of the cage. Each cage is linked to six tetrahedral anions (PO_4^{3-} or SO_4^{2-}) through 24 equivalent hydrogen bonds and each anion bridges a pair of cages through eight such hydrogen bonds. An unusual octahedral complex $M((CH_3)_2CO)_6^+$ (M = Rb or Cs), in which the M-C=O link is linear, appears to be a key structural component. A feature of this family of crystalline compounds is the presence of a range of complementary interactions which combine to generate materials that exhibit high crystallographic symmetry.

Keywords: complementarity; hydrogen bonding; crystal engineering; cyclotricatechylene

1. Introduction

The emergence of the field of supramolecular chemistry was recognised with the award of the 1987 Nobel Prize in Chemistry to Cram, Lehn and Pedersen "for their development and use of molecules with structure-specific interactions of high selectivity" [1]. In their work, each of these researchers developed molecular systems that exhibited selectivity in binding to alkali metal ions [2–8]. In the following decades, the field of metallosupramolecular chemistry proved to be a rich and fertile area with researchers such as Fujita producing a stunning array of cage-type compounds in which metal-centres are linked by bridging ligands [9,10]. In addition to metal-ligand assemblies, expansion of the field of supramolecular chemistry included the area of crystal engineering. Gautum Desiraju's pioneering work encouraged researchers to consider the crystal as a supramolecular entity in which a variety of complementary interactions control the arrangement of molecules [11,12]. In this context, Ward exploited complementary hydrogen bonding interactions between anionic sulphonates and guanidinium cations to generate an impressive series of network materials that represent an outstanding example of true crystal engineering [13,14]. All of the above examples rely upon different types of complementary interactions between components to yield supramolecular assemblies. In this current work, the serendipitous formation of a novel series of crystalline, supramolecular compounds is described, in which the tris (catechol) molecule, cyclotricatechylene (H_6ctc, **I**, Scheme 1), participates in all of the types of associations indicated in the examples listed above i.e., alkali metal binding, cage formation and hydrogen bonding within a highly symmetric extended network. The resulting compounds represent remarkable examples of high symmetry multi-component supramolecular assemblies in which a synergy exists between the different types of complementary interactions that govern their formation.

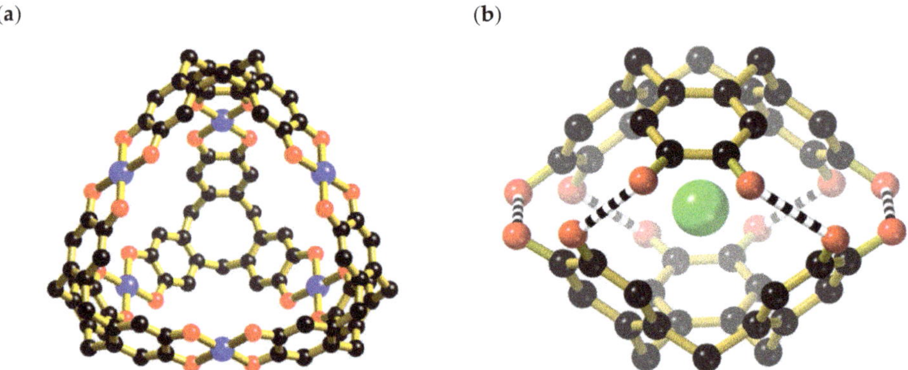

Scheme 1. Cyclotricatechylene (**I**) and cyclotriveratrylene (**II**).

Cyclotricatechylene possesses 3-fold symmetry and commonly adopts a bowl-shaped conformation with hydroxylic groups located on the rim of the bowl. The compound is closely related to its hexamethyl counterpart, cyclotriveratrylene (**II**, Scheme 1), which has received considerable attention with respect to its ability to host guest molecules within the bowl-shaped cavity [15,16]. In the last 12 years, interest in the supramolecular chemistry of cyclotricatechylene has grown and a variety of metal-based derivatives have been synthesised and structurally characterised. The structures of these assemblies depend upon the manner in which the metal ion interacts with the tris (catechol) ligand. For example, transition metal ions (M) such as V^{IV} (in the form of vanadyl) [17] and Cu^{II} [18] combine with ctc^{6-} to produce M_6ctc_4 anionic tetrahedral cages in which ctc hexaanions are located at the vertices of a tetrahedron. The catecholate arms extend along the tetrahedron edges and are linked by metal centres, each of which is linked to a pair of catecholate groups as indicated in Figure 1a. Recent work has demonstrated that 5-coordinate Si centres can fulfil the role of the metal centre to generate robust covalent cages [19,20].

(**a**)   (**b**)

Figure 1. Metallosupramolecular assemblies involving cyclotricatechylene, (**a**) the anionic cage $[Cu_6(ctc)_4]^{12-}$; C black, O red, Cu blue and (**b**) the anionic clam $[Cs(H_5ctc)(H_4ctc)]^{2-}$; C black, O red, Cs green, black and white striped connections represent hydrogen bonds, hydrogen atoms omitted for clarity.

In addition to the cages of the type described above, H_6ctc, in various stages of deprotonation, is able to form fascinating clam-like structures when it is combined with large alkali metal ions, in particular, Rb^+ and Cs^+ [21]. In these structures, pairs of partially deprotonated $H_nctc^{(6-n)-}$ units assemble through complementary hydrogen bonding interactions between hydroxyl groups which results in a rim-to-rim arrangement as shown in Figure 1b. A Rb^+ or Cs^+ ion occupies the internal cavity of the assembly and interacts only with the aromatic surfaces of the $H_nctc^{(6-n)-}$ units. A computational investigation of the stability of these clam-like structures, under different levels of protonation, was recently reported [22]. The affinity of ctc units for alkali metal ions is also apparent in the metal-like assemblies of the type depicted in Figure 1a with the large alkali metal cations associating with the internal aromatic surfaces of the cage.

In this current work, we report further investigations into supramolecular assemblies formed from the combination of alkali metal cations with H_6ctc, this time in the presence of the tetrahedral oxyanions, sulphate and phosphate. The original intention of the study was to investigate the interaction of large alkali metal cations (Rb^+ or Cs^+) with the internal aromatic surfaces of cyclotricatechylene, in various states of deprotonation, and determine how simple anions such as phosphate and sulphate impact upon the nature of the assembly. The high symmetry products obtained from this synthetic investigation were completely unexpected and display a remarkable level of complementarity involving the six components from which each of the compounds is assembled.

2. Materials and Methods

2.1. Synthetic Methods

All starting materials were obtained from commercial sources and were used without further purification. Cyclotricatechylene was prepared as previously described [23].

$[Cs((CH_3)_2CO)_6][K_4(H_6ctc)_4(H_2O)_8][Cs_4(H_2O)_6](PO_4)_3$ (**III**) was prepared by dissolving 24 mg (0.065 mmol) of H_6ctc in 3 mL of acetone and adding 14.8 mg (0.065 mmol) of K_2HPO_4 in 1 mL of water followed by 9.7 mg (0.065 mmol) of CsOH in 2 mL of water. Multi-faceted crystals formed and were collected by filtration; yield 30.5 mg (51%). Analysis: Found (%): C 39.6, H 4.5; calculated (%): C 39.8, 4.5.

$[Rb((CH_3)_2CO)_6][Rb_2K_2(H_6ctc)_4(H_2O)_6][Rb_4(H_2O)_6](PO_4)_3$ (**IV**) was prepared by dissolving 24 mg (0.065 mmol) of H_6ctc in 3 mL of acetone and adding 14.8 mg (0.065 mmol) of K_2HPO_4 in 1 mL of water followed by 6.7 mg (0.065 mmol) of RbOH in 2 mL of water. Multi-faceted crystals formed, which appeared, by visual inspection, to be homogenous in the mother liquor however the solid product underwent considerable deterioration during the attempted filtration of the crystals.

$[Cs((CH_3)_2CO)_6][K_4(H_6ctc)_4(H_2O)_8][Cs(H_2O)_9](SO_4)_3$ (**V**) was prepared by dissolving 24 mg (0.065 mmol) of H_6ctc in 3 mL of acetone and adding 8.9 mg (0.065 mmol) of $KHSO_4$ in 1 mL of water followed by 9.7 mg (0.065 mmol) of CsOH in 2 mL of water. Multi-faceted crystals formed and were collected by filtration; yield 19.0 mg (41%). Analysis: Found (%): C 45.4, H 4.5; calculated (%): C 44.8, 5.2.

$[Rb((CH_3)_2CO)_6][Rb_2K_2(H_6ctc)_4(H_2O)_6][Rb(H_2O)_9](SO_4)_3$ (**VI**) was prepared by dissolving 24 mg (0.065 mmol) of H_6ctc in 3 mL of acetone and adding 8.9 mg (0.065 mmol) of $KHSO_4$ in 1 mL of water followed by 6.7 mg (0.065 mmol) of RbOH in 2 mL of water. Multi-faceted crystals formed, however upon filtration significant deterioration of the solid was noted. A solid yield of 17 mg was recorded but analysis indicated the product was inconsistent with the single crystal X-ray analysis.

2.2. Crystal Structure Determinations

Single crystal X-ray analyses were undertaken on crystals transferred directly from the mother liquor to a protective oil before being mounted on the diffractometer in a stream of cooled N_2 gas. All data were collected on a Rigaku Oxford Diffraction XtaLAB Synergy dual source diffractometer. Data measurement and reduction was performed using CrysalisPro software [24]. The structure was solved using SHELXT [25] and refined using a full-matrix least squares method based on F^2 [26]. Crystallographic and refinement data are presented in Table 1. Twinning was apparent for compound **VI** and this resulted in elevated agreement values.

Table 1. Crystallographic and refinement data for compounds **III–VI**.

Compound	III	IV	V	VI
Formula	$C_{102}H_{136}Cs_5K_4O_{56}P_3$	$C_{102}H_{132}K_2O_{54}P_3Rb_7$	$C_{102}H_{142}Cs_2K_4O_{59}S_3$	$C_{102}H_{138}K_2O_{57}Rb_4S_3$
Crystal size (mm)	$0.12 \times 0.10 \times 0.05$	$0.26 \times 0.22 \times 0.14$	$0.14 \times 0.11 \times 0.10$	$0.34 \times 0.28 \times 0.21$
Crystal system	Cubic	Cubic	Cubic	Cubic
Space group	P-43m	P-43m	P-43m	P-43m
a (Å)	14.2581 (2)	14.2376 (2)	14.3955 (17)	14.39549 (3)
V (Å^3)	2898.58 (12)	2886.09 (12)	2983.2 (11)	2983.18 (2)
T (K)	100.0 (1)	100.0 (1)	100.0 (1)	100.0 (1)
Z	1	1	1	1
D_{calc} (g cm^{-3})	1.817	1.721	1.576	1.554
λ (Å)	0.71073 (Mo-Kα)	0.71073 (Mo-Kα)	1.54184 (Cu-Kα)	1.54184 (Cu-Kα)
μ (mm^{-1})	1.843	3.154	7.365	4.006
F (000)	1592	1518	1460	1440
θ range (°)	2.5–30.8	2.5–33.4	3.1–78.2	3.1–77.8
Reflections collected	7781	8768	64422	64233
Unique reflections	1481	1725	1261	1257
Observed reflections [$I > 2\sigma(I)$]	1399	1457	1256	1257
Parameters/restraints	82/18	85/17	85/20	84/29
Goodness of fit	1.112	1.089	1.051	1.056
Final R [$I > 2\sigma(I)$]	0.0356	0.0609	0.0674	0.0853
Final $wR2$ (all data)	0.0900	0.1740	0.1608	0.2107
Minimum, maximum electron density (eÅ^{-3})	1.01, −1.24	1.25, −2.59	2.04, −0.90	1.48, −1.81
CCDC Deposition number	2000019	2000020	2000021	2000022

3. Results and Discussion

The combination of H_6ctc with CsOH and K_2HPO_4 in aqueous acetone results in the overnight formation of large colourless crystals of composition, $[Cs((CH_3)_2CO)_6][K_4(H_6ctc)_4(H_2O)_8]$-$[Cs_4(H_2O)_6](PO_4)_3$ (**III**). The crystals possess cubic symmetry and adopt the space group P-43m. A key feature of this structure is the complementary interactions that exist between the various cationic, anionic and neutral species.

Within each unit cell there is a single $[K_4(H_6ctc)_4(H_2O)_8]^{4+}$ cage (Figure 2a) which is topologically different from the $M_6(ctc)_4$ anionic cages described in the introduction. All K^+ ions are symmetry related and exist in an 8-coordinate, slightly distorted hexagonal bipyramidal environment formed from six symmetry related co-planar catechol oxygen atoms and two *trans* water molecules. Each K^+ ion is located just above the plane of the six bound catechol oxygen atoms. The cage possesses the topology of a cube with four K^+ ions and four H_6ctc molecules units serving as the eight 3-connecting nodes and the catechol groups running along the cube edges. The topological relationship between the cube and the cage is apparent from inspection of Figure 2b. The formation of the cube rather than the tetrahedron, of the type depicted in Figure 1a, reflects the ability of K^+ ions to adopt an 8-coordinate hexagonal bipyramidal geometry.

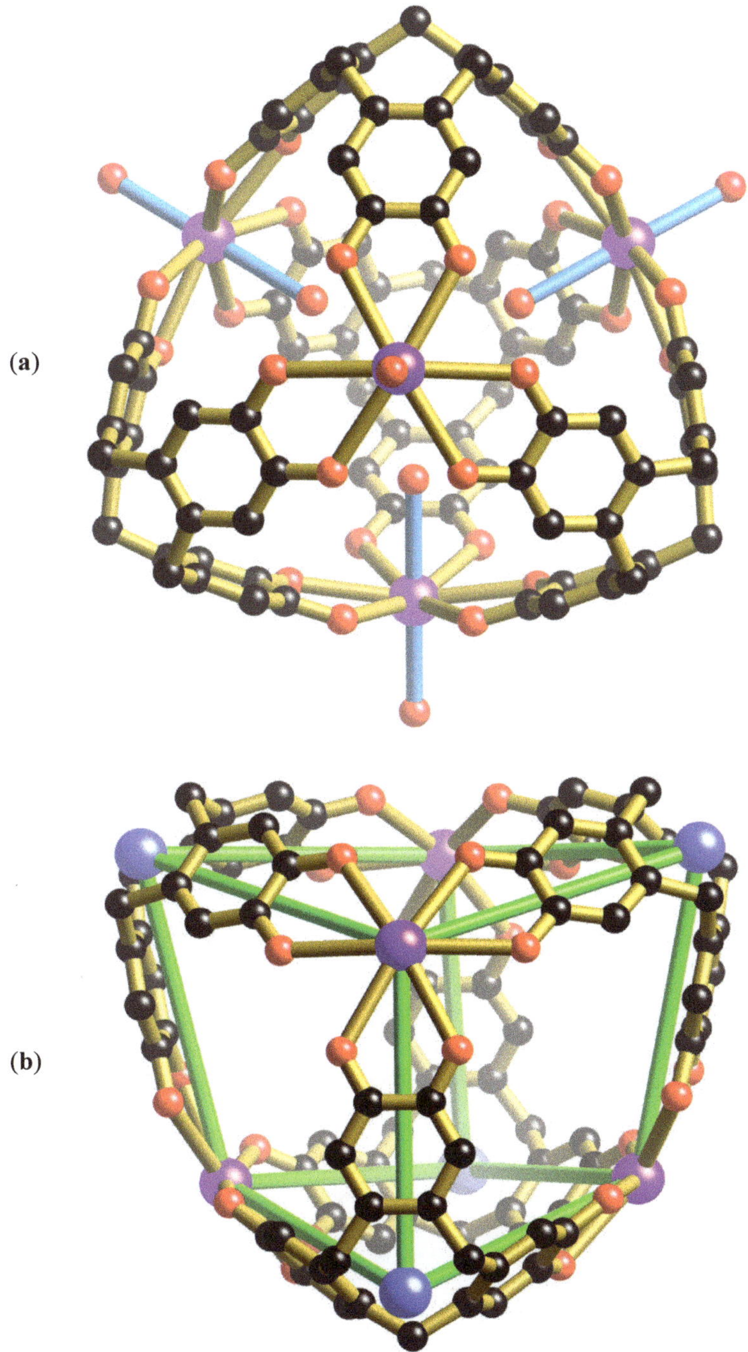

Figure 2. *Cont.*

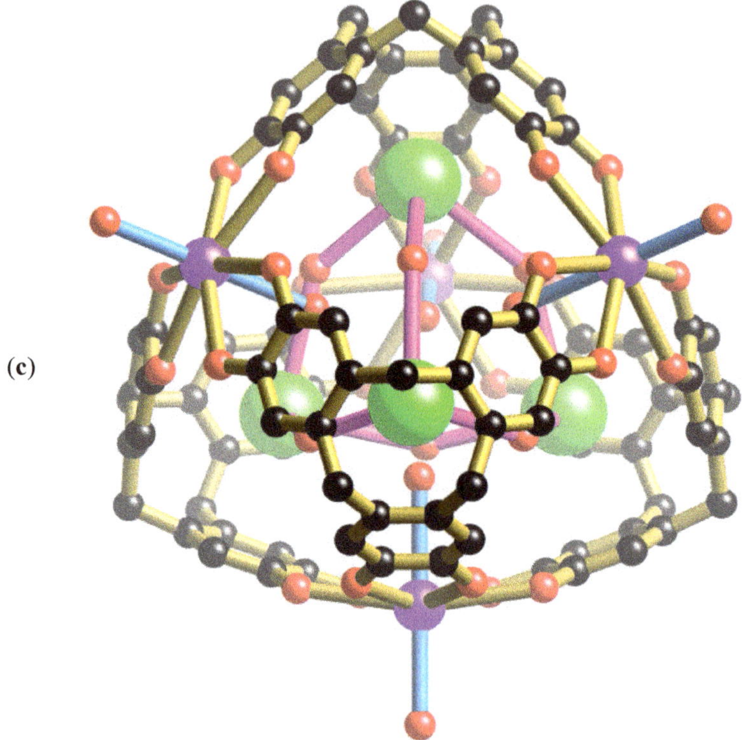

Figure 2. The structure of **III** showing (**a**) the $[K_4(H_6ctc)_4(H_2O)_8]^{4+}$ cage (**b**) the $[K_4(H_6ctc)_4(H_2O)_8]^{4+}$ cage indicating the topological relationship with a cube; each blue sphere represents a point on the 3-fold axis that passes through each of the H_6ctc molecules and (**c**) the $[K_4(H_6ctc)_4(H_2O)_8]^{4+}$ cage showing the inclusion of the four internal Cs^+ ions and six water molecules; C black, O red, K purple, Cs green.

In addition to the four water molecules that extend towards the centre of the cubic cage from the K^+ ions, there are four Cs^+ ions that associate with the internal aromatic surfaces of the $[K_4(H_6ctc)_4(H_2O)_8]^{4+}$ cage. Each of the Cs^+ ions is located on a 3-fold axis that passes through one of the K^+ ions, its two *trans* water molecules and along the central axis of a H_6ctc molecule with which it is in contact (Figure 2c). The location of the Cs^+ cation within the H_6ctc bowl is similar to the arrangement of Cs^+ ions found in the tetrahedral cages, $[(VO)_6(ctc)_4]^{12-}$ and $[(PhSi)_6(ctc)_4]^{6-}$. As with the tetrahedral cages, the closest contact is made with the aromatic C atoms bound to the methylene groups of the H_6ctc. As well as interacting with the internal surface of the H_6ctc, each Cs^+ ion is bound to a trio of symmetry-related water molecules each of which bridges a pair of Cs^+ ions. The result is a $[Cs_4(H_2O)_6]^{4+}$ cationic assembly located inside each cage. From a structural perspective, this cationic unit resembles an adamantane-type skeleton which is apparent in molecules such as P_4O_6.

The $[K_4(H_6ctc)_4(H_2O)_8]^{4+}$ cage has six windows corresponding to the faces of a cube. As indicated in Figure 3a, two oxygen atoms of a phosphate ion are involved in forming four equivalent hydrogen bonds with phenolic groups of the cage (O···O, 2.59 Å); the phosphorus atom sits above the mid-point of each window. The highly complementary nature of this interaction is apparent in the space-filling representation (Figure 3b) in which the phosphate ion neatly plugs each cage window. In addition to the four hydrogen bonds in a single window, each phosphate forms a quartet of hydrogen bonds to the phenolic hydrogen atoms lying in a window of a neighbouring cage. Thus, each phosphate anion participates, as an acceptor, in eight equivalent hydrogen bonds as shown in Figure 3c. Through the bridging phosphate ions, each $[K_4(H_6ctc)_4(H_2O)_8]^{4+}$ cage is linked to six equivalent cages each located in a neighbouring, face sharing, unit cell. The result is a three-dimensional primitive cubic network.

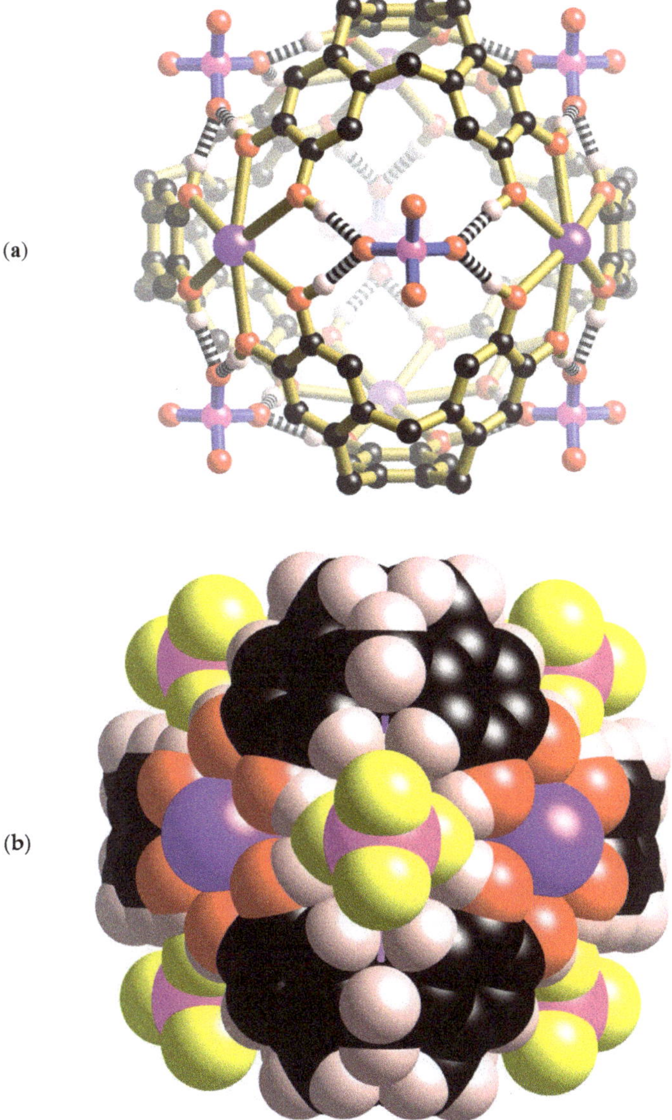

Figure 3. *Cont.*

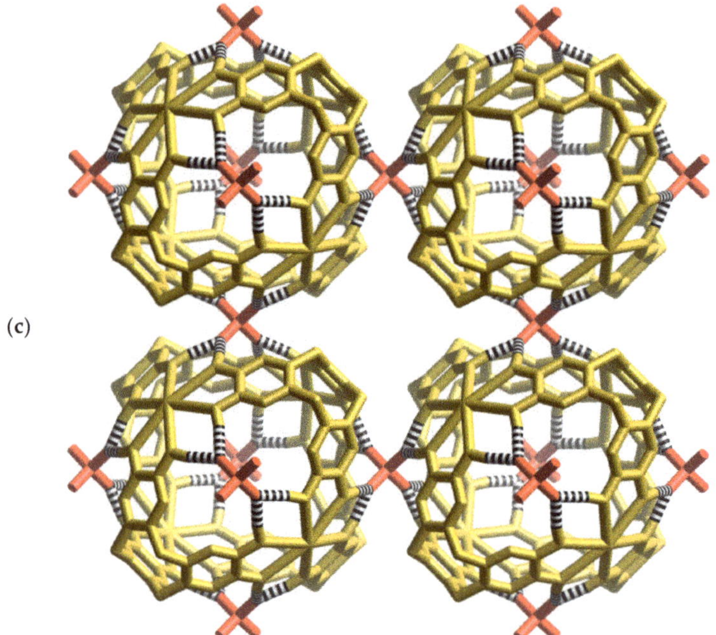

Figure 3. Hydrogen bonding interactions between the phosphate anions and the $[K_4(H_6ctc)_4(H_2O)_8]^{4+}$ cage (coordinated water molecules omitted for clarity) in **III**; (**a**) a single cage linked to six phosphate anions through 24 equivalent hydrogen bonds (H atoms bound to C atoms omitted for clarity) H pale pink, C black, O red, P pink, K purple; (**b**) a space filling representation; H pale pink, C black, O (H_6ctc) red, O ($PO_4{}^{3-}$) yellow; P pink, K purple and (**c**) part of the 3D hydrogen bonding network with each phosphate forming eight equivalent hydrogen bonds.

The hydrogen bonded network carries a net negative charge with the phosphate ions located in each face of the unit cell contributing 9- per unit cell whilst the $[K_4(H_6ctc)_4(H_2O)_8]^{4+}$ cage, along with the four internal Cs^+ ions, has a net charge of 8+. Charge balance is achieved by the incorporation of a metal complex in which a Cs^+ ion is coordinated by six symmetry related acetone molecules that lie along the unit cell axes (see Figure 4a). Although examples of alkali metal ions coordinated by six acetone molecules are known, they are part of aggregates involving more than one alkali metal ion e.g., cationic clusters such as $[Na_3((CH_3)_2CO)_9]^{3+}$ [27]. A search of the Cambridge Crystallographic Database reveals only three discrete $M[(CH_3)_2CO]_6{}^{n+}$ complexes, $[Mg((CH_3)_2CO)_6]^{2+}$ which serves as a counter cation for bis((μ2-iodo)-iodo-copper(I)) [28], $[Ni[(CH_3)_2CO]_6]^{2+}$ which accompanies a Ni-Cu carbide carbonyl cluster [29] and $[Fe((CH_3)_2CO)_6]^{2+}$ which is the countercation for $[FeCl_4]^-$ [30]. In each of these hexaacetone complexes there is a bend at the carbonyl oxygen atom whereas in **III** the Cs^+-O=C angle is 180° with the carbonyl oxygen group aligned along the unit cell axes. As indicated in Figure 4a the thermal ellipsoid of the carbonyl oxygen is somewhat elongated in a direction perpendicular to the Cs-O bond, however, attempts to model the oxygen atom over two symmetry related sites were unsuccessful.

A $Cs[(CH_3)_2CO]_6{}^+$ complex is located at each corner of the cubic unit cell as depicted in Figure 4b. The methyl hydrogen atoms make relatively close contact along the edges of the unit cell to create a cubic framework. The mean planes of the two acetone molecules that approach each other along each cell edge are rotated by 90° relative to each other.

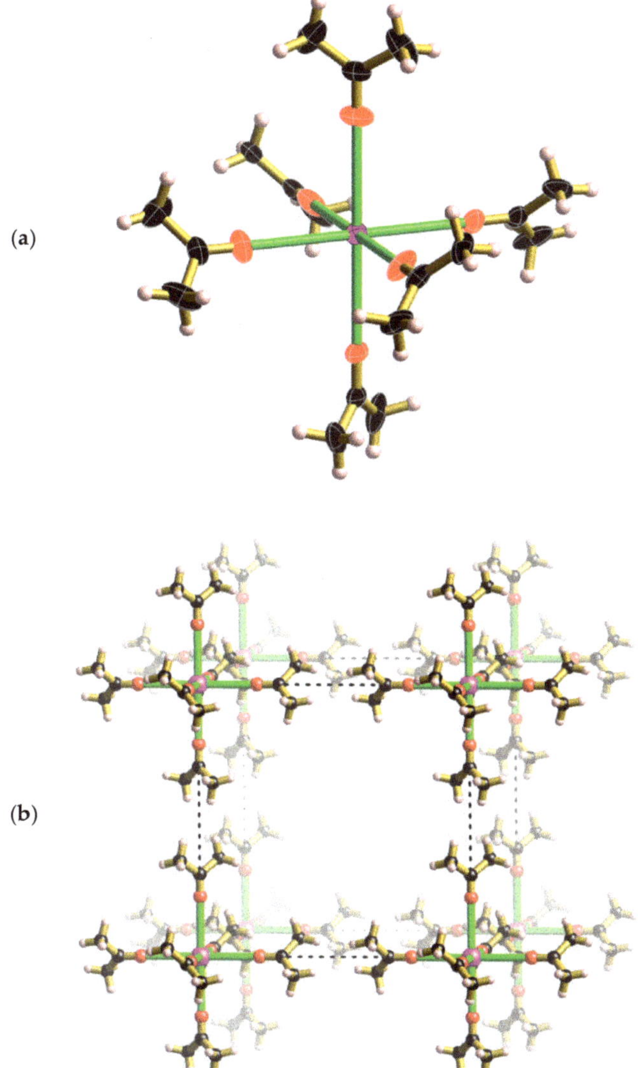

Figure 4. The Cs[(CH$_3$)$_2$CO]$_6$$^+$ complex in **III**; (**a**) a single complex with thermal ellipsoids shown at the 50% level, (**b**) eight Cs[(CH$_3$)$_2$CO]$_6$$^+$ complexes located at the vertices of the cubic unit cell; H pale pink, C black, O red, Cs pink.

From a crystallographic perspective one of the most appealing aspects of the structure is the high level of symmetry that arises from the assembly of the components. Figure 5 shows the relative positions of the Cs[(CH$_3$)$_2$CO]$_6$$^+$ complex, the [K$_4$(H$_6$ctc)$_4$(H$_2$O)$_8$]$^{4+}$ cage and the phosphate anions within a unit cell of **III**. The centre of the unit cell corresponds to the midpoint of the cage and is on a site of −43 m symmetry as is the Cs$^+$ ion of the Cs[(CH$_3$)$_2$CO]$_6$$^+$ complex. The phosphate ion, located in the middle of each face, is situated on a site of −42 m symmetry. The body diagonals of the unit cell are co-linear with 3-fold axes and pass through the K$^+$ ions and coordinated water molecules of the cage, the internal Cs$^+$ ion, the 9-membered ring of a H$_6$ctc in addition to the Cs$^+$ ion of the Cs[(CH$_3$)$_2$CO]$_6$$^+$ complex.

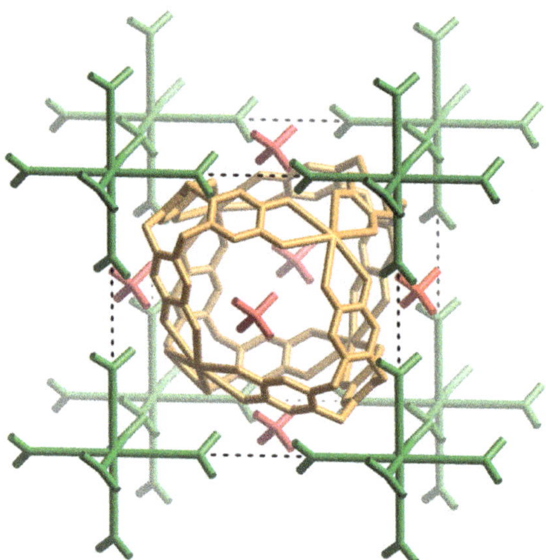

Figure 5. The unit cell of **III** showing the $Cs[(CH_3)_2CO]_6^+$ complexes at the unit cell vertices, the $[K_4(H_6ctc)_4(H_2O)_8]^{4+}$ cage and the phosphate anions located in the middle of each face; $[K_4(H_6ctc)_4(H_2O)_8]^{4+}$ cage gold connections; $Cs[(CH_3)_2CO]_6^+$ complexes green connections; phosphate red connections.

In investigations of the clam type structure of the type depicted in Figure 1b it was found that the clam would form both with Cs^+ and Rb^+ and it would thus appear that both ions are of an appropriate size to interact with the internal aromatic surfaces of the cyclotricatechylene. On this basis we attempted to synthesise compound **III** but with Rb^+ in place of Cs^+. Unfortunately, it was not possible to obtain a homogenous bulk product but we were successful in transferring a crystal from the mother liquor and into a protective oil for structure determination. The combination of H_6ctc with K_2HPO_4 and $RbOH$ in aqueous acetone yields crystals of composition $[Rb((CH_3)_2CO)_6][K_2Rb_2(H_6ctc)_4(H_2O)_6][Rb_4(H_2O)_6](PO_4)_3$ (**IV**). Apart from the inclusion of Rb^+ in place of Cs^+, the metal ions that form the $[M_4(H_6ctc)_4]^{4+}$ cages are an equal mixture of disordered K^+ and Rb^+ ions. Another difference relates to the occupation of the cage cavity by water molecules. There are only two coordinated water molecules inside the cage which are disordered over four symmetry related sites. In addition, the non-coordinated water molecules are now disordered over two sites. Presumably the smaller radius of the Rb^+ ion results in the water molecules being unable to symmetrically bridge the internal Rb^+ ions. The crystalline product in this case initially appeared to be homogeneous but crystallinity of the material was lost when the crystals were removed from the mother liquor.

The successful generation of the cubic compounds described above, prompted an investigation to determine whether other tetrahedral oxyanions employed in place of phosphate would yield similar structures. The combination of H_6ctc, $KHSO_4$ and $CsOH$ in aqueous acetone yields the compound $[Cs((CH_3)_2CO)_6][K_4(H_6ctc)_4(H_2O)_8][Cs(H_2O)_9](SO_4)_3$ (**V**). The structure closely resembles the phosphate analogue (**III**), except in regards to the contents of the $[K_4(H_6ctc)_4(H_2O)_8]^{4+}$ cage. Instead of four Cs^+ ions inside the cage, only one Cs^+ anion is present which is disordered over the four H_6ctc bowl sites. The decrease in the number of internal cations is required to maintain charge balance following the inclusion of sulphate in place of phosphate. In the crystallographic refinement of the structure, the best modelling of the disorder corresponded to refinement of the bowl sites with 25% occupancy by Cs^+ and 75% occupancy by the oxygen atom of a water molecule. The remaining six water molecules in the cage were each found to be disordered over a pair of sites.

The final compound considered in this series is $[Rb((CH_3)_2CO)_6][Rb_2K_2(H_6ctc)_4(H_2O)_6]$ $[Rb(H_2O)_9](SO_4)_3$ (**VI**) which is formed from the reaction of H_6ctc with $KHSO_4$ and $RbOH$, again in aqueous acetone. The internal contents of the cage are similar to that found in (**V**) but with Rb^+ in place of Cs^+ and only two coordinated water molecules instead of four. The cage itself is formed from a combination of H_6ctc with Rb^+ ions and K^+ ions as found for (**IV**).

A summary of the components from which compounds **III–VI** are formed is presented in Table 2.

Table 2. Summary of the components in compounds **III–VI**.

Compound	$M_4(H_6ctc)_4$ Cage, $M^+ =$	Internal Cage Cation/s	$M((CH_3)_2CO)_6$, $M^+ =$	Anion
III	$4 \times K^+$	$4 \times Cs^+$	Cs^+	PO_4^{3-}
IV	$2 \times K^+, 2 \times Rb^+$	$4 \times Rb^+$	Rb^+	PO_4^{3-}
V	$4 \times K^+$	$1 \times Cs^+$	Cs^+	SO_4^{2-}
VI	$2 \times K^+, 2 \times Rb^+$	$1 \times Rb^+$	Rb^+	SO_4^{2-}

The concept of complementarity, in which certain specific favourable interactions lead to the self-assembly of components into larger structures, is a key theme in both supramolecular chemistry and crystal engineering. In the examples considered in this current work, the complementary interactions involve three neutral molecules (cyclotricatechylene, acetone and water), two cations (K^+ and either Rb^+ or Cs^+) and an anion (either phosphate or sulphate) that participate in a variety of primary and secondary bonding interactions. Quite remarkably the six species combine to generate a series of crystalline materials exhibiting particularly high symmetry. It should be noted that whilst the initial formation of cubic crystals was serendipitous, the subsequent generation of related compounds was achieved through a recognition that some of the key interactions between components may persist upon the substitution of some components for structurally similar units.

Whilst prediction of the initial structure would seem most unlikely, the types of interactions between components are similar to those commonly encountered in the assembly of supramolecular species e.g., hydrogen bonding and metal coordination. Each of the compounds **III–VI** represents an additional example of the clear affinity of the larger alkali metal ions for the aromatic surfaces of cyclotricatechylene, although in contrast to previous examples the tris (catechol) is fully protonated.

Perhaps the most surprising aspect of the structures is the presence of the octahedral $M[(CH_3)_2CO]_6^+$ (M = Rb, Cs) complexes in each structure. At first inspection it would seem quite remarkable that such a species, with linear M-C=O links, would be generated particularly given that the solvent from which the crystals were obtained contains water, which would normally be expected to have a much greater affinity for the alkali metal ions than acetone. One can only speculate about the reason for the presence of the unusual $M[(CH_3)_2CO]_6^+$ octahedral complexes but it would seem likely that its size and shape neatly matches inter-cage spaces that are generated by the anionic hydrogen bonded network formed from the combination of the cages with the tetrahedral anions (PO_4^{3-} or SO_4^{2-}). As is apparent from inspection of Figure 6, the van der Waals contact between coordinated acetone molecules half-way along each edge of the unit cell leads to a M···M separation equal to the unit cell edge length. This distance matches the separation between the centres of oxyanions on opposite faces of the cubic unit cell which form hydrogen bonds with the cage at the centre of the unit cell. Close inspection of each of the structures reveals complementary interactions between the $M_4(H_2O)_8(H_6ctc)_4^{4+}$ cages and the surrounding $M[(CH_3)_2CO]_6^+$ octahedral complexes. Trios of acetone molecules orthogonal (*fac*) to one another in the octahedral $Cs[(CH_3)_2CO]_6^+$ complex are arranged in one of two ways. As is apparent from Figure 6, in one arrangement, the methyl hydrogen atoms are pointing towards catechol oxygen atoms which are bound to a single K^+ centre. The oxygen atom of the externally coordinated water molecule sits in the middle of a trio of methyl groups. The second case provides a contrast with three acetone molecules forming a concave cavity which acts as a bowl for a H_6ctc unit. In this arrangement, the mean plane of the acetone molecule is close to parallel with the aromatic surface to which it makes close contact. Through these two types of interaction each cage associates with eight $Cs[(CH_3)_2CO]_6^+$ complexes and each complex is surrounded by eight cages.

Inspection of Table 1 reveals that the sulphate containing compounds, **V** and **VI** have longer unit cell lengths than **III** and **IV**. This would appear to be a consequence of the difference in the hydrogen bond distances. In the case of the SO_4^{2-} structures the O⋯O separation between the sulphate oxygen atom and the oxygen atom of the H_6ctc is 2.73 Å for **V** and 2.72 Å for **VI**. This is significantly longer than the separation of 2.59 Å found in **III** and **IV** and perhaps reflects a stronger interaction between the H_6ctc and the anion when the charge on the anion is 3-.

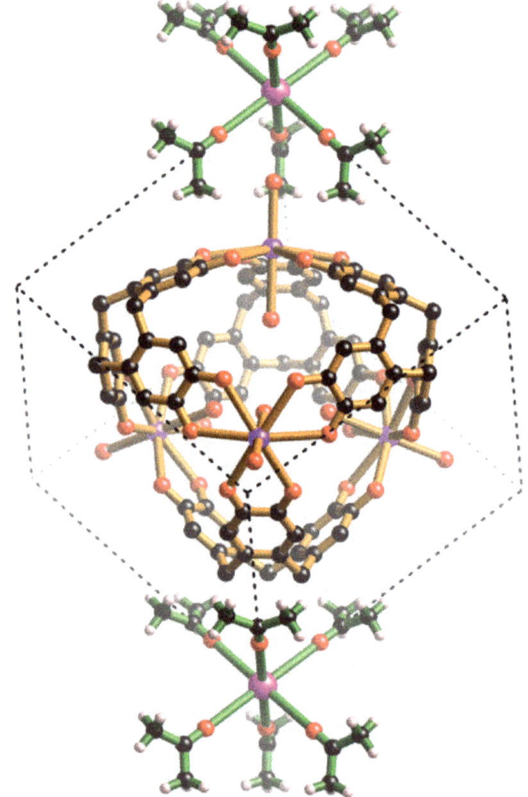

Figure 6. A representation of complementary interactions between the $Cs[(CH_3)_2CO]_6^+$ complex, the $[K_4(H_6ctc)_4(H_2O)_8]^{4+}$ cage and $Cs[(CH_3)_2CO]_6^+$ complexes; H pale pink, C black, O red, K purple, Cs pink.

4. Concluding Remarks

Complementary interactions provide a foundation for generating a wide range of discrete supramolecular assemblies or extended structures within crystals. In this work, we discovered an unexpected example of extreme complementarity in which six components are brought together in order to produce an aesthetically pleasing, highly symmetrical network structure. Through recognition of the nature of the complementary interactions that underpin this system, it was found that the same structural type could be replicated with different alkali metal ions and anions. It is clear that the reported system is amenable to at least some degree of structural tinkering. The insights provided by this work offer opportunities to generate further members of this family which may include the incorporation of a wider range of tetrahedral anions and the employment of other metal ions.

Whilst the relatively simple cations, anions and solvent molecules from which the crystals have been formed have all participated in the generation of the four supramolecular crystalline materials described above, the cyclotricatechylene has played the key structural role in this system. In its fully protonated neutral form, it has been able to coordinate to a trio of K^+ or Rb^+ ions through hydroxyl oxygen atoms to form the cage. In addition, each H_6ctc also serves as a hydrogen bond donor and forms six identical hydrogen bonds to three different anions by employing the same hydroxyl groups. Furthermore, its ability to interact with large alkali metal cations i.e., Rb^+ and Cs^+ through π-interactions allows the cage to accommodate the requisite number of alkali cations needed for charge balance. It is interesting to note that both cations and anions prefer to form bonds (hydrogen bonds or coordinate bonds) with the neutral H_6ctc molecule rather each other. Clearly, the symmetry of cyclotricatechylene (C_{3v} in its most symmetrical conformation) and versatility in being able to participate in a variety of different interactions makes it a particularly useful molecular building block for supramolecular assemblies and there would appear to be great scope for expanding the type of assemblies that may be formed in its combination with cations, anion and neutral species.

Author Contributions: Conceptualization, J.L.H., S.M.R., B.F.A.; methodology, J.L.H., S.M.R., B.F.A.; investigation, J.L.H., S.M.R., B.F.A., T.A.H., K.F.W.; writing—original draft preparation, J.L.H., B.F.A.; writing—review and editing, J.L.H., S.M.R., B.F.A., T.A.H., K.F.W.; funding acquisition, B.F.A. All authors have read and agreed to the published version of the manuscript.

Funding: The authors gratefully acknowledge the funding by the Australian Research Council (DP180101414).

Conflicts of Interest: The authors declare no conflicts of interest.

References

1. 1987 Nobel Prize in Chemistry. Available online: https://www.nobelprize.org/prizes/chemistry/1987/summary/ (accessed on 28 May 2020).
2. Christensen, J.J.; Eatough, D.J.; Izatt, R.M. The Synthesis and Ion Bindings of Synthetic Multidentate Macrocyclic Compounds. *Chem. Rev.* **1974**, *74*, 351–384. [CrossRef]
3. Cram, D.J. Preorganization-From Solvents to Spherands. *Angew. Chem. Int. Ed.* **1986**, *25*, 1039–1057. [CrossRef]
4. Izatt, R.M.; Bradshaw, J.S.; Nielsen, S.A.; Lamb, J.D.; Christensen, J.J.; Sen, D. Thermodynamic and kinetic data for cation-macrocycle interaction. *Chem. Rev.* **1985**, *85*, 271–339. [CrossRef]
5. Izatt, R.M.; Pawlak, K.; Bradshaw, J.S.; Bruening, R.L. Thermodynamic and kinetic data for macrocycle interactions with cations and anions. *Chem. Rev.* **1991**, *91*, 1721–2085. [CrossRef]
6. Kauffmann, E.; Lehn, J.-M.; Sauvage, J.-P. Enthalpy and Entropy of Formation of Alkali and Alkaline-Earth Macrobicyclic Cryptate Complexes [1]. *Helv. Chim. Acta* **1976**, *59*, 1099–1111. [CrossRef]
7. Lehn, J.M. Cryptates: The chemistry of macropolycyclic inclusion complexes. *Accounts Chem. Res.* **1978**, *11*, 49–57. [CrossRef]
8. Pedersen, C.J. The Discovery of Crown Ethers. *Science* **1988**, *241*, 536–540. [CrossRef]
9. Fujita, D.; Ueda, Y.; Sato, S.; Mizuno, N.; Kumasaka, T.; Fujita, M. Self-assembly of tetravalent Goldberg polyhedra from 144 small components. *Nature* **2016**, *540*, 563–566. [CrossRef]
10. Fujita, M.; Oguro, D.; Miyazawa, M.; Oka, H.; Yamaguchi, K.; Ogura, K. Self-assembly of ten molecules into nanometre-sized organic host frameworks. *Nature* **1995**, *378*, 469–471. [CrossRef]
11. Desiraju, G.R. Crystal engineering: Solid state supramolecular synthesis. *Curr. Opin. Solid State Mater. Sci.* **1997**, *2*, 451–454. [CrossRef]
12. Desiraju, G.R. Supramolecular Synthons in Crystal Engineering—A New Organic Synthesis. *Angew. Chem. Int. Ed.* **1995**, *34*, 2311–2327. [CrossRef]
13. Russell, V.A.; Etter, M.C.; Ward, M.D. Layered Materials by Molecular Design: Structural Enforcement by Hydrogen Bonding in Guanidinium Alkane- and Arenesulfonates. *J. Am. Chem. Soc.* **1994**, *116*, 1941–1952. [CrossRef]

14. Russell, V.A.; Evans, C.C.; Li, W.; Ward, M.D. Nanoporous Molecular Sandwiches: Pillared Two-Dimensional Hydrogen-Bonded Networks with Adjustable Porosity. *Science* **1997**, *276*, 575–579. [CrossRef]
15. Hardie, M.J.; Raston, C.L.; Wells, B. Altering the inclusion properties of CTV through crystal engineering: CTV, carborane, and DMF supramolecular assemblies. *Chem. Eur. J.* **2000**, *6*, 3293–3298. [CrossRef]
16. Hardie, M. Recent advances in the chemistry of cyclotriveratrylene. *Chem. Soc. Rev.* **2010**, *39*, 516–527. [CrossRef] [PubMed]
17. Abrahams, B.F.; Fitzgerald, N.J.; Robson, R. Cages with Tetrahedron-Like Topology Formed from the Combination of Cyclotricatechylene Ligands with Metal Cations†. *Angew. Chem. Int. Ed.* **2010**, *49*, 2896–2899. [CrossRef]
18. Abrahams, B.F.; Boughton, B.A.; Fitzgerald, N.J.; Holmes, J.; Robson, R. A highly symmetric diamond-like assembly of cyclotricatechylene-based tetrahedral cages. *Chem. Commun.* **2011**, *47*, 7404. [CrossRef]
19. Holmes, J.; Abrahams, B.F.; Ahveninen, A.; Boughton, B.A.; Hudson, T.A.; Robson, R.; Thinagaran, D. Self-assembly of a Si-based cage by the formation of 24 equivalent covalent bonds. *Chem. Commun.* **2018**, *54*, 11877–11880. [CrossRef]
20. Kawakami, Y.; Ogishima, T.; Kawara, T.; Yamauchi, S.; Okamoto, K.; Nikaido, S.; Souma, D.; Jin, R.-H.; Kabe, Y. Silane catecholates: Versatile tools for self-assembled dynamic covalent bond chemistry. *Chem. Commun.* **2019**, *55*, 6066–6069. [CrossRef]
21. Abrahams, B.F.; Fitzgerald, N.J.; Hudson, T.A.; Robson, R.; Waters, T. Closed and Open Clamlike Structures Formed by Hydrogen-Bonded Pairs of Cyclotricatechylene Anions that Contain Cationic "Meat". *Angew. Chem. Int. Ed.* **2009**, *48*, 3129–3132. [CrossRef]
22. Mehta, N.; Abrahams, B.F.; Goerigk, L. Clam-like Cyclotricatechylene-based Capsules: Identifying the Roles of Protonation State and Guests as well as the Drivers for Stability and (Anti-)Cooperativity. *Chem. Asian J.* **2020**, *15*, 1301–1314. [CrossRef] [PubMed]
23. Hyatt, J.A. Octopus molecules in the cyclotriveratrylene series. *J. Org. Chem.* **1978**, *43*, 1808–1811. [CrossRef]
24. *CrysAlis PRO Software System*, Rigaku Oxford Diffraction: Oxfordshire, UK, 2018.
25. Sheldrick, G.M. SHELXT - integrated space-group and crystal-structure determination. *Acta Crystallogr. Sect. A Found. Adv.* **2015**, *71*, 3–8. [CrossRef] [PubMed]
26. Sheldrick, G.M. Crystal structure refinement with SHELXL. *Acta Crystallogr. Sect. C Struct. Chem.* **2015**, *71*, 3–8. [CrossRef] [PubMed]
27. Adonin, S.; Peresypkina, E.; Sokolov, M.N.; Fedin, V.P. Trinuclear iodobismuthate complex $[Na_3(Me_2CO)_{12}][Bi_3I_{12}]$: Synthesis and crystal structure. *Russ. J. Coord. Chem.* **2014**, *40*, 867–870. [CrossRef]
28. Hoyer, M.; Hartl, H. Synthese und Strukturuntersuchungen von Iodocupraten(I). XV Iodocuprate(I) mit solvatisierten Kationen: $[Li(CH_3CN)_4][Cu_2I_3]$ und $[Mg\{(CH_3)_2CO\}_6][Cu_2I_4]$. *Z. Anorg. Allg. Chem.* **1992**, *612*, 45–50. [CrossRef]
29. Bernardi, A.; Ciabatti, I.; Femoni, C.; Iapalucci, M.C.; Longoni, G.; Zacchini, S. Ni–Cu tetracarbide carbonyls with vacant Ni(CO) fragments as borderline compounds between molecular and quasi-molecular clusters. *Dalton Trans.* **2013**, *42*, 407–421. [CrossRef] [PubMed]
30. Hanson, C.S.; Psaltakis, M.C.; Cortes, J.J.; Devery, J.J.; Devery, I.J.J. Catalyst Behavior in Metal-Catalyzed Carbonyl-Olefin Metathesis. *J. Am. Chem. Soc.* **2019**, *141*, 11870–11880. [CrossRef]

 *chemistry*

Article

Metallosupramolecular Compounds Based on Copper(II/I) Chloride and Two Bis-Tetrazole Organosulfur Ligands

Olaya Gómez-Paz [1], Rosa Carballo [1,*], Ana Belén Lago [2,*] and Ezequiel M. Vázquez-López [1]

[1] Departamento de Química Inorgánica, Instituto de Investigación Sanitaria Galicia Sur (IISGS)-Universidade de Vigo, 36310 Vigo, Galicia, Spain; olgomez@uvigo.es (O.G.-P.); ezequiel@uvigo.es (E.M.V.-L.)

[2] Departamento de Química, Facultad de Ciencias, Sección Química Inorgánica, Universidad de la Laguna, 38206 La Laguna, Spain

* Correspondence: rcrial@uvigo.es (R.C.); alagobla@ull.edu.es (A.B.L.)

Received: 13 January 2020; Accepted: 29 January 2020; Published: 4 February 2020

Abstract: The present work assesses the ability of two flexible bis-tetrazole organosulfur ligands to build up different metallosupramolecular compounds based on the analysis of the different supramolecular interactions. The reaction of copper(II) chloride dihydrate with the *N,N'*-donor dithioether ligands bis(1-methyl-1*H*-tetrazole-5-ylthio)methane (BMTTM) and 1,2-bis(1-methyl-1*H*-tetrazole-5-ylthio)ethane (BMTTE) was investigated using different synthetic methods. Four compounds have been obtained as single crystals: two pseudopolymorphic 1D Cu(II) coordination polymers with the ligand BMTTM, a 2D Cu(II) coordination polymer and a discrete Cu(I) tetramer with the BMTTE ligand. The effects of the weak interactions on the crystal packing and the Hirshfeld surfaces of the structures were analyzed to clarify the nature of the intermolecular interactions.

Keywords: copper chloride complexes; H-bonding pattern; tetrazole ligands; X-ray diffraction; Hirshfeld surfaces

1. Introduction

Structurally well-defined supramolecular architectures produced by the formation of ordered crystalline materials have attracted considerable attention in recent years due to their different novel chemical properties and their possible new applications. Coordination chemistry has played a central role in the blossoming of this fast-evolving field. In this way, metallosupramolecular chemistry, which concerns non-covalent interactions between discrete or polymeric coordination compounds, has become an interdisciplinary research area that has provided insights into and spurred developments across biology, chemistry, nanotechnology, materials science and physics [1].

Amongst the metallosupramolecular compounds, supramolecular metal–organic frameworks (SMOFs) are materials that can be considered as analogs to metal–organic frameworks (MOFs) in the sense that some coordination bonds are replaced by hydrogen bonds as directional interactions to build the final crystal. In SMOFs, the coordination bonds are released from guiding the crystal structure and supramolecular interactions play this role instead. The strategy for the preparation of SMOFs is based on the synthesis of coordination compounds, with the choice of metal center made on the basis of its coordinative preferences and the ligand or a combination of ligands with the ability to coordinate the metal cation and also facilitate several weak interactions between the diverse rigid molecular units [2].

The final goal of supramolecular chemistry is to understand the inherent complexities of the association mechanisms of molecular and ionic building blocks organized through multiple noncovalent interactions [3,4]. The relatively greater strength of ligand–metal coordination bonds when compared with other noncovalent interactions allows the stabilization of a range of different structures, whereas the weak and reversible forces are key to understanding these self-assembling systems. In terms of the weak intermolecular noncovalent interactions, the analysis of C–H$\cdots$X (X = Cl, S, N) hydrogen bonds in metallosupramolecular systems has received less attention despite its proven participation in several biological systems [5,6]. In this way, the hydrogen bond acceptor capability of halogens has attracted attention on a number of fronts. In the context of metallosupramolecular chemistry, halide ligands (M–X) have been used to drive the self-assembly of coordination compounds due to their directionality and versatility [7,8].

In order to facilitate the hydrogen bond acceptor role of a halide, such as chloride in a SMOF, copper(II) chloride is used as the metal source and, although the Cu(II) cation is stable under ambient conditions, Cu(II)-to-Cu(I) transformations at room temperature can be produced by reaction conditions such as temperature, pH value, solvents and pressure due to the low standard electrode potential between Cu(II) and Cu(I).

The tetrazole organosulfur derivatives and their transition metal complexes are important in medicinal chemistry and drug design [9] and also as industrial materials [10]. In the field of metallosupramolecular chemistry, these compounds are particularly interesting since the tetrazole moiety contains several nitrogen atoms that can facilitate simultaneously the coordination to one or more metal centers and the formation of hydrogen bonds acting as acceptors.

With the aim of studying the role of C–H$\cdots$X (X = Cl, S, N) hydrogen bonds in the crystalline supramolecular networks based on copper(II/I) chloride/bis-tetrazole organosulfur systems, we report here the crystal structures of four compounds resulting from reactions under different synthetic conditions between the ligands bis(1-methyl-1H-tetrazole-5-ylthio)methane (BMTTM) and 1,2-bis(1-methyl-1H-tetrazole-5-ylthio)ethane (BMTTE) and copper(II) chloride. These ligands have attractive features, such as the multiple heteroatomic potential coordination sites, six N donors and two S atoms, which also contribute to the flexibility of the ligands. In both ligands, there are three adjacent N donors in each methyl-tetrazole group, and these may promote the construction of multinuclear clusters.

In the literature there is structural information on several compounds based on copper(II/I) polyoxometalates (POMs = $H_2Mo_8O_{26}^{2-}$, $PMo_{12}O_{40}^{3-}$, $SiW_{12}O_{40}^{3-}$, $PW_{12}O_{40}^{3-}$, $SiMo_{12}O_{40}^{4-}$, $SiW_{12}O_{40}^{4-}$, $HSiMo_{12}O_{40}^{3-}$) and these two ligands [11–14]. These compounds are 1D, 2D or 3D coordination polymers in which the ligands are able to coordinate two, three or four copper cations by chelating and/or bridging coordination modes. However, a study of the weak interactions responsible for the supramolecular organization has not been undertaken for any of these compounds.

2. Results and Discussion

2.1. Synthesis of the Complexes

The study of the reactivity in the $CuCl_2$/BMTTM or BMTTE system was performed using different stoichiometric ratios and four synthetic methods: stirring at room temperature, diffusion, reflux under microwave irradiation and hydro/solvothermal techniques. Conventional synthesis at room temperature afforded the compounds in good yields. However, hydrothermal and microwave methods, providing the same compounds in lower yields, were used to synthesized compounds as pure materials and as single crystals. Acetonitrile was used as the solvent for the reactions due to the better solubility of the ligands in this solvent. The synthetic conditions allowed the preparation of the crystalline complexes **1**, **1·solv**, **2** and **3** (Scheme 1).

In the reactions of copper(II) chloride with BMTTM, compound **1** was obtained in good yield with a 1:1 metal/ligand ratio by diffusion and by stirring at room temperature. However, when the reaction was performed under microwave irradiation, in addition to **1**, some crystals of the pseudopolymorph **1·solv** ($4CH_3CN$) were also obtained. The hydro/solvothermal methods tested at different metal/BMTTM ligand ratios at different temperatures between 70 and 160 °C in all cases yielded the copper(I) 2D coordination polymer **Cu(mtS)** (mtS = 1-methyl-1*H*-tetrazole-5-thiolato) [15], which indicates that cleavage of the BMTTM ligand had taken place under these synthetic conditions.

Reactions with BMTTE were conditioned by its low solubility in all of the solvents tested and, as a consequence, a copper complex could not be isolated by diffusion due to the rapid crystallization of BMTTE [16,17]. Copper(II) compound **3** was obtained as an orange crystalline powder upon stirring at room temperature using a 1:1 molar ratio, and as orange single crystals by reaction under microwave irradiation using a 4:1 metal/ligand molar ratio. The copper(I) complex **2** was obtained under hydrothermal conditions at 160 °C with a 3:1 metal/ligand ratio. This hydrothermal reduction of copper(II) to copper(I) was previously observed in the preparation of copper polyoxometalates of BMTTM and BMTTE from copper(II) acetate, although the authors indicated that metal/ligand ratios greater than 10:1 are required for this reduction to occur [11–14].

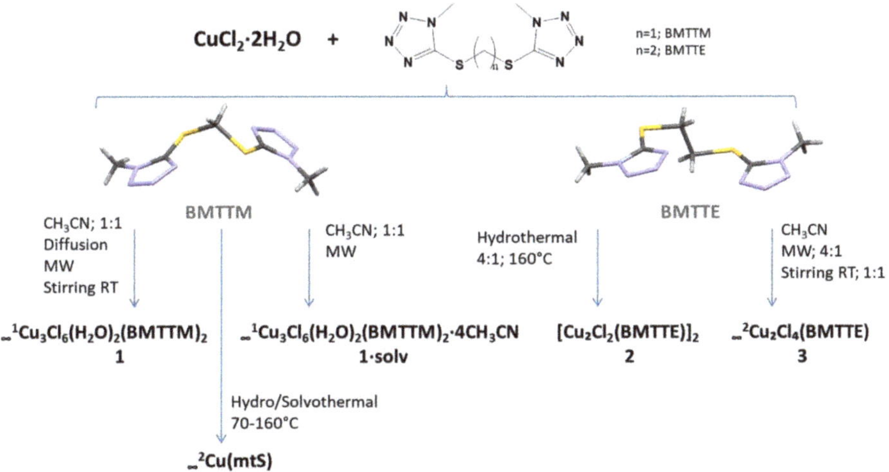

Scheme 1. Summary of the synthetic routes for the complexes and the Cu(II):L stoichiometric ratios, temperature and solvent used where relevant (RT: room temperature; MW: microwave).

2.2. Structural Studies: General Features

All of the compounds described here were isolated as single crystals and their structures were elucidated by X-ray diffraction. The combination of the flexible polydentate ligands with different chlorocuprate clusters resulted in the formation of 1D coordination polymer chains (**1** and **1·solv**) or a 2D coordination layer (**3**). Moreover, the reduction of copper(II) to copper(I) produced a stable discrete tetrameric Cu(I) coordination compound (**2**). The main structural features for each compound are provided in Table 1. The structures can be deconstructed into two components: trinuclear units {Cu₃Cl₆N₆O₂} in **1** and **1·solv**, tetranuclear units {Cu₄Cl₄N₆} in **2**, and dinuclear units {Cu₂Cl₄N} in **3** as inorganic chlorocuprate building clusters, and the corresponding flexible BMTTM and BMTTE tetrazole bridging ligands.

The two methyl tetrazole groups in BMTTM and BMTTE are separated by a flexible organosulfur spacer that allows rotation around the C–S and C–C bonds to adjust the direction of the coordination nitrogen atoms. It is therefore apparent that the rigid and geometrically well-defined structures of the inorganic units with the flexibility and the potential N-hexacoordination ligands are essential to achieve the structural diversity observed in these systems.

Compounds **1**, **1·solv** and **3** are polymeric coordination compounds. The layers of **3** are formed by chloride-bridged Cu(μ-Cl)₂ chains and bridging N-donor ligands in a second dimension that act as cross-linking ligands. The copper-chloride chain is based on the repetition of the dinuclear Cu₂Cl₂ unit, as shown in Table 1. The polymeric chains of **1** and **1·solv** incorporate trinuclear Cu_3Cl_6 units and organic subunits in an alternate manner.

Table 1. Main structural features.

Compound Dimensionality	Coordination Mode of Ligand	Inorganic Units	Coordination Geometry
$^1_\infty Cu_3Cl_6(H_2O)_2(BMTTM)_2$ **(1)** 1D $^1_\infty Cu_3Cl_6(H_2O)_2(BMTTM)_2 \cdot 4CH_3CN$ **(1·solv)** 1D $\mu_3\text{-}1\kappa N3{:}2\kappa N4{:}3\kappa N4'$	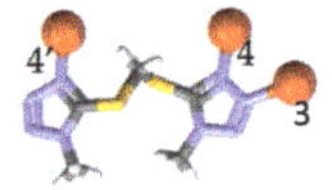	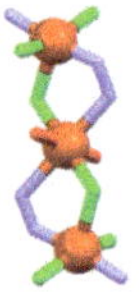 **{Cu₃Cl₆N₆O₂}**	 Cu2  Cu1
$[Cu_2Cl_2(BMTTE)]_2$ **(2)** 0D, tetramer $\mu\text{-}1\kappa^2 N4,N4'{:}2\kappa N3$	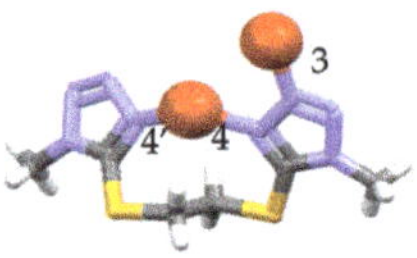	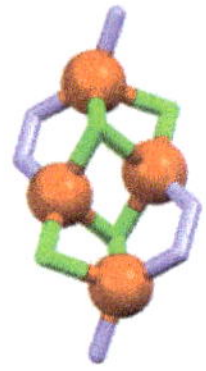 **{Cu₄Cl₄N₆}**	  Cu1 Cu2
$^2_\infty Cu_2Cl_4(BMTTE)$ **(3)** 2D $\mu\text{-}\kappa N2$		 **{Cu₂Cl₄N}**	

Compound **2** is a discrete tetrameric compound based on a stair-step Cu_4Cl_4 cluster coordinated by two BMTTE ligands in an octahedral {$Cu_4Cl_4N_6$} motif, which has been less widely studied than other copper halide clusters such as cubane organizations. In this octahedral motif, a tetranuclear copper core defines the basal plane of an octahedron with two capping μ_3-chloride atoms in the apical positions and bridging μ_2-chloride and nitrogen BMTTE atoms along the meridian positions [18].

2.3. Crystal Structures of 1 and 1·solv

Complex **1** and the solvate **1·solv** crystallize in the monoclinic $P2_1/n$ and triclinic $P\text{-}1$ space groups, respectively, and they could be considered pseudopolymorphs [19] since they have the same crystalline form, although **1·solv** crystallizes with acetonitrile molecules trapped within the crystal network. Significant structural parameters for compounds **1–1.solv** are listed in Table 2 and crystal structure and refinement data are listed in Table S1. Both structures are 1D chains based on a {Cu_3Cl_6} cluster and the BMTTM ligand, as shown in Figure 1. The three copper atoms in the {Cu_3Cl_6} unit are aligned with a Cu1–Cu2–Cu1 angle of 180° [20]. The conformation of the outer two coppers is identical. The Cu1$\cdots$Cu2 distance is 3.903 Å in **1** and slightly shorter (3.842 Å) in **1·solv**.

Table 2. Selected bond lengths/Å and angles/°.

Atoms	1	1·solv	Atoms	1	1·solv
Cu1–Cl1	2.5477(12)	2.5119(4)	N4–Cu1–N8#1	171.83(15)	174.19(4)
Cu1–Cl2	2.2895(11)	2.2660(4)	N8#1–Cu1–Cl1	93.88(11)	93.94(3)
Cu1–Cl3	2.2744(12)	2.2927(3)	N8#1–Cu1–Cl2	90.83(11)	90.23(3)
Cu2–N3	2.561(4)	2.4137(11)	N8#1–Cu1–Cl3	90.22(11)	87.76(3)
Cu1–N4	2.038(4)	2.0500(11)	Cl1#2–Cu2–Cl1	180.00(5)	180.000(14)
Cu1–N8#1	2.052(4)	2.0346(11)	Cl1–Cu2–N3	83.54(9)	85.98(3)
Cu2–Cl1	2.2981(10)	2.2716(3)	Cl1–Cu2–N3#2	96.46(9)	94.02(3)
Cu2–Cl1#2	2.2980(10)	2.2716(3)	N3#2–Cu2–N3	180.0	180.0
Cu2–O1	1.959(3)	2.0140(9)	O1–Cu2–Cl1	90.18(10)	88.06(3)
Cl2–Cu1–Cl1	97.16(4)	102.660(12)	O1–Cu2–Cl1#2	89.82(10)	91.94(3)
Cl3–Cu1–Cl1	105.98(4)	100.253(12)	O1–Cu2–N3	89.96(14)	93.80(4)
Cl3–Cu1–Cl2	156.72(4)	157.082(13)	O1#2–Cu2–N3	90.03(13)	86.20(4)
N4–Cu1–Cl1	94.23(11)	91.57(3)	O1–Cu2–O1#2	180.0	180.00(5)
N4–Cu1–Cl2	87.23(11)	90.35(3)	Cu2–Cl1–Cu1	107.19(4)	106.772(12)
N4–Cu1–Cl3	88.45(11)	89.45(3)			

Symmetry code: (1) #1 −x + 2, −y + 1, −z + 1; #2 −x + 1, −y + 2, −z + 1; (1·solv) #1 −x + 1, −y + 1, −z + 2; #2 −x + 1, −y + 2, −z + 2.

The two bridges of chloride and N-N tetrazole atoms between the Cu1 and Cu2 lead to the formation of an unusual planar {Cu$_3$Cl$_6$N$_6$O$_2$} cluster. These units are connected by two BMTTM ligands to form infinite tapes, as shown in Figure 1.

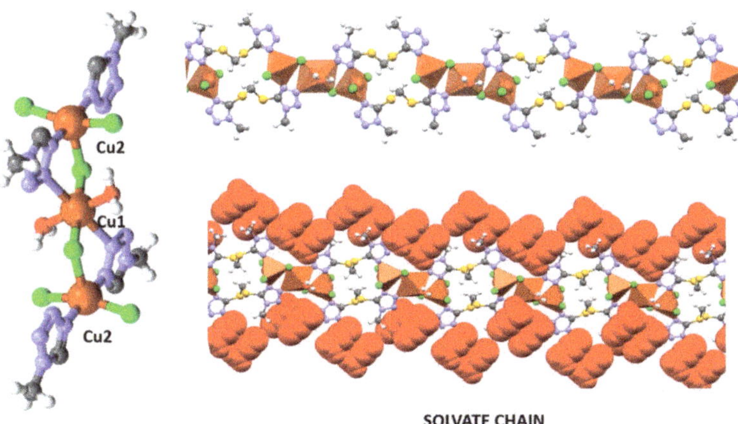

Figure 1. Coordination environments and polymeric chains of **1** and **1·solv**.

The Cu1 center is pentacoordinated and the value of the Addison parameter [21] τ of 0.25 (0.28 in **1·solv**) indicates a square pyramidal coordination geometry around this copper atom, with the two nitrogen atoms and the two terminal chloride atoms in the *trans* positions of the pyramid base and a chloride atom acting as a bridge between Cu1 and Cu2 at the apex position. The Cu–Cl bond lengths with the chloride terminal groups are in the range 2.27–2.29 Å and the bridging chloride atom in the apical position is located at a longer distance, with Cu–Cl distances of 2.5477(12) Å in **1** and 2.5119(4) Å in **1·solv**.

The Cu2 atom is in an elongated octahedral environment and it exhibits the expected Jahn–Teller distortion, with four short metal–ligand bonds (Cu–Ow and Cu–Cl) and two long bonds with nitrogen atoms of BMTTM bridging ligands [Cu2–N distances of 2.561(4) and 2.4137(11) Å in **1** and **1·solv**, respectively]. Here, it is worth highlighting the equatorial coordination position of the water molecule, which contrasts with the typical axial position in similar systems.

The BMTTM ligand uses three nitrogen atoms of the two tetrazole rings to coordinate three metal centers, as shown in Table 2, with a bridging role that results in the formation of a tape, as shown in Figure 1, with 16-membered macrocycles ($Cu_2N_4S_4C_6$). This macrocyclic motif has previously been observed in complexes with other thioether N-donor ligands [22]. This coordination mode (μ_3-1κN3:2κN4:3κN4′) of BMTTM was not observed in the copper-POM complexes, in which the ligand had a bidentate chelating coordination mode with one, two or three copper cations [12,14]. The intermetallic distances through the BMTTM-bridge are 6.97 Å for compound **1** and 6.69 Å for **1·solv**. Moreover, the intramolecular N···N distance is shorter in **1·solv** (6.48 Å) than in **1** (6.70 Å). On comparing the two structures, it is evident that BMTTM acts as a semirigid ligand, which shrinks due to the presence of the guest molecules in **1·solv** but retains the same connectivity and geometric environment in both cases. In accordance with this situation, the value of the C–S–C–S torsion angles are 81.78° and 83.20° in **1** and 77.14° and 84.45° in **1·solv**. The chains are achiral and incorporate molecules of BMTTM with opposite helicity in each macrocycle unit.

2.4. Crystal Structure of **2**

The Cu(I) compound **2** crystallizes in the $P2_1/n$ space group with an inversion center located between the Cu2 central atoms. Significant structural parameters are listed in Table 3, and crystal structure and refinement data are listed in Table S1. Four Cu(I) atoms and four chloride atoms form a stair-like [Cu_4Cl_4] cluster core with each of the two BMTTE ligands anchored at each end of the stair through two Cu(I) centers, as shown in Figure 2 [23,24]. In the cluster, the Cu2···Cu2 distance of 2.7651(5) Å is slightly shorter than the sum of the van der Waals radii of two Cu atoms (2.80 Å), thus indicating attractive Cu···Cu interactions, but the Cu1···Cu2 distance of 2.9556(3) Å suggests that copper–copper interactions are not present [25].

Table 3. Selected bond lengths/Å and angles/°.

Atoms	2	3	Atoms	2
Cu1–Cu2#1	2.9556(3)		N4–Cu1–Cu2#1	124.81(4)
Cu1–Cl1	2.4039(4)	2.3132(4)	N4–Cu1–Cl1	106.22(4)
Cu1–Cl1#2		2.3163(4)	N4–Cu1–Cl2#1	96.70(4)
Cu1–Cl2		2.2987(4)	N4–Cu1–N8	137.74(6)
Cu1–Cl2#1	2.4794(4)	2.2979(4)	N8–Cu1–Cu2#1	96.92(4)
Cu1–N2		2.3981(13)	N8–Cu1–Cl1	100.74(4)
Cu1–N4	2.0097(14)		N8–Cu1–Cl2#1	108.89(4)
Cu1–N8	2.0268(14)		Cu2#1–Cu2–Cu1#1	72.140(10)
Cu2–Cu2#1	2.7651(5)		Cl1–Cu2–Cu1#1	111.431(14)
Cu2–Cl1	2.3556(5)		Cl1–Cu2– Cl1#1	113.629(13)
Cu2–Cl1#1	2.6728(5)		Cl1#1–Cu2– Cl1#1	50.261(10)
Cu2–Cl2	2.2330(4)		Cl1#1–Cu2–Cu2#1	51.302(12)
Cu2–N7	2.0013(14)		Cl1–Cu2–Cu2#1	62.326(13)
Cl1–Cu1–Cu2#1	58.759(12)		Cl2–Cu2–Cu1#1	54.982(12)
Cl2#1–Cu1–Cu2#1	47.525(11)		Cl2–Cu2–Cu2#1	121.487(16)
Cl1–Cu1–Cl1#2		176.292(10)	Cl2–Cu2–Cl1	113.376(17)
Cl2–Cu1–Cl1		93.746(16)	Cl2–Cu2–Cl1#1	100.991(16)
Cl2#1–Cu1–Cl1	102.006(16)	85.599(15)	N7–Cu2–Cu1#1	137.43(4)
Cl2#1–Cu1–Cl1#2		94.895(15)	N7–Cu2–Cu2#1	99.24(4)
Cl2–Cu1–Cl1#2		85.511(15)	N7–Cu2–Cl2	135.88(4)
Cl2#1–Cu1–Cl2		176.067(11)	N7–Cu2–Cl1	99.50(4)
Cl1–Cu1–N2		96.33(3)	N7–Cu2–Cl1#1	91.18(4)
Cl1#2–Cu1–N2		87.29(3)	Cu2–Cl2–Cu1#1	77.492(15)
Cl2–Cu1–N2		88.39(3)	Cu1–Cl1–Cu2#1	70.981(13)
Cl2#1–Cu1–N2		95.53(3)	Cu2–Cl1–Cu1	90.204(15)
Cu1#2–Cu1–N2		90.896(14)	Cu2–Cl1–Cu2#1	66.371(14)
Cu1–Cl1–Cu1#1		90.072(14)		

Symmetry code: (**2**) #1 −x + 1, −y + 2, −z + 1; (**3**) #1 −x + 1/2, y + 1/2, −z + 3/2; #2 −x + 1/2, y − 1/2, −z + 3/2.

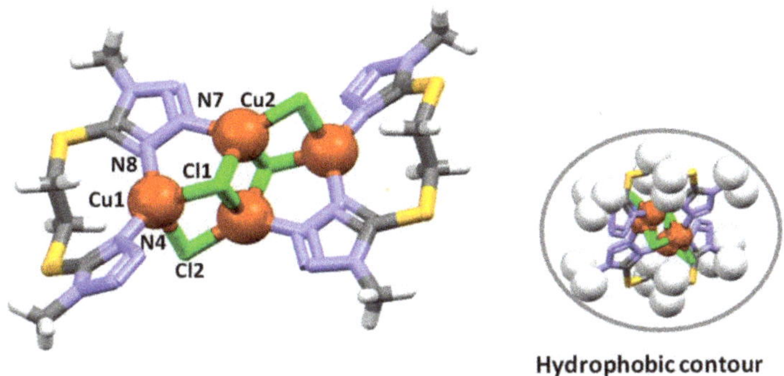

Figure 2. Molecular structure of **2** showing the hydrophobic contour of the tetrameric unit.

The chloride ligands display μ_3-Cl1 and μ-Cl2 bridging modes. Each Cu(I) center is tetrahedrally coordinated—Cu1 by two BMTTE nitrogen atoms and Cl1 and Cl2 atoms, and Cu2 by one BMTTE nitrogen atom, two μ_3-Cl1 and one μ-Cl2 atoms, so the values of the τ_4 parameters [26] of 0.80 for Cu1 and 0.77 for Cu2 indicate a fairly regular tetrahedral geometry. Cl2, which is bonded to adjacent copper atoms with Cu–Cl distances of 2.4794(4) Å and 2.2330(4) Å, resides 0.813 Å above and below the planar copper core. The triply bonded chloride ligand connects Cu1 and Cu2 atoms with variable distances [2.4039(4) Å, 2.3556(5) Å and 2.6728(5) Å].

In compound **2**, each BMTTE molecule provides three N atoms to coordinate two copper ions, as shown in Figure 2. Thus, BMTTE acts as a bis-monodentate bridging ligand to link Cu1 and Cu2 through the tetrazole fragment and also as a bidentate chelating ligand with Cu1 to give a nine-membered ring ($CuN_2S_2C_4$). This coordination mode of BMTTE has also been observed in a copper-POM 1D coordination polymer [13]. The S–C–C–S torsion angle is 168.93° and the two ligands in the tetramer have opposite axial chirality.

2.5. Crystal Structure of 3

The Cu(II) polymeric compound **3** crystallizes in the *C*2/*c* space group with one crystallographically independent copper atom. Significant structural parameters are listed in Table 3, and crystal structure and refinement data are listed in Table S1. A dimeric Cu_2Cl_4 motif with two metal cations linked by two bridging chloride atoms leads to the formation of Cu_2Cl_4-based inorganic chains along the *b* axis, as shown in Figure 3. The nonbonding Cu···Cu distance through the halide bridge is 3.276 Å. These chains are bridged by bis-monodentate BMTTE ligands to yield the final 2D coordination compound. The 2D structure contains 32-membered macrometallocycles formed by (Cu_3Cl_4) inorganic units and two BMTTE ligands, as shown in Figure 3, with the methyl groups of each ligand molecule orientated to opposite sides.

The copper(II) cation is pentacoordinated, as shown in Figure 3, and the value of the Addison parameter τ [21] of 0.04 indicates that the coordination sphere is almost an ideal square pyramid, with four chloride atoms in basal positions (Cu—Cl distances between 2.313 and 2.298 Å) and a nitrogen atom belonging to a BMTTE molecule at the apex (Cu—N2 = 2.3981(13) Å). At a longer distance (2.854 Å) in the sixth coordination position is a nitrogen atom (N3) of an opposite ligand molecule, so that each tetrazole group coordinates a metal center and establishes a weak contact with a neighboring metal center.

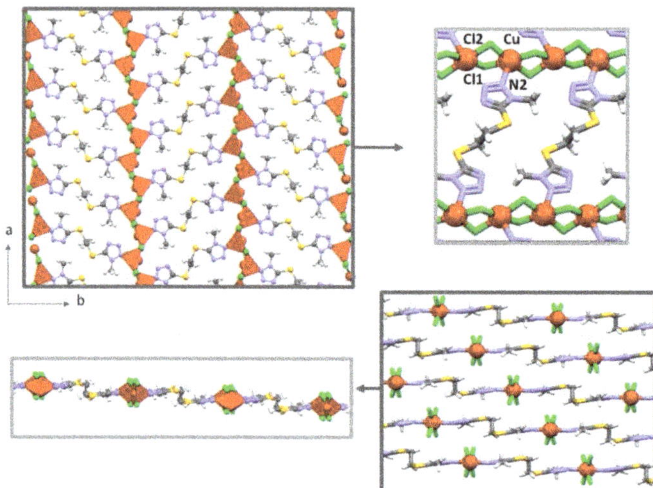

Figure 3. Top; 2D structure of **3** showing in detail the macrometallocycle. Bottom; supramolecular arrangement of the layers.

The BMTTE acts as a bis-monodentate bridging (μ-κN2) ligand. This coordination mode is different to the bis-monodentate bridging mode observed in some copper-POM complexes, where the nitrogen atom involved in coordination is N3 [11,12] and not N2. The S–C–C–S torsion angle is 180° and each metallocycle contains two ligands of opposite chirality to yield an achiral layer. The distance between two neighboring copper(II) ions through the BMTTE ligand is 14.585 Å and the intramolecular N···N distance is 9.794 Å.

2.6. Supramolecular Arrangements

The strategic use of supramolecular synthons to control crystal structure allows the design of new solids with interesting physicochemical properties. However, to achieve this objective, it is necessary to understand the intermolecular interactions in the context of crystal packing. In the current systems, common structural and compositional characteristics allow a deeper study of this aspect. In this sense, chloride ligands have shown through crystallographic data analysis that the weak C–H···Cl–M interactions exhibit the characteristics of conventional hydrogen bonds and hence are very significant in molecular recognition and crystal engineering. Another interesting feature in these structures is the presence of coordinatively unsaturated metal centers (UMCs), which can interact at longer distances with other atoms. Besides, it should be noted in these compounds that not all of the nitrogen atoms of each tetrazole take part in coordination, so free N-atoms are still available to act as hydrogen acceptors in different interactions, thus playing a significant role in the final supramolecular arrangements. Moreover, the flexible organosulfur chains between the tetrazole groups offer the possibility of establishing noncovalent interactions involving the sulfur atom as an acceptor or the organic fragment as a donor. In addition, the methyl substituent in the tetrazole moiety increases the hydrophobicity and the stability of the molecules, but the most remarkable feature observed in these structures is its role as a donor in multiple interactions.

In this way, chains of **1** are organized into a supramolecular 2D network along the *ab* plane by means of O–H···Cl hydrogen bonds that involve coordinated water molecules and chloride atoms of neighboring chains. Nonpolar hydrophobic methyl groups are oriented out of the layers and establish C–H···Cl interactions with other chains to give the final 3D supramolecular array, as shown in Figure 4. An interchain Cl···π interaction [Cl2···centroid of N1/N2/N3/N4/C1; d(Cl···centroid = 3.355 Å; 1 − x,

2 − y, 1 − z)] and $CH_{Me}\cdots N$ and $CH_{Me}\cdots S$ interactions, as shown in Table 4, also participate in the formation of the crystal network, which has a Kitaigorodskii packing index [27] of 75%.

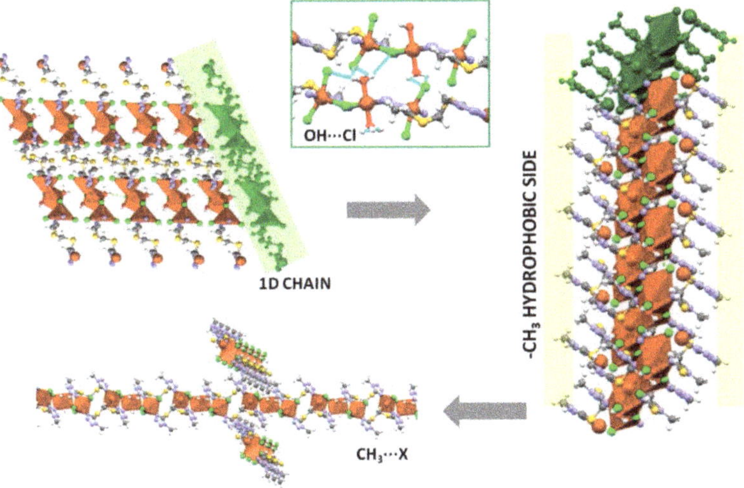

Figure 4. Supramolecular arrangement in the crystal structure of **1**.

The supramolecular arrangement in **1·solv** is mainly due to $C-H\cdots X$ interactions (X = Cl and N) but, as observed in similar compounds [28], solvent molecules are involved in different $C-H\cdots X$ (X = Cl, O, N; Table 4) interactions with the metal–organic framework, so the polymeric chains are extended along the *b* axis and arranged in a sinusoidal manner, with solvent molecules hosted between them, as shown in Figure 5. These hydrogen-bonding interactions are established between the methyl groups of tetrazole or acetonitrile as donors with the available nitrogen atoms of both fragments as acceptors. The metallorganic chains are linked through solvent molecules that act as a bridge in the *ac* plane, as shown in Figure 5 (top, left), to give a shifted parallel arrangement. Furthermore, several weak intra- and inter-molecular $C-H\cdots S$ interactions reinforce the zig-zag disposition of the chains. The solvent volume makes up 30% of the unit cell volume, as calculated using Mercury [29], and the Kitaigorodskii packing index [27] of 75% is slightly lower than that in **1**.

In the crystal packing of **2**, the nonpolar methyl $-CH_3$ and methylene $-CH_2-$ spacer groups, which are oriented outwards from the tetrameric molecule to define a hydrophobic contour, as shown in Figure 2, play an important role as H-donors towards the chloride atoms to establish several intermolecular $C-H\cdots Cl$ interactions with $C\cdots Cl$ distances in the range 3.32–3.47 Å, as shown in Table 4 and Figure 6. This kind of interaction is also established with the nitrogen atoms of tetrazole units. Furthermore, a $\pi\cdots\pi$ interaction between the tetrazole groups N1/N2/N3/N4/C1 and N5/N6/N7/N8/C4 (intermolecular centroid–centroid distance of 3.6976 Å and interplanar dihedral angle of 33.40°; symmetry code: 1 − x, 1 − y, 1 − z) contributes to the crystal packing, with a Kitaigorodskii index [27] of 70%.

The metallorganic layer of **3** has a stair-step disposition along the *ab* plane, as shown in Figure 3, and each layer stacks with adjacent ones through $CH_{methylene}\cdots Cl$ interactions, as shown in Table 4, to give a 3D supramolecular array reinforced by the contribution of an $S\cdots\pi$ interaction with the tetrazole ring ($S\cdots$ centroid distance = 3.357 Å; symmetry: 1 − x, y, 1.5 − z). The resulting Kitaigorodskii packing index [27] is 77%.

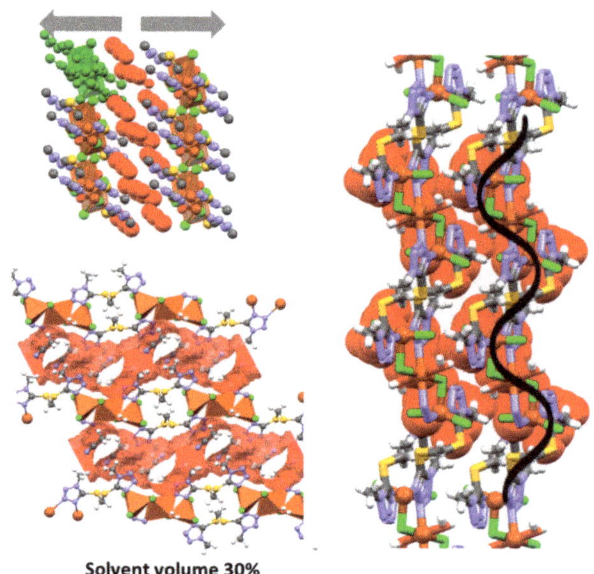

Figure 5. Supramolecular arrangement in the crystal structure of **1·solv**.

Table 4. Main hydrogen bonds (Å,°).

Structure	D–H···A	d(D–H)	d(H···A)	d(D···A)	∠(DHA)
1	O1–HA … Cl3#1	0.84(2)	2.82(6)	3.250(4)	114(5)
	O1–HA … Cl1#1	0.84(2)	2.93(4)	3.685(3)	152(6)
	C12–H12C … Cl2#4	0.98	2.49	3.418(5)	158.6
	C12–H12B … N7#3	0.98	2.86	3.207(6)	102.0
	C12–H12A … N6#2	0.98	2.68	3.327(6)	124.0
	C12–H12A … S2#1	0.98	2.90	3.717(5)	141.2
1·solv	O1–H1C … Cl1#6	0.89	2.99	3.6713(10)	135.4
	O1–H1C … Cl2#6	0.89	2.72	3.4096(10)	135.1
	C12–H12C … N41#7	0.98	2.95	3.221(2)	96.9
	C12–H12B … N31#4	0.98	2.68	3.650(2)	170.0
	C22–H22C … Cl3#5	0.98	2.95	3.5901(15)	123.9
	C1–H1B … Cl2#3	0.99	2.95	3.5478(13)	119.9
	C1–H1B … S2#2	0.99	2.95	3.5065(13)	116.3
	C31–H31A … N6#5	0.98	2.64	3.559(2)	156.5
	C31–H31B … N7#1	0.98	2.59	3.473(2)	150.1
	C31–H31C … O1#7	0.98	2.71	3.536(2)	141.7
	C41–H12A … Cl1#6	0.98	2.98	3.3693(15)	104.8
2	C2–H2B … Cl2#1	0.99	2.89	3.3201(17)	107.2
	C1–H1B … Cl2#1	0.99	2.84	3.4698(17)	122.1
	C12–H12C … Cl2#3	0.98	2.87	3.4037(19)	115.0
	C22–H22C … Cl1#4	0.98	3.05	3.4498(17)	106.1
	C12–H12B … S2#2	0.98	2.95	3.7522(18)	139.5
3	C1–H1A···Cl2#1	0.99	2.78	3.6987(16)	153.7

Symmetry code: (**1**) #1 x −1, y, z; #2 −x + 1/2, y + 1/2, −z + 1/2; #3 −x + 3/2, y + 1/2, −z + 1/2; #4 x − 1/2, −y + 3/2, z − 1/2; (**1·solv**) #1 −x + 1, −y + 1, −z + 2; #2 −x + 1, −y + 2, −z + 2; #3 x + 1, y, z; #4 x, y − 1, z + 1; #5 −x + 2, −y + 1, −z + 2; #6 −x, −y + 2, −z + 2; #7 1 − x, 1 − y, 3 − z; (**2**) #1 x + 1/2, −y + 3/2, z + 1/2; #2 −x + 3/2, y − 1/2, −z + 3/2; #3 x − 1/2, −y + 3/2, z + 1/2; #4 1 + x, y, z; (**3**) #1 −x + 1, y, −z + 3/2.

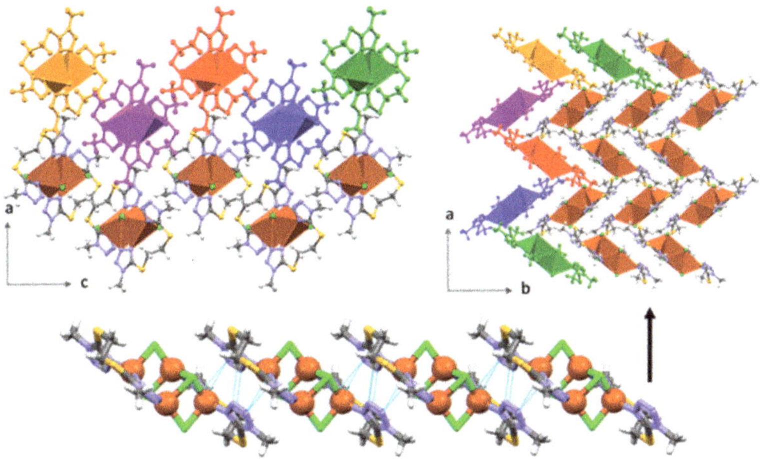

Figure 6. Supramolecular organization in the crystal structure of **2**.

2.7. Hirshfeld Surface Study

A Hirshfeld surface study was carried out to gain a fuller appreciation of the nature and quantitative contributions of intermolecular interactions to the supramolecular assembly of complexes **1**, **2** and **3**. The decomposition of contributions from different interaction types, which overlap in the full fingerprint, proved to be helpful to highlight graphically the surface regions that are involved in a specific type of intermolecular contact. The contributions to the Hirshfeld surface area from the various close intermolecular contacts are presented in the histogram in Figure 7.

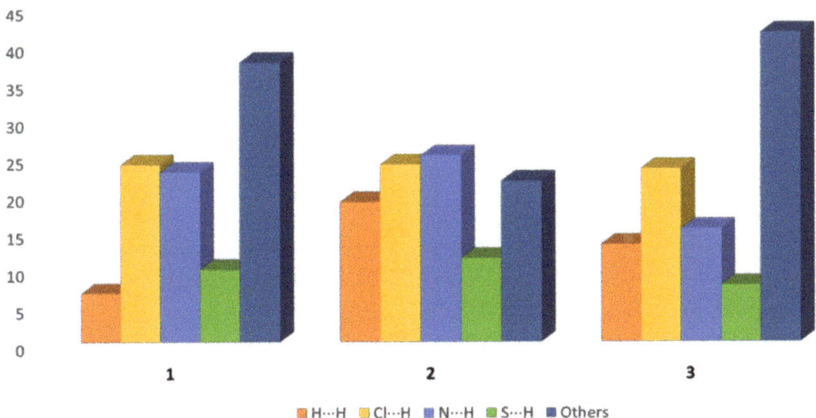

Figure 7. Contributions to the Hirshfeld surface area from the various close intermolecular contacts.

The analysis shows that in compounds **1** and **3**, the Cl$\cdots$H interactions have the highest priority (the highest contribution to the Hirshfeld surface) and the N$\cdots$H interactions have the highest priority in **2**.

In this approach to assess and visualize the contribution of polar and non-polar interactions to the crystal packing forces, the two-dimensional fingerprint plots for **1**, **2** and **3** were delineated for different types of contacts, such as H$\cdots$H, Cl$\cdots$H, N$\cdots$H and S$\cdots$H, as shown in Figure 8. The decomposition of contributions from different interaction types proved to be helpful to highlight graphically the surface regions that are involved in a specific type of intermolecular contact. The visual analysis of the different fingerprint plots show that the molecular environments are clearly different in each compound.

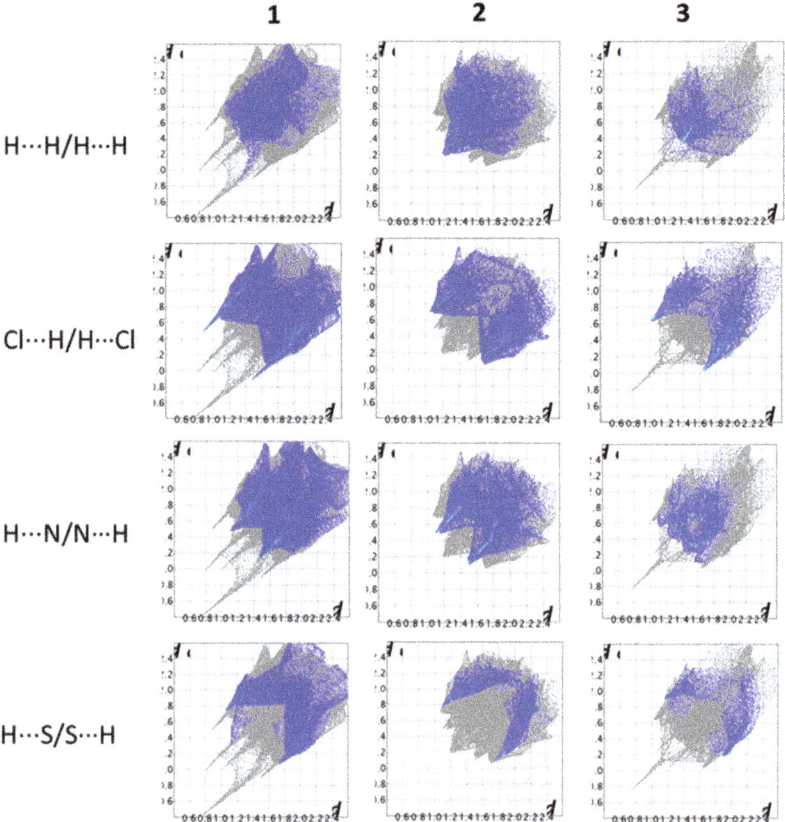

Figure 8. Fingerprint plots for H···H, Cl···H/H···Cl, H···N/N···H and H···S/S···H contacts. The outline of the full fingerprint is shown in gray.

Despite the different crystal packing arrangements, in general the Hirshfeld analysis revealed that 50%–60% of the total surface areas can be identified with Cl/H, N/H and S/H contacts, which correspond to CH···Cl, CH···N or CH···S hydrogen bonds.

The proportion of Cl···H/H···Cl interactions is almost the same in all structures and they contribute around 24% of the Hirshfeld surfaces for each compound. Furthermore, in all structures H···N contacts are present; with the highest amount in **2** (24.9%) and decreasing gradually in **1** (22.7%) and **3** (15.1%).

The H/H contacts, which are less directed than H-bonds, are present on 2D fingerprint plots as bulk central areas. These contributions correspond to the van der Waals interactions and they represent less than 20% of the Hirshfeld surfaces (barely 6.5% in **1**), thus showing that these lattices are mostly stabilized by H-bonds rather than dispersion forces. The high percentage of other interactions, as shown in Figure 7, is consistent with the number of non-classical π···π, Cl···π or S···π interactions present in the structures. These interactions are associated with C···N, N···N, Cl···C/N and S···C/N contacts and show that tetrazole units enhance the stacking interactions in the structures. Thus, as expected, another common feature in all of the structures is the relatively low area associated with C···C interactions.

The presence of other interactions is particularly significant in the structures of **1** and **3**. The presence of water oxygen atoms in **1** produces O/H contacts that represent around 5% of the surface in this structure. Furthermore, the unsaturated coordination environment of the copper metal center in **3** gives rise to weak $Cu \cdots Cl$ interactions that represent around 10.2% of the total surface. These contacts are almost negligible in the other structures due to their more saturated copper coordination environments.

2.8. Thermal Analysis

The thermal stability of compounds **1**, **2** and **3** was evaluated by TGA experiments performed from room temperature to 900 °C. All of the complexes showed good thermal stability, and this indicated the strength of their corresponding networks. The least thermally stable compound was the 1D polymer **1**, which lost its coordinated water molecules (exp. 3.02%, calcd. 3.88%) at 110 °C. From 190 to 650 °C, the consecutive loss of the organic ligand and the anion occurred. Despite their different covalent dimensionality, compounds **2** and **3** were stable up to 180 °C and then began to decompose in a process that ended at 650 °C, as in compound **1**.

3. Conclusions

In the copper/chloride compounds described here, the flexible bis-tetrazole organosulfur ligands BMTTM and BMTTE have proven to be able to adopt different coordinative modes to produce different coordination geometries (tetrahedral, octahedral and square-pyramidal) around the copper ions and different covalent dimensionality (from 0D to 2D) in the resulting crystalline solids. With BMTTM, all of the synthetic methods used (except the hydro/solvothermal method) led to the same 1D coordination polymer. However, with BMTTE, the softest methods led to a 2D coordination polymer and the reaction under hydrothermal conditions led to the reduction of copper(II) to copper(I) and the crystallization of a discrete tetramer.

The supramolecular organizations in these thermally stable compounds are mainly determined by $C–H \cdots X$ (X = Cl, N) interactions, with a significant role of the coordinated chloride and the uncoordinated nitrogen atom of the tretrazole as acceptors and of the methyl and methylene groups of ligands as donors. Other, less common interactions, such as $Cl \cdots \pi$ (**1**), $\pi \cdots \pi$ (**2**) and $S \cdots \pi$ (**3**) contribute to the stabilization of the corresponding supramolecular networks, which achieve a very efficient packing. The Hirshfeld surface analysis corroborated the main geometrical observations about the crystal packing and highlight the importance of the methyl groups in establishing $C–H \cdots X$ (X = Cl, N or S) interactions.

4. Experimental

4.1. Materials and Physical Measurements

Solvents and reagents were obtained commercially and were used as supplied. BMTTM [30] and BMTTE [16,17] ligands were synthesized by ourselves. Elemental organic analysis (C, H, N) was carried out on a Carlo Erba 1108/Chromatographic combustion elemental analyzer. FT-IR spectra in the 4000–400 cm^{-1} region were recorded on a Jasco FT/IR-6100 spectrophotometer. Thermogravimetric analysis (TGA) and Differential Scanning Calorimetry analysis (DSC) profiles were obtained with a TGA-ATD/DSC, SETSYS Evolution 1750 (Setaram) thermal analyzer.

4.2. Synthesis and Crystallization of the Complexes

4.2.1. $_\infty{}^1Cu_3Cl_6(H_2O)_2(BMTTM)_2$ (**1**) and $_\infty{}^1Cu_3Cl_6(H_2O)_2(BMTTM)_2 \cdot 4CH_3CN$ (**1·solv**)

The pseudopolymorphs **1** and **1·solv** were obtained as crystalline materials using different synthetic methodologies:

Method 1: Diffusion

To a solution of $CuCl_2 \cdot 2H_2O$ (0.25 mmol) in CH_3CN (10 mL) in a test-tube was slowly added a solution of ligand (0.25 mmol) in CH_3CN (10 mL). Blue single crystals of **1** mixed with some colorless crystals of ligand were obtained after two days. Yield: 84%.

Method 2: Microwave Irradiation

To a solution of $CuCl_2 \cdot 2H_2O$ (0.9 mmol) in CH_3CN (10 mL) was slowly added a solution of ligand (0.9 mmol) in CH_3CN (10 mL). This mixture was irradiated for 10 min in a modified conventional microwave oven [31]. After slow evaporation at room temperature, we first isolated a few prismatic blue single crystals of **1·solv** and then blue plate single crystals of **1**. Yield: 69%.

Method 3: Stirring at Room Temperature

To a solution of $CuCl_2 \cdot 2H_2O$ (0.9 mmol) in CH_3CN (10 mL) was slowly added a solution of ligand (0.9 mmol) in CH_3CN (10 mL). The mixture was stirred for 24 h at room temperature. The blue crystalline powder of **1** was filtered off and dried over $CaCl_2$. Yield: 87%.

Data for **1**: MP: 170–175 °C. Anal. Calc. for $C_{10}H_{20}Cl_6N_{16}O_2S_4Cu_3$: N 24.1%, C 13.0%, H 2.2%; Found: N 24.1%, C 13.2%, H 2.4%. IR (cm^{-1}): 1480m, 1414m, 1393m, 1365m, ν(ring); 1177m, δ(CH); 1087m, 1039m, 1003m, δ(ring); 708s, ν(C–S).

4.2.2. $[Cu_2Cl_2(BMTTE)]_2$ (2) and $_\infty^2Cu_2Cl_4(BMTTE)$ (3)

Synthesis of $[Cu_2Cl_2(BMTTE)]_2$ (2)

A mixture of $CuCl_2 \cdot 2H_2O$ (2.7 mmol) and BMTTE (0.9 mmol) in H_2O (5–15 mL) was sealed in a 20-mL Teflon-lined autoclave and heated to 160 °C over 100 min. The autoclave was kept at 160 °C for 3 days and then slowly cooled to room temperature at a rate of 10 °C/h. Yellow single crystals of **2** were obtained.

Data for **2**: Yield: 32%. MP: 200–202 °C. Anal. calc. for $C_{12}H_{20}Cl_4N_{16}S_4Cu_4$ N 24.7%, C 15.9%, H 2.2%; Found: N 24.3%, C 15.7%, H 2.1%. IR (cm^{-1}): 1467m, 1406m, 1382m, ν(ring); 1296m ω(CH,CH$_2$); 1173m δ(CH); 1147w, 1097w, 1078w, δ(ring); 728m, γ(CH); 704vs ν(C–S).

Synthesis of $_\infty^2Cu_2Cl_4(BMTTE)$ (3)

Method 1: Microwave Radiation.

To a suspension of $CuCl_2 \cdot 2H_2O$ (3.56 mmol) in CH_3CN (10 mL) was slowly added a solution of ligand (0.89 mmol) in CH_3CN (10 mL). The mixture was irradiated for 10 min in a modified conventional microwave oven. After cooling (30 min), orange single crystals of **3** were obtained. Yield: 35%.

Method 2: Stirring at Room Temperature.

To a solution of $CuCl_2 \cdot 2H_2O$ (0.9 mmol) in CH_3CN (10 mL) was slowly added a solution of ligand (0.9 mmol) in CH_3CN (10 mL). After stirring at room temperature for 24 h, an orange crystalline powder was filtered off under vacuum and dried. Yield: 47%.

Data for **3**: Yield: 35%. MP: 225–230 °C. Anal. calc. for $C_6H_{10}Cl_4N_8S_2Cu_2$ N 20.40%, C 13.75%, H 1.92%; Found: N 20.80%, C 13.70%, H 1.91%. IR (cm^{-1}): 1473m, 1403m, 1384m, ν(ring); 1240m ω(CH,CH$_2$); 1183m δ(CH); 1137w, 1090w, 1052w, δ(ring); 739m, γ(CH); 708vs ν(C–S).

4.3. X-ray Structure Determination

Crystallographic data were collected at 293 K on a Bruker Smart 1000 CCD diffractometer using graphite-monochromated Mo-Kα radiation (λ = 0.71073 Å). The software SMART [32] was used to collect frames of data, index reflections, and determine lattice parameters. SAINT [33] was used for integration of intensity of reflections and SADABS [34] for scaling and empirical absorption correction. The structure was solved by a dual-space algorithm using the program SHELXT [35]. All non-hydrogen atoms were refined with anisotropic thermal parameters by full-matrix least-squares calculations on F^2 using the program SHELXL [35] with OLEX2 [36]. Hydrogen atoms were inserted at calculated positions and constrained with isotropic thermal parameters. Drawings were produced with Mercury [28]. Special computations for the crystal structure discussions were carried out with PLATON [37]. Crystal data and structure refinement parameters are reported in Table S1.

X-ray powder diffraction (XRPD) was performed using a Siemens D-5000 diffractometer with Cu-Kα radiation (λ = 1.5418 Å) over the range 5.0–60.0° in steps of 0.20° (2θ) with a count time per step of 5.0 s. The program FULLPROF [38] was used to perform profile matching between powder diffraction data and that calculated from the single-crystal structures.

4.4. Hirshfeld Surface Study

Hirshfeld surfaces and their respective 2D fingerprint plots for all complexes were calculated with CRYSTALEXPLORER 3.1 software [39]. The d_{norm} surface and the breakdown of two-dimensional fingerprint plots were used to analyze intermolecular interactions in the different crystal lattices. The sizes and shapes of the fingerprints illustrate the significant differences between the intermolecular interaction patterns.

Supplementary Materials: The following are available online at http://www.mdpi.com/2624-8549/2/1/5/s1, Figure S1: Infrared Spectra, Figure S2: Thermogravimetric analysis, Figure S3: Powder X-Ray Diffraction patterns, Table S1: Crystal data and structure refinement. CCDC 1971817-1971820 contains the supplementary crystallographic data for **1**, **1.solv**, **2** and **3**.

Author Contributions: Conceptualization, R.C. and A.B.L.; methodology, R.C. and O.G.-P.; synthesis, characterization, thermal analysis, O.G.-P.; formal analysis, R.C. and O.G.-P.; investigation, O.G.-P.; resources, R.C. and O.G.-P.; writing—original draft preparation, R.C. and A.B.L.; writing—review and editing, R.C., A.B.L. and E.M.V.-L.; visualization, R.C. and A.B.L.; supervision, E.M.V.-L.; project administration, R.C. and E.M.V.-L.; funding acquisition, R.C. and E.M.V.-L. All authors have read and agreed to the published version of the manuscript.

Funding: This research was funded by ERDF (EU), MEC (Spain) and Xunta de Galicia (Spain) (research projects CTQ2015-71211-REDT, CTQ2015-7091-R and ED431D 2017/01).

Acknowledgments: SC-XDR measurements were performed at the Unidade de Difracción de Raios X de Monocristal (CACTI-Universidade de Vigo), Spain. O.G.P. thanks the Xunta de Galicia for a predoctoral contract.

Conflicts of Interest: The authors declare no conflict of interest. The funders had no role in the design of the study; in the collection, analyses, or interpretation of data; in the writing of the manuscript, or in the decision to publish the results.

References

1. Hirao, T.; Kim, D.S.; Chi, X.; Lynch, V.M.; Ohara, K.; Park, J.S.; Yamaguchi, K.; Sessler, J.L. Control over multiple molecular states with directional changes driven by molecular recognition. *Nat. Commun.* **2018**, *9*, 823. [CrossRef]
2. Thomas-Gipson, J.; Pérez-Aguirre, R.; Beobide, G.; Castillo, O.; Luque, A.; Pérez-Yañez, S.; Román, P. Unravelling the Growth of Supramolecular Metal–Organic Frameworks Based on Metal-Nucleobase Entities. *Cryst. Growth Des.* **2015**, *15*, 975–983. [CrossRef]
3. Steed, J.W.; Atwood, J.L. *Supramolecular Chemistry*; John Wiley & Sons: London, UK, 2013.
4. Lehn, J.-M. *Supramolecular Chemistry*; VCH: Weinheim, Germany, 1995.
5. Taylor, R.; Kennard, O. Crystallographic Evidence for the Existence of C−H$\cdots$O, C−H$\cdots$N, and C−H$\cdots$Cl Hydrogen Bonds. *J. Am. Chem. Soc.* **1982**, *104*, 5063–5070. [CrossRef]
6. Desiraju, G.R.; Steinter, T. *The Weak Hydrogen Bond in Structural Chemistry and Biology*; Oxford University Press: Oxford, UK, 1999.
7. Brammer, L.; Bruton, E.A.; Sherwood, P. Understanding the Behavior of Halogens as Hydrogen Bond Acceptors. *Cryst. Growth Des.* **2001**, *1*, 277–290. [CrossRef]
8. Thallapally, P.K.; Nangia, A. A Cambridge Structural Database analysis of the C−H$\cdots$Cl interaction: C−H$\cdots$Cl$^-$ and C−H$\cdots$Cl−M often behave as hydrogen bonds but C−H$\cdots$Cl−C is generally a van der Waals interaction. *CrystEngComm* **2001**, *3*, 114–119. [CrossRef]

9. Popova, E.A.; Trifonov, R.E.; Ostrovskii, R.A. Tetrazoles for biomedicine. *Russ. Chem. Rev.* **2019**, *88*, 644–676. [CrossRef]

10. Neochoritis, C.G.; Zhao, T.; Dömling, A. Tetrazoles via Multicomponent Reactions. *Chem. Rev.* **2019**, *119*, 1970–2042. [CrossRef]

11. Wang, X.L.; Hu, H.L.; Tian, A.X. Influence of Transition Metal Coordination Nature on the Assembly of Multinuclear Subunits in Polyoxometalates-Based Compounds. *Cryst. Growth Des.* **2010**, *10*, 4786–4794. [CrossRef]

12. Wang, X.; Hu, H.; Tian, A.; Lin, H.; Li, J. Application of Tetrazole-Functionalized Thioethers with Different Spacer Lengths in the Self-Assembly of Polyoxometalate-Based Hybrid Compounds. *Inorg. Chem.* **2010**, *49*, 10299–10306. [CrossRef]

13. Wang, X.; Wang, Y.; Liu, G.; Hu, H.; Tian, A. Polyoxometalates-Directed Assembly of Inorganic-Organic Hybrid Compounds with Copper Multinuclear Nano-Cluster Based on Flexible Double Tetrazole-Based Thioether. *J. Clust. Sci.* **2011**, *22*, 211–223. [CrossRef]

14. Wang, X.; Gao, Q.; Tian, A.; Hu, H.; Liu, G. Assembly of a Series of Keggin-Based Multi- and Mono-Nuclear Structures by Tuning the Bis (Tetrazole)-Functionalized Thioether Ligands. *Inorg. Chim. Acta* **2012**, *384*, 62–68. [CrossRef]

15. Liu, G.; Zhang, S.Z. Crystal structure of poly[(μ_4-1-methyl-1*H*tetrazole-5-thiolato-$\kappa^3 S$:*S*:*N*:*N*') copper(I)], $C_2H_3CuN_4S$. *Kristallogr. NCS* **2016**, *231*, 791–792.

16. Li, C.-R.; Chen, T.; Xia, Z.-Q. Bis[(1-methyl-1H-tetrazol-5-yl)sulfanyl]-ethane. *Acta Crystallogr.* **2011**, *E67*, o769. [CrossRef]

17. Argibay-Otero, S.; Gómez-Paz, O.; Carballo, R. A Monoclinic Polymorph of 1,2-Bis[(1-Methyl-1*H*-Tetrazol-5-yl)Sulfanyl]ethane (BMTTE). *Acta Crystallog.* **2017**, *E73*, 1523–1525. [CrossRef]

18. Chen, K.; Shearer, J.; Catalano, V.J. Subtle Modulation of $Cu_4X_4L_2$ Phosphine Cluster Cores Leads to Changes in Luminescence. *Inorg. Chem.* **2015**, *54*, 6245–6256. [CrossRef] [PubMed]

19. Robin, A.Y.; Fromm, K.M. Coordination polymer networks with O- and N-donors: What they are, why and how they are made. *Coord. Chem. Rev.* **2006**, *250*, 2127–2157. [CrossRef]

20. Xue, K.; Chai, W.X.; Wu, Y.W.; Ling, C.; Song, L. Syntheses, Structures and Properties of Two Coordination Polymers Built Upon Copper(I, II) Halide Clusters and a New Thiodiazole Ligand. *J. Clust. Sci.* **2014**, *25*, 1005–1017. [CrossRef]

21. Addison, A.W.; Rao, T.N.; Reedij, J.; van Rijn, J.; Verschoor, G.C. Synthesis, structure, and spectroscopic properties of copper(II) compounds containing nitrogen–sulphur donor ligands; the crystal and molecular structure of aqua [1,7-bis(*N*-methylbenzimidazol-2'yl)-2,6-dithiaheptane] copper(II) perchlorate. *J. Chem. Soc. Dalton Trans.* **1984**, *7*, 1349–1356. [CrossRef]

22. Amoedo, A.; Carballo, R.; García-Martínez, E.; Lago, A.B.; Vázquez López, E.M. Molecular metallocycles, acyclic metallodimers and 2D coordination polymers containing the twisted ligand bis(pyrimidin-2-ylthio)methane. *Dalton Trans.* **2010**, *39*, 2385–2394. [CrossRef]

23. Li, L.; Li, H.; Ren, Z.; Lang, J. Unique Deca- and Tetranuclear Halocuprate(I) Clusters of a Clamplike Ligand: Isolation, Structure, and Luminescence Properties. *Eur. J. Inorg. Chem.* **2014**, *5*, 824–830. [CrossRef]

24. Kwon, H.; Lee, E. Static and dynamic coordination behaviours of copper(I) ions in hexa(2-pyridyl)benzene ligand systems. *Dalton Trans.* **2018**, *47*, 8448–8455. [CrossRef] [PubMed]

25. Mehrotra, P.K.; Hoffmann, R. Copper(I)-copper(I) interactions. Bonding relationships in d^{10}-d^{10} systems. *Inorg. Chem.* **1978**, *17*, 2187–2189. [CrossRef]

26. Yang, L.; Powell, D.R.; Houser, R.P. Structural Variation in Copper(I) Complexes with Pyridylmethylamide Ligands: Structural Analysis with a New Four-Coordinate Geometry Index, $\tau(4)$. *Dalton Trans.* **2007**, 955–964. [CrossRef] [PubMed]

27. Kitaigorodskii, A.I. *Molecular Crystals and Molecules*; Academic Press: New York, NY, USA, 1973.

28. Lago, A.B.; Carballo, R.; Rodríguez-Hermida, S.; Vázquez-López, E.M. Copper(II) Acetate/Bis(4-pyridylthio)methane System: Synthesis, Structural Diversity, and Single-Crystal to Single-Crystal Transformation. *Cryst. Growth Des.* **2014**, *14*, 3096–3109. [CrossRef]

29. Macrae, C.F.; Edgington, P.R.; McCabe, P.; Pidcock, E.; Shields, G.P.; Taylor, R.; Towler, M.; van de Streek, J. Mercury: Visualization and analysis of crystal structures. *J. Appl. Cryst.* **2006**, *39*, 453–457. [CrossRef]

30. Wei, W.; Xia, Z.-Q.; Chen, A.-P.; Gao, S.-L. Bis[(1-methyl-1H-tetrazol-5-yl)sulfanyl]-methane. *Acta Crystallogr.* **2011**, *E67*, o999. [CrossRef]

31. Ardon, M.; Hayes, P.D.; Hogarth, G. Microwave-Assisted Reflux in Organometallic Chemistry: Synthesis and Structural Determination of Molybdenum Carbonyl Complexes. An Intermediate-Level Organometallic-Inorganic Experiment. *J. Chem. Educ.* **2002**, *79*, 1249–1251. [CrossRef]

32. *SMART, Version 5.054*; Bruker AXS: Madison, WI, USA, 1997.

33. *SAINT, Version 8.38A*; Bruker AXS: Madison, WI, USA, 2017.

34. Sheldrick, G.M. *SADABS, Version 2016/2*; Program for Absorption Corrections; Göttingen University: Göttingen, Germany, 1996.

35. Sheldrick, G.M. Crystal Structure Refinement with SHELXL. *Acta Crystallogr.* **2015**, *C71*, 3.

36. Dolomanov, O.V.; Bourhis, L.J.; Gildea, R.J.; Howard, J.A.K.; Puschmann, H. OLEX2: A complete structure solution, refinement and analysis program. *J. Appl. Cryst.* **2009**, *42*, 339–341. [CrossRef]

37. Spek, A.L. Structure validation in chemical crystallography. *Acta Crystallogr.* **2009**, *D65*, 148–155. [CrossRef]

38. Rodriguez-Carvajal, J. Fullprof: A Program for Rietveld Refinement and Pattern Matching Analysis. In Proceedings of the Satellite Meeting on Powder Diffraction of the XV Congress of the IUCr, Toulouse, France, 19–28 July 1990; p. 127.

39. Wolff, S.K.; Grimwood, D.J.; McKinnon, J.J.; Turner, M.J.; Jayatilaka, D.; Spackman, M.A. *CrystalExplorer, Version 3.1*; University of Western Australia: Crawley, Australia, 2012.

 *chemistry*

Article

Solvent Effect on the Regulation of Urea Hydrolysis Reactions by Copper Complexes

Caio B. Castro [1], Rafael G. Silveira [2], Felippe M. Colombari [3], André Farias de Moura [4], Otaciro R. Nascimento [5] and Caterina G. C. Marques Netto [1,*]

[1] Laboratório de Metaloenzimas e Biomiméticos, Departamento de Química, Universidade Federal de São Carlos (UFSCar), Rod. Washington Luiz, km 235, São Carlos, São Paulo 13565-905, Brazil; caiobezerradecastro@gmail.com

[2] Instituto Federal Goiano, Rod. GO, km 154, s/n 03, Ceres, Goiás 76300-000, Brazil; rafael.silveira@ifgoiano.edu.br

[3] Laboratório Nacional de Nanotecnologia, Rua Giuseppe Máximo Scolfaro, 10.000, Polo II de Alta Tecnologia de Campinas, Campinas, SP 13083-100, Brazil; colombarifm@hotmail.com

[4] Laboratório de Química Teórica, Departamento de Química, Universidade Federal de São Carlos (UFSCar), Rod. Washington Luiz, km 235, São Carlos, São Paulo 13565-905, Brazil; moura@ufscar.br

[5] Instituto de Física de São Carlos, Universidade de São Paulo (USP), Av. João Dagnone, 1100, Jardim Santa Angelina, São Carlos, São Paulo 13563-120, Brazil; ciro@ifsc.usp.br

* Correspondence: caterina@ufscar.br; Tel.: +55-16-3306-6653

Received: 14 February 2020; Accepted: 20 May 2020; Published: 2 June 2020

Abstract: Abiotic allosterism is most commonly observed in hetero-bimetallic supramolecular complexes and less frequently in homo-bimetallic complexes. The use of hemilabile ligands with high synthetic complexity enables the catalytic center by the addition or removal of allosteric effectors and simplicity is unusually seen in these systems. Here we describe a simpler approach to achieve kinetic regulation by the use of dimeric Schiff base copper complexes connected by a chlorido ligand bridge. The chlorido ligand acts as a weak link between monomers, generating homo-bimetallic self-aggregating supramolecular complexes that generate monomeric species in different reaction rates depending on the solvent and on the radical moiety of the ligand. The ligand exchange was observed by electron paramagnetic resonance (EPR) and conductivity measurements, indicating that complexes with ligands bearing methoxyl ($Cu^{II}L2$) and ethoxyl ($Cu^{II}L5$) radicals were more prone to form dimeric complexes in comparison to ligands bearing hydrogen ($Cu^{II}L1$), methyl ($Cu^{II}L3$), or t-butyl ($Cu^{II}L4$) radicals. The equilibrium between dimer and monomer afforded different reactivities of the complexes in acetonitrile/water and methanol/water mixtures toward urea hydrolysis as a model reaction. It was evident that the dimeric species were inactive and that by increasing the water concentration in the reaction medium, the dimeric structures dissociated to form the active monomeric structures. This behavior was more pronounced when methanol/water mixtures were employed due to a slower displacement of the chlorido bridge in this medium than in the acetonitrile/water mixtures, enabling the reaction kinetics to be evaluated. This effect was attributed to the preferential solvation shell by the organic solvents and in essence, an upregulation behavior was observed due to the intrinsic nature of the complexes to form dimeric structures in solution that could be dismantled in the presence of water, indicating their possible use as water-sensors in organic solvents.

Keywords: copper complexes; chlorido ligand displacement; catalysis regulation; Schiff base ligands; urea hydrolysis; supramolecular chemistry

---

1. Introduction

Allosterism is commonly observed in proteins that suffer conformational changes induced by ligand binding to an orthosteric site, producing an activation, inhibition, or regulation of the enzymatic activity [1]. Allosteric enzymes optimize the interactions between the ligand and host to tune the populations of active and inactive states for a specific metabolic function [2]. Therefore, allosterism control leads to a high specificity of these enzymes [3,4], which can inspire the design and development of supramolecular devices with high selectivity toward a substrate or analyte. Therein, abiotic allosteric catalysts have shown that metal ions can induce the control of conformation and reactivity of dinuclear catalytic sites [5–7]. The use of redox switching [8] and anion binding affinity [9–11] has also been explored in allosteric coordination chemistry, in which the weak-link approach (WLA) is afforded by the employment of hemilabile ligands to obtain systems that are stimuli-responsive [10].

The WLA approach allows for reversibility in enzyme mimics, leading to allosteric responses [12] as well as to the development of small-molecule sensors such as enzyme-linked immunosorbent assays (ELISA) [13,14] and polymerase chain reaction (PCR) [15]. In the WLA approach, systems with ditopic ligands are commonly used since they enable the interaction of the complex with other molecular species such as anions and cations, causing the change in the overall conformation of the complex [16]. This change in conformation induces an alteration in the properties of one of the metal centers. Classical examples of systems employing WLA in an allosteric conformational manner can be found in the literature, in which the molecular structure [17], binding specificity [18], and catalytic activity [19,20] are modulated upon the interaction with a regulator.

Ligands based on Schiff bases are common moieties in complexes bearing allosteric behaviors [12,21,22] and homo-bimetallic complexes have been shown to be effective in several catalytic reactions [23–26], with special attention to the supramolecular assemblies forming dimeric structures that exhibit significant rate acceleration when compared to the corresponding monomeric catalyst [27,28]. However, these systems are based on coordination compounds with rather sophisticated structures and it would be valuable to find simpler structural complexes bearing abiotic allosterism or catalytic regulation. In this aspect, dimeric or polymeric structures can be easily achieved with chlorido bridges [29–33], which can suffer ligand substitution reactions to form monomeric species in solution, enabling an easy approach to obtain different reactivities of the complex in an allosteric manner.

In this work, a simple coordination system was designed to achieve allosteric behavior by the regulation of the equilibrium between monomeric and dimeric species. For this purpose, Schiff base ligands based on *L*-proline were designed and coordinated to Cu^{II}, as shown in Figure 1. The equilibrium between monomeric and dimeric species was dependent on the solvent mixture used in the reaction, enabling their use in a model reaction. As a proof-of-concept, urea hydrolysis was performed by these complexes, and water was shown to act as an allosteric regulator, since it induced the formation of active monomeric structures. Hence, we demonstrate that coordination systems with less complexity can also be used in regulatory reactions, serving as an inspiration to the development of cheaper sensors.

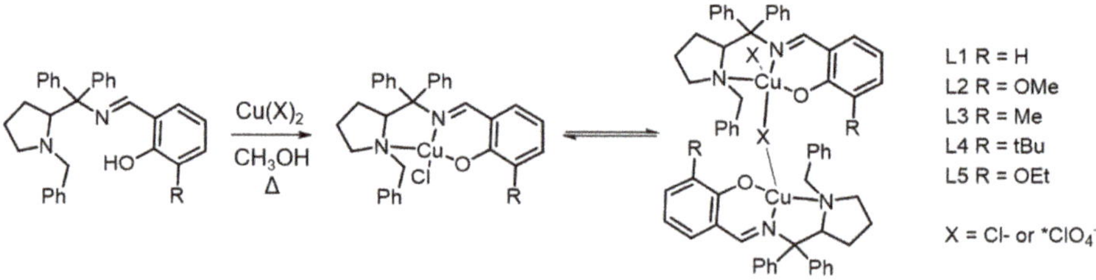

Figure 1. Synthesis of Cu^{II}L1-L5(X) complexes. * Note: The perchlorate anion was used only for ligand L2.

2. Materials and Instruments

The reagents were of analytical grade and were used without prior purification. Thionylchloride ($SOCl_{2(l)}$) used in the esterification of *L*-proline, as described in the literature, was previously distilled under argon. The solvents used in the synthesis and experiments were previously distilled. High Resolution Mass Spectra (HRMS) were obtained with a MICROTOF–Bruker Daltonics (Billerica, MA, USA) in the positive mode. Nebulizer: 0.3 bar, Dry gas: 4 mL/min, temperature: 180 °C, High Voltage: 4500 V. Nuclear Magnetic Resonance (NMR) spectra were recorded on a BRUKER DRX 400 MHz using $CDCl_3$ as the solvent. X-ray diffraction data were obtained on a Rigaku XtaLAB Mini diffractometer (Rigaku, Tokio, Japan) with an X-ray generator operating at 50 kV and 12 mA with a graphite monochromatic Mo-K (λ = 0.71073 Å) using the Olex2 program (version Olex2 v1.3© OlexSys Ltd. 2004–2020, Chemistry Department, Durham University, Durham, UK). EPR spectra were obtained on a Varian E109 EPR X-Band (Varian, Palo Alto, CA, USA) using the rectangular cavity field modulation at 100 kMz. Parameters: microwave power of 20 mW, modulation amplitude of 0.4 mT peak to peak, gain adjustable for each sample, field scan of 160 mT, time constant of 0.064 s, scan time of 3 min. For the measurements of liquid, N_2 was used in the Dewar immersion method. An EPR standard was used to calibrate the magnetic field (MgO crystal: Cr^{III} g = 1.9797) and the resonance frequency was measured with a microwave frequency meter. FTIR spectra were recorded on a Bomen-Michelson FT model MB-102 spectrometer (ABB BOMEM, Quebec, QC, Canada) in the 4000–200 cm^{-1} region. In situ FTIR were recorded on a Nicolet 6700 FTIR spectrometer equipped with a Mercury-Cadmium-Telluride (MCT) detector using p-polarized light and employed a 60 CaF_2 prism located in the bottom of a gas cell in the configuration described. Each spectrum consisted of 32 interferograms, recorded with a spectral resolution of 4 cm^{-1}. UV–Vis spectra were recorded on a HP-Hewlett Packard 8452 A spectrophotometer (Hewlett Packard, Palo Alto, CA, USA) in the 190–800 nm range.

2.1. Synthesis of Ligands (L1–L5)

The ligands were synthesized starting from L-proline. Five synthetic steps are necessary, according to the method reported in the literature [34–36] for the synthesis of L1. Some minor modifications were performed. Essentially, 205 mg (0.60 mmol) of ((S)-1-benzylpyrrolidin-2-yl)diphenylmethanamine (6) was added in a reaction flask containing 2.0 mL anhydrous methanol. Then, 1.05 eq of the salicylaldehyde derivative was added (63.94 µL (0.63 mmol) of 2-hydroxybenzaldehyde (L1), 96.00 mg (0.63 mmol) of 2-hydroxy-3-methoxybenzaldehyde (L2), 76.58 µL (0.63 mmol) of 2-hydroxy-3-methylbenzaldehyde (L3), 107.60 µL (0.63 mmol) of 3-tert-butyl-2-hydroxybenzaldehyde (L4), and 105.00 mg (0.63 mmol) of 3-ethoxy-2-hydroxybenzaldehyde (L5)), in addition to 20.00 mg (0.23 eq) of anhydrous $Na_2SO_4(s)$. The reactions were left under stirring at 40 °C and followed by thin layer chromatography until no change in reagent consumption was observed. L1, L2, and L3 were filtered after 20 h of reaction and washed extensively with methanol to obtain yellowish solids. The solids were dissolved in dichloromethane and methanol was added to obtain yellowish crystals by slow evaporation for 24 h. L4 and L5 were left in reaction for 48 and 10 h, respectively, with the formation of a viscous yellowish solid. The solid was purified by column chromatography on silica gel with ethyl acetate/hexane (5:95) and (20:80) mixture as the eluent, respectively, for ligands L4 and L5.

(*E*)-2-((((1-benzylpyrrolidin-2-yl)diphenylmethyl)imino)methyl)phenol (L1). Yield: 75%. ^{1}HNMR (400 MHz, $CDCl_3$, 298 K): δ 14.71 (s, 1H), 7.97 (d, 2H), 7.33–7.20 (m, 7H), 7.18–7.01 (m, 8H), 6.96 (d, 1H), 6.75 (t, 1H), 3.96 (dd, 1H), 3.34 (d, 1H), 3.10 (d, 1H), 2.75–2.63 (m, 1H), 2.18–2.02 (m, 2H), 1.72–1.62 (m, 1H), 1.43–1.32 (m, 1H), 0.84–0.72 (m, 1H) ppm. ^{13}C NMR (400 MHz, CDCl3, 298 K): δ 164.55, 162.06, 144.55, 142.85, 140.44, 132.60, 132.09, 130.19, 129.10, 128.48, 128.12, 127.98, 127.83, 127.18, 126.86, 126.44, 118.87, 118.22, 117.51, 77.76, 71.96, 62.03, 55.10, 30.72, 24.00 ppm. HRMS (ESI$^+$, CH_3OH) *m/z* calculated for $C_{31}H_{31}N_2O$ 447.2436 [M+H]$^+$; found 447.2413. IR (KBr): 3347 (ν OH), 3056 (ν Csp_2H), 2969 (ν Csp_3H), 2818 (ν Csp_3H), 1620 (ν C=N), 1490 (ν C=C), 1280 (ν C–O) cm^{-1}. UV–Vis ε(L mol^{-1} cm^{-1}) in CH_2Cl_2: 240 (8104), 260 (9513), 320 (3729), and 414 (540) nm.

(*E*)-2-(((((1-benzylpyrrolidin-2-yl)diphenylmethyl)imino)methyl)-6-methoxyphenol (L2) Yield: 69%. ^{1}HNMR (400 MHz, $CDCl_3$, 298 K): δ 15.23 (s, 1H), 7.91 (s, 1H), 7.45–7.40 (m, 2H), 7.32–7.20 (m, 6H), 7.17–7.00 (m, 7H), 6.82 (dd, 1H), 6.65–6.58 (m, 2H), 3.96 (dd, 1H), 3.31 (d, 1H), 3.08 (d, 1H),

2.76–2.69 (m, 1H), 2.18–2.06 (m, 2H), 1.72–1.63 (m, 1H), 1.39 (dd, 1H), 1.09 (dd, 1H) ppm. ^{13}C NMR (400 MHz, CDCl$_3$, 298 K): δ 164.31, 155.95, 149.95, 144.01, 142.30, 139.91, 130.05, 128.87, 128.58, 128.30, 128.00, 127.95, 127.43, 127.00, 126.53, 123.74, 117.63, 116.63, 113.61, 77.22, 71.75, 62.00, 55.05, 30.70, 24.03 ppm. HRMS (ESI$^+$, CH$_3$OH) m/z calculated 477.2542 [M+H]$^+$; found 477.2518. IR (KBr): 3417 (ν OH), 3020 (ν Csp$_2$H), 2966 (ν Csp$_3$H), 2805 (ν Csp$_3$H), 1623 (ν C=N), 1491 (ν C=C), 1253 (ν C–O) cm^{-1}. UV–Vis (L mol^{-1} cm^{-1}) in CH$_2$Cl$_2$: 232 (27,271), 266 (15,775), 324 (3046), and 432 (872) nm.

(*E*)-2-((((1-benzylpyrrolidin-2-yl)diphenylmethyl)imino)methyl)-6-methylphenol (L3). Yield: 72%. ^{1}H NMR (400 MHz, CDCl$_3$, 298 K): δ 14.63 (s, 1H), 7.95 (s, 1H), 7.44 (dt, 2H), 7.32–7.02 (m, 14H), 6.87 (dd, 1H), 6.65 (t, 1H), 3.98 (dd, 1H), 3.49 (d, 1H), 3.15 (d, 1H), 2.64 (m, 1H), 2.26 (s, 3H), 2.15 (m, 1H), 2.07 (m, 1H), 1.72 (m, 1H), 1.35 (m, 1H), 0.97 (m, 1H) ppm. ^{13}C NMR (400 MHz, CDCl$_3$, 298 K): δ 165.42, 160.18, 144.30, 143.29, 140.51, 133.45, 130.11, 129.78, 129.38, 128.57, 128.01, 127.73, 127.05, 126.79, 126.41, 126.28, 118.16, 117.74, 77.64, 72.13, 62.08, 54.98, 30.61, 23.89, 15.66 ppm. HRMS (ESI$^+$, CH$_3$OH) m/z calculated for C$_{32}$H$_{33}$N$_2$O 461.2592 [M+H]$^+$; found 461.2569. IR (KBr): 3412 (ν OH), 3052 (ν Csp$_2$H), 2976 (ν Csp$_3$H), 2813 (ν Csp$_3$H), 1619 (ν C=N), 1491 (ν C=C), 1264 (ν C–O) cm^{-1}. UV–Vis ε(L mol^{-1} cm^{-1}) in CH$_2$Cl$_2$: 232 (24,009), 262 (18,426), 326 (5085), and 420 (358) nm.

(*E*)-2-((((1-benzylpyrrolidin-2-yl)diphenylmethyl)imino)methyl)-6-(tert-butyl)phenol (L4). Yield: 45%. ^{1}H NMR (400 MHz, CDCl$_3$, 298 K): δ 15.13 (s, 1H), 7.96 (s, 1H), 7.43 (dd, 2H), 7.35–7.06 (m, 14H), 6.88 (dd, 1H), 6.67 (dd, 1H), 3.96 (dd, 1H), 3.49 (d, 1H), 3.16 (d, 1H), 2.65 (ddd, 1H), 2.16–2.03 (m, 2H), 1.75–1.66 (m, 1H), 1.39 (s, 9H), 0.92–0.85 (m, 2H) ppm. ^{13}C NMR (400 MHz, CDCl$_3$, 298 K): δ 165.45, 161.15, 144.44, 143.08, 140.63, 137.75, 130.28, 130.23, 129.40, 128.47, 127.89, 127.77, 126.99, 126.81, 126.33, 118.80, 117.30, 77.66, 72.23, 62.04, 54.97, 34.93, 30.64, 29.33, 23.81 ppm. HRMS (ESI$^+$, CH$_3$OH) m/z calculated for C$_{35}$H$_{39}$N$_2$O 503.3062 [M+H]$^+$; found 503.3035. IR (KBr): 3413 (ν OH), 3058 (ν Csp$_2$H), 2954 (ν Csp$_3$H), 1619 (νC=N), 1264 (ν C–O) cm^{-1}. UV–Vis ε(L mol^{-1} cm^{-1}) in CH$_2$Cl$_2$: 232 (22,025), 264 (16,192), 330 (5320), and 400 (269) nm.

(*E*)-2-((((1-benzylpyrrolidin-2-yl)diphenylmethyl)imino)methyl)-6 ethoxy phenol (L5). Yield: 38%. ^{1}H NMR (400 MHz, CDCl$_3$, 298 K): δ 14.92 (s, 1H), 7.93 (s, 1H), 7.44 (dd, 2H), 7.32–7.05 (m, 13H), 7.01 (dd, 2H), 6.84 (t, 1H), 6.63 (d, 2H), 4.08 (q, 2H), 3.99 (dd, 1H), 3.42 (d, 1H), 3.12 (d, 1H), 2.66 (m, 1H), 2.10 (m, 2H), 1.70 (m, 1H), 1.45 (t, 3H), 1.35 (m, 1H), 1.03 (m, 1H) ppm. ^{13}C NMR (400 MHz, CDCl$_3$, 298 K): δ 165.11, 154.02, 148.19, 144.16, 143.05, 140.33, 132.44, 130.08, 130.00, 129.19, 128.59, 128.15, 127.96, 127.79, 127.18, 126.85, 126.45, 123.76, 118.48, 117.15, 115.08, 77.50, 71.86, 64.38, 62.06, 54.99, 30.63, 23.95, 14.88 ppm. HRMS (ESI$^+$, CH$_3$OH) m/z calculated for C$_{32}$H$_{33}$N$_2$O$_2$ 477.2537 [M+H]$^+$; found 477.2503. IR (KBr): 3421 (ν OH), 3055 (νCsp$_2$H), 2965 (ν Csp$_3$H), 1621 (ν C=N), 1492 (ν C=C), 1272 (ν C–O) cm^{-1}. UV–Vis ε (L mol^{-1} cm^{-1}) in CH$_2$Cl$_2$: 232 (15,984), 264 (9288), 332 (2716), and 430 (622) nm.

2.2. General Procedure of Synthesis of Cu^{II} Chlorido Complexes

In a reaction flask, 50.00 mg of CuCl$_2$ (0.37 mmol, 1.1 eq) wasadded to 3.0 mL of anhydrous methanol. The methanolic solution was heated at reflux temperature for 10 min, followed by the addition of 1.0 eq of the ligands (150.00 mg (0.33 mmol) of HL1, 157.00 mg (0.33 mmol) of HL2, 152.00 mg (0.33 mmol) of HL3, 166.00 mg (0.33 mmol) of HL4 and 162.00 mg (0.33 mmol) of HL5). After 4 h, the reaction mixture was cooled to room temperature and filtered. The filtrate was evaporated to dryness and suspended in dichloromethane. The mixture was centrifuged and the supernatant was removed. The complexes were obtained after removal of the solvents by rotatory evaporation under vacuum.

Spectroscopic data of Cu^{II}L1. Dark green powder, yield 75%. HRMS (ESI$^+$, CH$_2$Cl$_2$/CH$_3$CN) m/z 566.1162 calculated for [M+Na]$^+$, found 566.1122; IR (KBr): 3546, 3472, 3412 (ν OH), 3080, 3057 (ν Csp$_2$H), 3026 (ν N=Csp$_2$H), 2959, 2922, 2851 (ν Csp$_3$H), 1654, 1637, 1617(ν C=N), 1598, 1580 (ν C=C), 1455, 1445 (δ CH2), 1317, 1276, 1261 (ν C–O), 1089, 1074, 1028 (ν C–N), 760, 704 (γ Csp$_2$H), 638 (ν Cu–O), 474 (ν Cu–N) cm^{-1}. UV–Vis Vis ε L mol^{-1} cm^{-1} (CH$_2$Cl$_2$): 248 (18,246), 276 (16,002), 380 (4383), 636 (265) nm. C$_{31.5}$H$_{30.5}$Cl$_{11.5}$CuN$_2$O$_{1.25}$[Cu(L1)Cl]·(0.25CH$_3$OH)·(0.25CH$_2$Cl$_2$) calculated C, 65.93; H, 5.36; N, 4.88. Found: C, 65.95; H, 5.23; N, 4.95.

Spectroscopic data of Cu^{II}L2. Dark brown powder, yield 91%. HRMS (ESI$^+$, CH$_2$Cl$_2$/CH$_3$CN) m/z calculated for [M–Cl]$^+$ 538.1676, found 538.1669; HRMS (ESI$^+$, CH$_2$Cl$_2$/CH$_3$OH) m/z calculated 1169.2638 for [2M + Na$^+$]$^+$, found m/z 1169.2410; IR (KBr): 3458, 3410 (ν OH), 3055 (ν Csp$_2$H), 3026 (ν N=Csp$_2$H), 2959, 2926 (ν Csp$_3$H), 1619 (ν C=N), 1577, 1544 (ν C=C), 1469, 1444 (δ CH2), 1316, 1276 (ν C–O), 1245, 1218 (ν C–O–C), 1081, 1004 (ν C–N), 748, 704 (γ Csp$_2$H), 638 (ν Cu–O), 557 (ν C–N) cm^{-1}. UV–Vis Vis ε(L mol^{-1} cm^{-1}) in CH$_2$Cl$_2$: 234 (15,660), 284 (13,953), 362 (2479), ~600 (–) nm. C$_{32.6}$H$_{32.2}$Cl$_{2.2}$CuN$_2$O$_2$[Cu(L2)Cl]·(0.66CH$_2$Cl$_2$) calculated C, 62.16; H, 5.16; N, 4.44. Found: C, 62.14; H, 4.82; N, 4.71

Spectroscopic data of Cu^{II}L3. Green brownish powder, yield 91%. HRMS (ESI$^+$, CH$_2$Cl$_2$/CH$_3$OH) m/z calculated for [M–HCl]$^+$ 522.1727, found 522.1691. IR (KBr): 3549, 3450, 3410 (ν OH), 3082, 3057 (ν Csp$_2$H), 3026 (ν N=Csp$_2$H), 2949, 2920 (ν Csp$_3$H), 1654, 1615 (ν C=N), 1600, 1577, 1544 (ν C=C), 1467, 1446, 1421 (δ CH2), 1317, 1276 (ν C–O), 1087, 1028 (ν C–N), 748, 704 (γ Csp^{2}H), 638 (ν Cu–O), 567 (νCu–N) cm^{-1}. UV–Vis Vis ε(L mol^{-1} cm^{-1}) in CH$_2$Cl$_2$:252 (19,765), 280 (12,308), 378 (3053), 600–700 (–) nm. C$_{33}$H$_{34}$Cl$_2$CuN$_2$O$_{1.5}$ [Cu(L3)Cl]·(0.5CH$_3$OH)·(0.5CH$_2$Cl$_2$) calculated C, 64.23; H, 5.55; N, 4.54. Found: C, 64.00; H, 5.78; N, 4.48.

Spectroscopic data of Cu^{II}L4. Dark green, yield 72%. HRMS (ESI$^+$, CH$_2$Cl$_2$/CH$_3$OH) m/z calculated for [M–HCl]$^+$ 564.2196, found 564.2168. IR (KBr): 3545, 3472, 3412 (νOH), 3084, 3054 (ν Csp^{2}H), 3026 (ν N=Csp^{2}H), 2949, 2920 (ν Csp^{3}H), 1654, 1615 (ν C=N), 1596, 1534, 1492 (ν C=C), 1465, 1443, 1415 (δ CH$_2$), 1336, 1326 (ν C–O), 1143, 1085 (ν C–N), 748, 702 (γ Csp^{2}H), 567 (ν Cu–N) cm^{-1}. UV–Vis Vis ε(L mol^{-1} cm^{-1}) in CH$_2$Cl$_2$:250 (18,922), 278 (12,694), 332 (3959), 388 (4499) e 650 (294) nm. C$_{36}$H$_{39.8}$Cl$_{2.2}$CuN$_2$O$_{1.4}$[Cu(L4)Cl]·(0.4CH$_3$OH)·(0.6CH$_2$Cl$_2$) calculated C, 65.07; H, 6.04; N, 4.22. Found: C, 64.84; H, 5.97; N, 4.71.

Spectroscopic data of Cu^{II}L5. Dark brown, yield 96%. HRMS (ESI$^+$, CH$_2$Cl$_2$/CH$_3$OH) m/z calculated for [M–HCl]$^+$ 552.1833, found 552.1784; [M–CH$_4$–HCl]$^+$ 536.1519, found 536.1691. IR (KBr): 3458, 3414(ν OH), 3080, 3057 (ν Csp^{2}H), 3026 (ν N=Csp^{2}H), 2974, 2924, 2853 (ν Csp^{3}H), 1654, 1615 (ν C=N), 1602, 1577, 1560 (ν C=C), 1465, 1448 (δ CH$_2$), 1317, 1278 (ν C–O), 1245, 1216 (ν C–O–C), 1073, 1028 (ν C–N), 763, 740, 704 (γ Csp^{2}H), 638 (ν Cu–O) cm^{-1}. UV–Vis Vis ε(L mol^{-1} cm^{-1}) in CH$_2$Cl$_2$: 236 (10,803), 252 (15,770) 356 (2348), ~600 (n.d.) nm. C$_{34.1}$H$_{35.7}$Cl$_{2.7}$CuN$_2$O$_{2.25}$ [2Cu(L5)Cl]·(1.6CH$_2$Cl$_2$) calculated C, 61.57; H, 5.21; N, 4.31. Found: C, 61.64; H, 4.87; N, 4.44.

2.3. Synthesis of Cu^{II}Perchlorate Complex

The perchlorate complex was synthesized similarly to the chlorido complex, but using the precursor Cu(ClO$_4$)·6H$_2$O. After 4 h of reaction, cold distilled water was added and the obtained solid was centrifuged, filtered, and washed with cold distilled water. The solid was left in a desiccator at high vacuum. Dark green. Yield: 89%. HRMS (ESI$^+$, CH$_2$Cl$_2$/CH$_3$OH) m/z calculated for [2M–ClO$_4^-$]$^+$ 1175.2848, found 1175.2834. IR (KBr): 3530, 3446, 3317 (ν OH), 3060 (ν Csp^{2}H), 3033 (ν N=Csp^{2}H), 2967, 2841 (ν Csp^{3}H), 1655, 1623 (ν C=N), 1606, 1577, 1545 (ν C=C), 1493, 1470, 1440 (δ CH$_2$), 1319, 1279 (ν C–O), 1246, 1220 (ν C–O–C), 1120, 1108, 1087 (ν$_3$ ClO$_4^-$), 1005 (ν C–N), 943, 921 (ν$_4$ ClO$_4^-$) 748, 706 (γ Csp^{2}H), 638, 624 (ν Cu–O), 556, 522 (ν Cu–N)cm^{-1}. UV–Vis (CH$_2$Cl$_2$): 244, 286, 392, 592 nm. C$_{65.3}$H$_{66.6}$Cl$_{2.6}$Cu$_2$N$_4$O$_{13}$ [2Cu^{II}L2ClO$_4$·(CH$_3$OH)](0.3CH$_2$Cl$_2$) calculated C, 58.76; H, 5.03; N, 4.20. Found: C, 58.87; H, 5.15; N, 4.16. Conductivity: 13 μS cm^{-1} in dichloromethane.

2.4. Catalysis Protocol

For the catalytic assays, fresh solutions of urea (60 mM in water) and of the complexes (1.6 mM in dichloromethane (DCM)) were prepared.

The catalytic assays were performed by diluting 50 μL of the solution containing the complex into 850 μL of the reaction solvent (methanol, acetonitrile, and dimethylsulfoxide, tetrahydrofuran, and ethanol). The reaction typically started by the addition of the urea solution (100 μL). Vigorous stirring was maintained over the course of the reaction. Aliquots of 100 μL of the reaction were taken after 5, 10, 20, 30, 40, 60, 120, 240 and 480 s of reaction. These aliquots were analyzed by the Berthelot method to quantify the ammonia content. The reaction temperature was kept at 36 °C. All reactions were performed in triplicate.

For each of the catalytic protocols, a control experiment was also performed without the complexes and the blank was subtracted from the catalytic measurements.

The amount of water in the medium, when increasing amount of urea concentrations were employed, were 10, 20, 30 and 40%, respectively, to the urea concentrations of 5.2, 10.4, 15.6 and 20.4 mM (except in acetonitrile, in which urea 20.4 mM was also employed).

2.5. Ammonia Quantification

Fresh solutions of sodium hypochlorite (2.5% in water), sodium citrate (0.38 mM in 0.46 mM aqueous solution of NaOH), and sodium salicylate (2.75 mM in water containing 9.39 μmol of sodium nitroprussiate) were prepared prior to the ammonia quantification. To a microtube containing 250 μL of the hypochlorite solution and 250 μL of citrate/NaOH solution, the 100 μL aliquot from the reaction was added, followed by the addition of 300 μL of the sodium salicylate solution. After mixing, the reaction was allowed to proceed for 15 min at room temperature. The reaction was analyzed by UV–Vis spectroscopy at 654 nm. Quantification of ammonia was performed by using a calibration curve with seven measurements and $R^2 = 0.99$, using ammonium chloride as standard.

2.6. Computer Simulations

Geometries for the $Cu^{II}L1$ monomer and the dimer were optimized at the semi-empirical GFN1-xTB level in vacuum using the xtb software [37–39], with convergence criteria for the self-consistent charge (SCC) iterations of $2 \cdot 10^{-7}$ E_h for the energy change and $4 \cdot 10^{-6}$ for the charge change between cycles and for the geometry optimization of $1 \cdot 10^{-6}$ E_h for the energy change between steps and a maximum gradient of $8 \cdot 10^{-4}$ E_h/α. All calculations were performed with an electronic temperature of 500 K to allow some degree of Fermi smearing of nearly-degenerate energy levels and to take static correlation into account. All model systems were considered neutral with each Cu(II) cation in the $Ar[3d^9]$ configuration, yielding a doublet state for the monomer. Regarding the dimer, both singlet and triplet states were considered for the sake of completeness. The same protocols were applied to acetonitrile, water, and methanol molecules in vacuum. After full geometry optimization of the monomer and the two spin states of the dimer, further geometric relaxation was performed by means of 10 ps-long molecular dynamics simulations performed in the canonical, constant-NVT ensemble (meaning amount of substance (N), volume (V) and temperature (T) are conserved), using the Berendsen weak-coupling scheme to control the temperature around 300 K. Equations of motion were integrated using a 0.5 fs time step, and recording structures and energies each of 0.5 ps. The last structure for each system was subjected to further geometry optimization and these optimized structures were considered as the lowest lying reference states for the thermochemical analyses. The next step of the modeling consisted of searching the thermodynamically most probable position and relative orientation of each solvent around each complex, as described in detail in the Supplementary Materials. This systematic search amounted to ca. 1 million quantum chemical calculations and the most probable solvent-$Cu^{II}L1$ structures were subjected to further geometry optimization as described above.

3. Results and Discussion

3.1. Probing Copper Coordination

Amine 6 was obtained after a series of reactions starting from *L*-proline (Figure 2). The reaction between amine 6 with the corresponding aldehyde afforded ligands L1–L5. After purification, these compounds were characterized by ^1H NMR, ^{13}C NMR, 2D NMR, mass spectrometry, infrared spectroscopy, and X-ray diffractometry (see support information). The coordination of L1–L5 to $CuCl_2$ in methanol resulted in the formation of complexes $Cu^{II}L1$–L5 in high yields (71–91%). All complexes were characterized by proper microanalysis, EPR, infrared spectroscopy, mass spectrometry, cyclic voltammetry, and electronic spectroscopy at the UV–Vis region (Supplementary Materials). Some features of the characterization of the complexes deserve to be described since they indicate the coordination mode and the extent of dimerization. For instance, through FTIR spectroscopy (Figures S25–S29), it was evident that the coordination occurred via the nitrogen of azomethine [40,41], since the ν C=N vibration mode was red shifted by 4–7 cm^{-1}. An increase of the νC–O energy indicated that the coordination was also occurring via a phenolate [34]. Hence, the presence of a band in the 640–470 cm^{-1} range corroborated that coordination occurred through oxygen and nitrogen, consistent with the ν M–O and M–N modes [42].

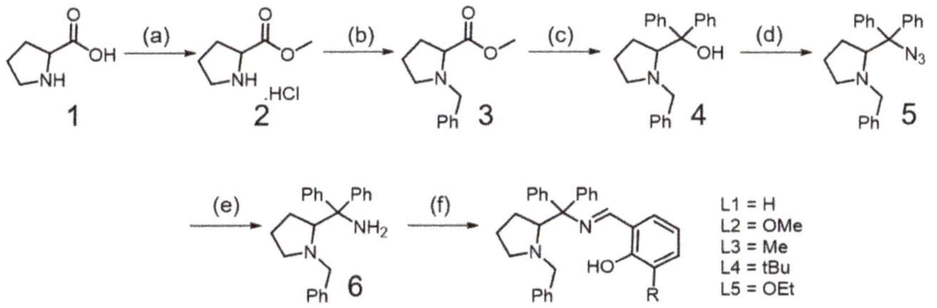

Figure 2. Synthesis of L1–L5 ligands from L-proline. Reaction conditions: (**a**) SOCl$_2$(l), CH$_3$OH; (**b**) BrCH$_2$(Ph), K$_2$CO$_3$(s), CH$_3$CN; (**c**) MgBr(Ph), THF; (**d**) NaN$_3$(s), H$_2$SO$_4$ 70%, CHCl$_3$; (**e**) LiAlH$_4$(s), THF; (**f**) salicylaldehyde and derivatives, Na$_2$SO$_4$(s), CH$_3$OH.

Corroborating to the FTIR spectroscopy, a bathocromic shift was observed for the $\pi\rightarrow\pi^*$ bands of azomethine in the electronic spectroscopy, characterizing the coordination through this moiety [43]. In addition, the disappearance of the $n\rightarrow\pi^*$ azomethine transition indicated that the Cu^{II} coordination occurred on that position of the Schiff base [40]. Moreover, a broadband in the 550–770 nm region was consistent with the d–d transition of the metal center [43–45] of complexes Cu^{II}L1 and Cu^{II}L4 (636 and 650 nm, respectively). However, the other complexes exhibited a less evident d–d band, possibly due to a higher planar geometry in comparison to Cu^{II}L1 and Cu^{II}L4 (Figures S30–S33) [46].

In the cyclic voltammetry, for all ligands, the oxidation of phenol to quinone and the formation of radical cations was observed. These observations are in agreement with the redox behavior of other Schiff bases [47,48]. After coordination, these redox processes were displaced to higher potentials as an effect of electron depletion upon coordination by the metal [49].

The possibility of the formation of monomeric and dimeric structures was first inspected by high resolution mass spectrometry (HRMS) in which, for example, a peak corresponding to the monomer was present at *m/z* 538.1669, whereas the dimer was observed at *m/z* 1169.2410 for the Cu^{II}L2 complex (Figure S58). However, since dimers can be formed in the gas phase depending on the solution concentration [50], we performed EPR analysis in solid and in solution. With these analyses, we evaluated the existence of equilibrium between monomers and dimers in solution that could be controlled by ligand substitution reactions.

3.2. Ligand Substitution and Electron Paramagnetic Resonance Measurements

First, ligand substitution was followed by conductivity measurements to give us insights on the natural dissociation of the complexes. In dichloromethane (DCM) at 298 K, the conductivities were consistent with neutral compounds (values in the 2.00–6.00 μS cm^{-1} range). However, in acetonitrile (ACN) at 298 K, the conductivity was shown to constantly increase over time, somewhat reaching a plateau after half an hour (Figure 3A). The quasi-stabilized values for complexes Cu^{II}L2, CuIIL3, and Cu^{II}L4 were 63, 43, and 23 μS cm^{-1}, respectively. These values indicate that the labilization of the chlorido ligand by acetonitrile was more pronounced for Cu^{II}L2. The solvolysis of the chlorido ligand was facilitated when water was present in solution, as shown in Figure 3B, which shows that in 2 min, the conductivity was already stabilized in limiting values of the 1:1 electrolyte range (55.0–90.0 µS cm^{-1}) [51]. Interestingly, the conductivity of these complexes was more stable both in methanol and the methanol/water mixture (80/20 *v/v*), revealing a slower rate of ligand substitution in these solvents (Figures S53–S55). The observed values ranged from 45.0 to 53.0 µS cm^{-1} and 47.0 to 50.0 µS cm^{-1} in methanol and the methanol/water mixture, respectively, which were lower than that expected for the 1:1 electrolyte in methanol. The lower values might indicate the presence of a mixture of charged and neutral species in solution. The displaced chloride could be detected by the addition of silver nitrate solution as a white precipitate of AgCl.

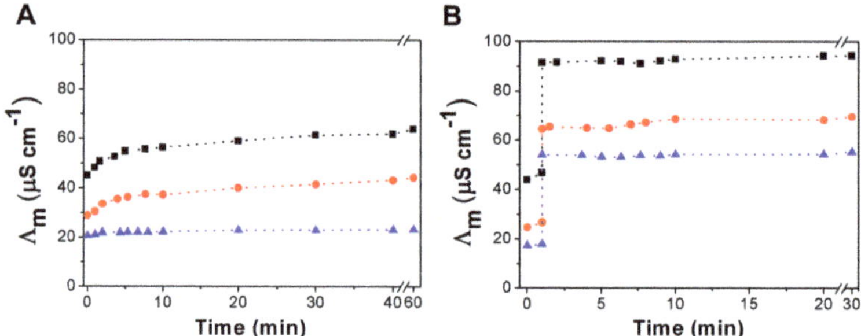

Figure 3. Influence of the time in conductivity measurements of the complexes in (**A**) acetonitrile and (**B**) acetonitrile/water (80/20) mixture. Conductivity of Cu^{II}L2 is shown as black squares, Cu^{II}L3 is shown as red circles, and Cu^{II}L4 is shown as blue triangles.

Considering the existence of a mixture of species in solution, EPR measurements were performed in the same solvents of the conductivity analyses to enable a better structural comprehension (Figure 4). In dichloromethane, all complexes presented values of g$_z$ higher than g$_x$ and g$_y$ (Table 1 and Table S7), suggestive of an axial symmetry [52–54], due to the presence of an unpaired electron in the dx^2–y^2 orbital. The axial symmetry supports the proposal of a square planar geometry of the complexes [42,55,56]. However, as shown in Table 1, the differences between the values of g$_x$ and g$_y$ are indicative of a distortion of the plane. In addition, the highest observed g$_z$ for Cu^{II}L4 could mean that this complex has a tetrahedral distortion. In contrast, the A$_z$ is smaller for the Cu^{II}L5 complex as a result of a greater distortion, indicating that aside from the difference between substituents in the ligands, another factor might be affecting the tetrahedral distortion of complexes Cu^{II}L4 and Cu^{II}L5.

Table 1. Electron Paramagnetic Resonance parameters for the Cu^{II} complexes of this work in dichloromethane at 298 K.

Compounds	G			t$_{corr}$, ps	A, cm^{-1} ($\times10^{-4}$)		
	g$_x$	g$_y$	g$_z$		A$_x$	A$_y$	A$_z$
Cu^{II}L1	2.0501	2.0932	2.1626	71.9	14.53	11.51	200.2
Cu^{II}L2	2.0584	2.0932	2.1626	114.5	11.88	11.51	198.8
Cu^{II}L3	2.0515	2.0932	2.1626	75.3	11.88	11.51	201.0
Cu^{II}L4	2.0494	2.0518	2.2063	44.2	14.53	14.06	199.5
Cu^{II}L5	2.0504	2.0932	2.1626	116.7	11.88	11.51	188.6

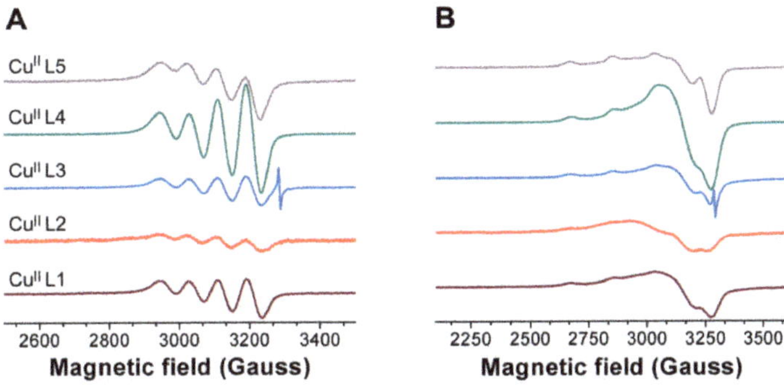

Figure 4. Comparison of the experimental EPR spectra of the complexes in dichloromethane at (**A**) 298 and (**B**) 77 K. Complex Cu^{II}L1 is shown in purple, Cu^{II}L2 is shown in red, Cu^{II}L3 is shown in blue, Cu^{II}L4 is shown in green, and Cu^{II}L4 is shown in grey lines. The narrow line in the high magnetic field for Cu^{II}L3 are a standard signal of Cr^{III}:MgO sample (g = 1.9797).

In acetonitrile and acetonitrile/water (80:20) mixture, the values of g_x and g_y were more similar to each other, as an effect of ligand substitution, in which the nitrogen atom of ACN or the oxygen atom of water are coordinated to copper, forming a plane with higher symmetry than the previous N, N, O, Cl coordination plane. The increase in symmetry observed in the EPR measurements in ACN and ACN/H$_2$O is in agreement with the chlorido displacement observed in the conductivity measurements. However, there was still no evidence of dimeric and monomeric species in equilibrium in these solutions.

Hence, the spinning radiuses of the molecules in solution were obtained from the spectral simulations using the EasySpin program [57] to obtain the rotational time correlation t_{corr} (Table 2). These values are associated withthe spinning velocity of a molecule in solution, expressing large values of t_{corr} when effective intramolecular interactions are generated. Curiously, the increasing order of t_{corr} was Cu^{II}L4 > Cu^{II}L1 > Cu^{II}L3 > Cu^{II}L2 > Cu^{II}L5, revealing that complexes bearing a substituent group with oxygen (Cu^{II}L2 and Cu^{II}L5) are more effectively interacting in solution than Cu^{II}L4. With the t_{corr} values, we were able to calculate the radiuses and volumes of rotation of the complexes using the Stokes–Einstein–Debye (SED) equation ($t_{corr} = 4\pi\eta a^3/3K_BT$), where a is the molecular radius of rotation (Table 2). In general, all compounds have reduced their radius of rotation in acetonitrile and the acetonitrile/water (80:20) mixture. Therefore, the complexes were probably arranged mostly as dimeric structures in dichloromethane, and when coordinating solvents were present such as acetonitrile and water, the equilibrium between dimeric and monomeric species shifted to monomeric ones.

Table 2. Comparison of the values of the radiuses of rotation and approximate volumes of the complexes Cu^{II}L1–Cu^{II}L5 considering a spherical model.

Compounds	a, Å			V, Å^3 ($\times 10^2$)		
	DCM	**ACN**	**ACN/H$_2$O**	**DCM**	**ACN**	**ACN/H$_2$O**
Cu^{II}L1	5.55	4.20	4.38	7.16	3.10	3.51
Cu^{II}L2	6.49	5.14	4.19	11.4	5.68	3.08
Cu^{II}L3	5.64	4.69	4.14	7.51	4.32	2.97
Cu^{II}L4	4.72	4.69	4.34	4.40	4.32	3.42
Cu^{II}L5	6.53	4.68	4.48	11.7	4.29	3.76

In fact, when frozen solutions of the complexes in DCM were analyzed by EPR spectra and compared to simulated spectra (Figure 5), it was evident that a component attributed to molecular aggregates needed to be introduced for a better fit. These molecular aggregates had a magnetic interaction, indicating that two Cu^{II} centers probably interacted with each other, causing the lines to broaden. This broadening was more pronounced for Cu^{II}L2 and Cu^{II}L5, which were the most effective complexes to form dimeric structures, as observed by the t$_{corr}$ values. A similar trend was observed for the measurements performed in acetonitrile and in the acetonitrile/water (80:20) mixture, indicating equilibrium between monomeric and dimeric species in frozen solution. The stronger interaction observed for complexes Cu^{II}L2 and Cu^{II}L5 might indicate the presence of halogen-bonds between the oxygen from methoxy and ethoxy radicals with the chloride. Halogen bonds are more sensitive to steric effects and could be the reason for the higher volume observed for Cu^{II}L2 in ACN than Cu^{II}L5 [58]. Moreover, the dimerization of copper complexes in different solvents has already been observed by EPR measurements by other groups [56], corroborating our observations.

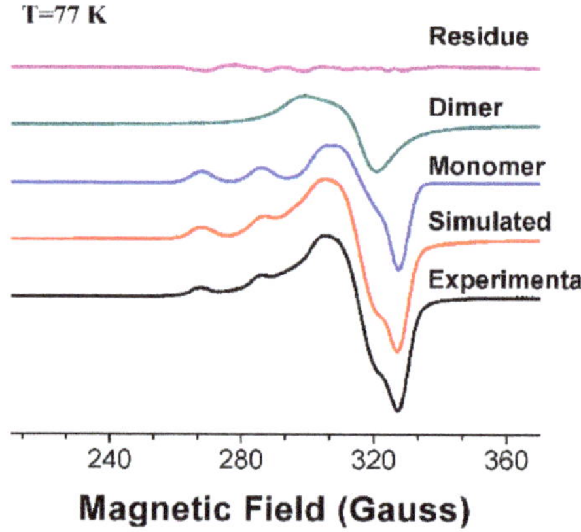

Figure 5. Comparison between the experimental and simulated EPR spectrum of Cu^{II}L4 complex at 77 K in dichloromethane. The experimental EPR spectrum is shown as a black line, whereas the simulated spectrum is shown as a red line. Simulation spectra are composed by a monomeric and dimeric species that are shown in blue and green, respectively.

The difference in the g$_z$ observed for the different complexes may be a result of distinct aggregation structures. For instance, Cu^{II}L1, Cu^{II}L2, and Cu^{II}L5 present values of g$_z$0 higher than g$_x$0 and g$_y$0, which are similar to monomeric species, possibly due to the maintenance of an axial geometry even after aggregation. An opposite behavior was observed for complexes Cu^{II}L3 and Cu^{II}L4, which hadg$_z$0 values lower than g$_x$0 and g$_y$0 in the aggregate species. Therefore, these complexes (Cu^{II}L3 and Cu^{II}L4) may have formed aggregates with nonaxial geometry, suggesting that the unpaired Cu^{II} electron is not of the dx^2–y^2 orbital, possibly due to a trigonal bipyramidal geometry, as shown in Figure 6. Structural features behind thermodynamic differences for the interactions between one solvent molecule and the Cu^{II}L1 monomer and dimer were evaluated by quantum chemical calculations. The dimers had very distinctive geometries in each electronic spin state, with the lowest-lying singlet states having a single Cl bridge between the two monomers (Figure 6B–D), which renders each monomer structurally different from the other, while the higher energy triplet state has two Cl bridging the two Cu(II) atoms (Figure S76). It was evidenced by the simulations that for all of the solvents considered as well as for the bare dimer in vacuum, the singlet state was always lower lying than the triplet state, corroborating the proposition of the structures in Figure 6.

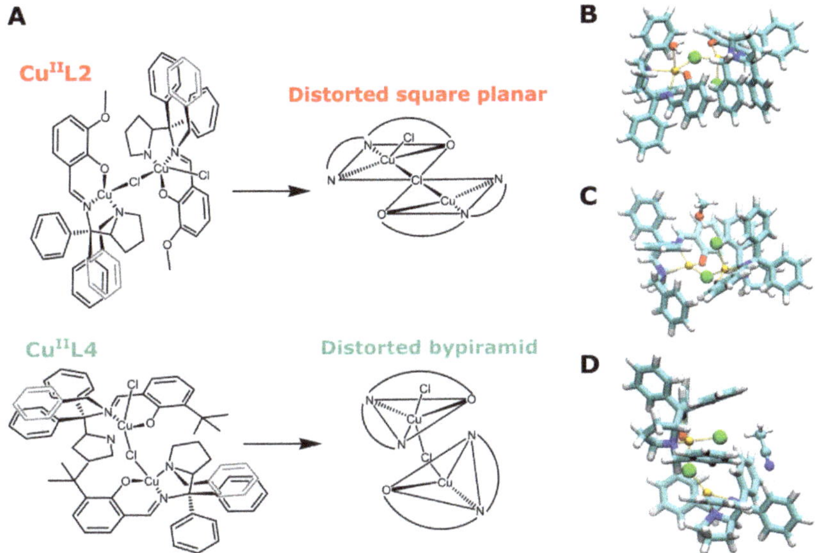

Figure 6. Proposition of the aggregate structures of complexes $Cu^{II}L2$ and $Cu^{II}L4$ in frozen solutions of dichloromethane (**A**). Optimized structures for the $Cu^{II}L1$dimer in its singlet state interacting with H_2O (**B**), MeOH (**C**), and ACN (**D**).

Since all complexes behaved similarly in the EPR measurements in dichloromethane, acetonitrile, and the acetonitrile/water mixture at 298 K, only $Cu^{II}L2$ was evaluated in methanol and the methanol/water in EPR measurements at 298 K. To provide a comparison between dimeric and monomeric species in solution, an analogous of complex $Cu^{II}L2$ was synthesized with perchlorate as a counterion ([$Cu^{II}L2ClO_4$] Figure 1). The perchlorate ion is known for its high volume and would be expected to generate only monomeric species in solution. Therefore, this complex was compared with $Cu^{II}L2$ in EPR measurements performed at 298 K in methanol and the methanol/water (80/20, *v/v*) mixture. Interestingly, the EPR spectrum of $Cu^{II}L2$ in methanol (Figure 7) is visibly a mixture between two species. However, unexpectedly, [$Cu^{II}L2ClO_4$], despite presenting a profile of monomeric species in solution, exhibited a three times higher t_{corr} than $Cu^{II}L2$, which could indicate that [$Cu^{II}L2ClO_4$] is in fact, dimeric. Indeed, the HRMS spectra exhibited *m/z* peaks corresponding to dimeric structures (1175.2834, Figure S59) and in the FTIR spectra, three bands associated with monodentated ClO_4^- species were observed at 1121, 1108 and 1027 cm^{-1}, corroborating the hypothesis of [$Cu^{II}L2ClO_4$] complex dimerization. The addition of 20% water keeps the equilibrium in solution, as expected, due to the similarity of the conductivities of the complexes in methanol and the methanol/water mixtures. Therefore, it can be assumed that methanol and water do not fully displace the chloride. The unexpected dimerization of [$Cu^{II}L2ClO_4$] might strengthen the proposition of halogen bond formation in the $Cu^{II}L2$ complex, suggesting a supramolecular structure with the possibility of use in dynamic catalysis [59].

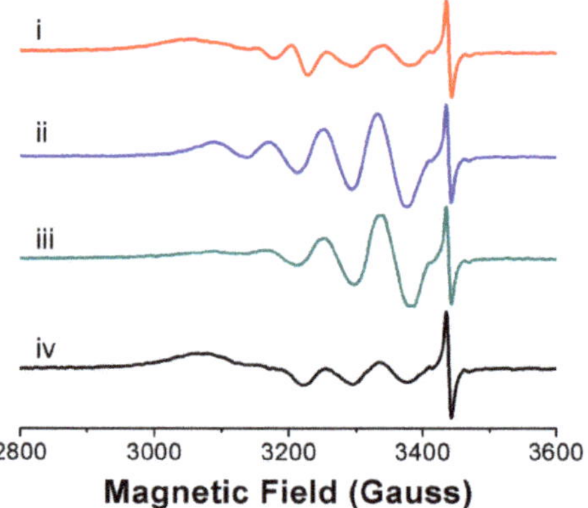

Figure 7. EPR spectra of Cu^{II}L2 and [Cu^{II}L2ClO$_4$] in methanol (i and ii) and Cu^{II}L2 and [Cu^{II}L2(ClO$_4$)]in the methanol/water (80:20) mixture (iii and iv). The narrow line in the high magnetic field for Cu^{II}L3 isa standard signal of the Cr^{III}:MgO sample (g = 1.9797).

3.3. Urea Hydrolysis as a Model Reaction: Kinetics of NH$_3$ Formation

The self-organization of these complexes into dimeric or monomeric structures was shown to be dependent on the solvent and ligand exchange reactions, as shown in Figure 8. Due to the distinct behaviors of the complexes in acetonitrile/water and methanol/water, we suspected that hydrolytic catalysis could be tuned by influencing the equilibrium between monomeric and dimeric species. Hence, we decided to evaluate their potential as catalysts to hydrolyze urea, as a model reaction. Catalysis was performed primarily in the acetonitrile/water and methanol/water mixtures. In this reaction, ammonia is expected to be formed and aliquots of the reaction were analyzed by the Berthelot method [60] over the reaction times (5, 10, 20, 30, 40, 50, 60, 120, 240, and 480 s).

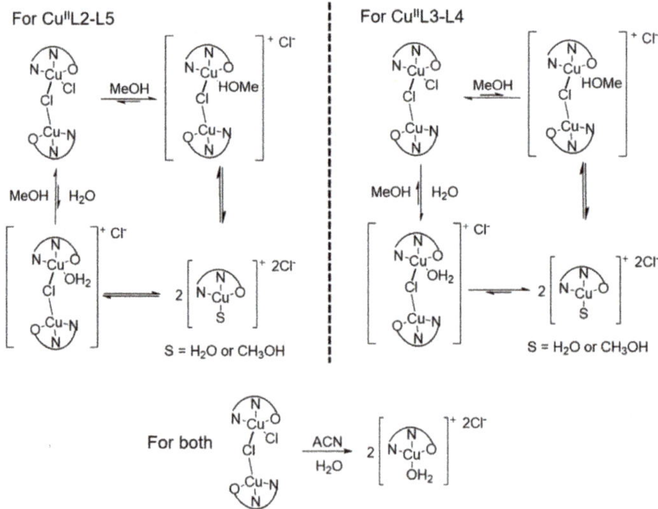

Figure 8. Equilibrium of dimeric and monomeric species dependent on water-association solvent.

Complexes Cu^{II}L2, Cu^{II}L3, and Cu^{II}L4 were chosen to evaluate the influence of the complexes' aggregation in the reaction. In the acetonitrile/water mixture, the reaction was faster than the employed method to detect ammonia formation (minimum reaction time: 5 s) and we could only observe the decrease in ammonia concentration over the reaction time (Figure 9A and Figures S63–S66). In contrast, in the methanol/water mixtures, the reaction profile changed, in which the increase in ammonia concentration was observed until a saturation level was reached (Figure 9B). The lability of the chlorido was smaller in methanol and the methanol/water (80/20) mixture when compared to the acetonitrile system, and indicates that the labilization of chlorido affects the path of the reaction. For instance, in acetonitrile/water, all complexes exhibited a decrease of the volume, possibly due to the formation of monomeric species in solution. Therefore, the equilibrium dimer/monomer is still present in methanol/water and it can be inferred that the presence of dimeric structures in solution is possibly slowing the reaction.

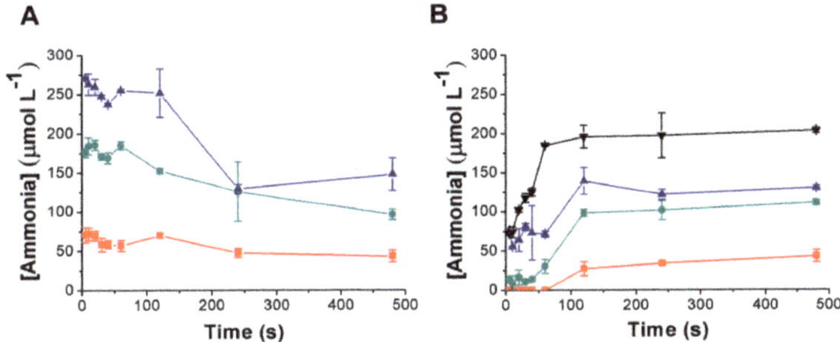

Figure 9. Ammonia quantification produced by $Cu^{II}L2$ complex in (**A**) acetonitrile/water and (**B**) methanol/water mixtures up to 480 s at 308 K and at different urea concentrations: 5.2 mmol L^{-1} (brown line, squares), 10.4 mmol L^{-1} (green line, circles), 15.6 mmol L^{-1} (blue line, triangles), and 20.8 mmol L^{-1} (black line, inverted triangles).

Considering that complex $[Cu^{II}L2ClO_4]$ was mostly dimeric in solution, it would be expected to observe a slower reaction rate of urea hydrolysis by this complex. In general, the behavior of $[Cu^{II}L2ClO_4]$ was similar to the chloride complex (faster in acetonitrile/water and slower in methanol/water mixtures), but indeed, a four times lower conversion was observed (Figure 10). Moreover, the reaction in methanol/water only started to produce ammonia after 5 min of reaction, whereas the $Cu^{II}L2$ complex was able to produce it after 1 min of reaction. Hence, it may indicate that the dimeric structure is not active toward urea hydrolysis.

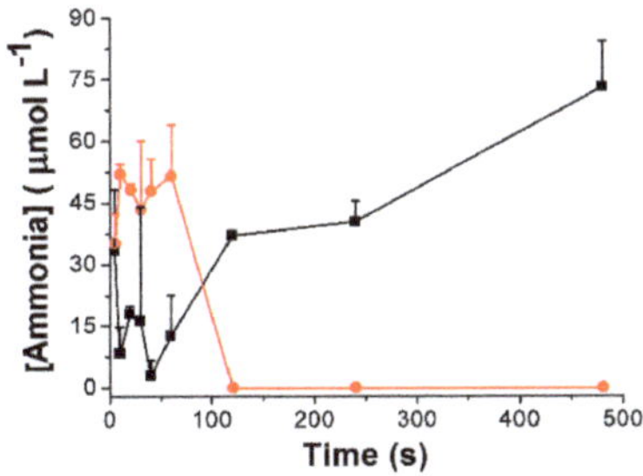

Figure 10. Ammonia quantification produced by the $[Cu^{II}L2ClO_4]$ complex in the acetonitrile/water (red line) and methanol/water mixture (black line) at 308 K using a 10.4 mmol L^{-1} urea concentration. Only the positive portion of the error bars is shown in the graphic.

In order to verify how the solvent was affecting the equilibrium monomer/dimer, we performed the reaction of urea hydrolysis in other solvents (DMSO, THF, and ethanol), Figures S70–S72. We observed that a high reaction rate of urea hydrolysis was achieved in solvents that exhibited less pronounced hydrogen bonds with water (ACN, DMSO, and THF) than the organic solvents methanol and ethanol [61]. Therefore, we reasoned that the effect of preferential solvation [62] by the organic solvents in the aquation reaction of our complexes could be tuning the equilibrium dimer/monomer. The interaction between solvent molecules and the complex is probably occurring via the apolar sites of the solvent, since the complex has a neutral charge. This mode of solvation results in a secondary sphere organized in a way that the dipoles of the solvents are oriented to the bulk solution, and water can interact with these sites via hydrogen bonds (Figure 11A). The stronger hydrogen bond between methanol and water stabilizes the initial state of reaction (dimer–solvent), leading to a slower rate of ligand substitution due to the higher activation energy of the reaction in methanol/water (Figure 11B), thus forming less monomers than in ACN (or the DMSO, THF/H_2O mixtures). These results are in contrast to the increased reaction rate observed in methanol and ethanol by the other groups [63] due to the stabilization of the transition state, which strengthens our supposition. Moreover, the reaction rate in the ethanol/water mixtures was even slower than in the methanol/water mixtures due to the longer chain of ethanol, resulting in a higher stabilization of the hydrogen bonds with water in the tertiary coordination sphere. Therefore, our analysis of the equilibrium monomer/dimer in solution verified the occurrence of only the monomer species in ACN/H_2O by EPR assays, whereas in the MeOH/H_2O mixtures, we detected the presence of dimeric structures by HRMS and EPR experiments, corroborating this hypothesis. The thermochemical data obtained from the DFT calculations of the $Cu^{II}L1$ monomer or dimer interacting with a single solvent molecule support that the interaction between the dimeric structure is more stabilized in methanol and water than in acetonitrile. The enthalpic difference between methanol and water was less than 1 kJ/mol, but an eight times higher enthalpy difference was observed between acetonitrile and water (8.5 kJ/mol). Thus, the stabilization of the ground state in strong hydrogen-bond solvents and the competition between solvents is more pronounced in methanol/water systems, which could result in a lower substitution rate of the chloride in methanol/water mixtures in comparison to the acetonitrile/water mixtures, reinforcing our experimental data.

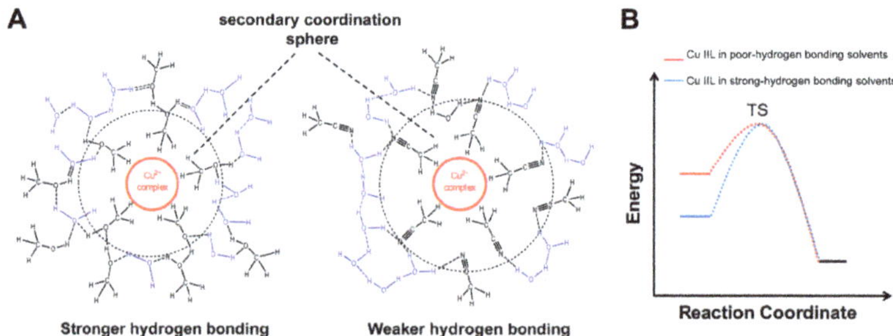

Figure 11. Preferential solvation shell of the complexes in methanol/water and acetonitrile/water mixtures (**A**) and its effect in the stabilization of the ground state (dimeric species) (**B**).

Noticing the strong effect of the solvent in the equilibrium monomer/dimer, we suspected that this could enable the tuning of the catalytic behavior by an allosterism (or upregulation).In order to check this possibility, we evaluated complexes $Cu^{II}L2$, $Cu^{II}L3$, and $Cu^{II}L4$ in different urea concentrations (Figure 9 and Figure S68). It should be noted, however, that the increase in urea concentration also increased the water content of the mixture, which could influence the monomer/dimer equilibrium. Noticeably, the complexes presented different behaviors upon an increase in urea (and water) concentrations. For instance: $Cu^{II}L3$ had a sigmoidal behavior, whereas the behavior for $Cu^{II}L2$ was linear (Figure 12). By keeping the water constant at 20%, the water effect in the monomer/dimer equilibrium decreased, and essentially, this effect was more pronounced for $Cu^{II}L2$, observing a 2-fold decrease of the reaction rate when the water concentration was constant. This result indicates that water has a positive effect in catalysis for the $Cu^{II}L2$ complex due to the shift of inactive dimeric species into active monomeric species. In contrast, the $Cu^{II}L3$ complex and $Cu^{II}L4$, already have their equilibrium shifted to monomeric species and therefore, do not present a strong influence of water in catalysis, even though a slight positive allosteric effect of water is also observed.

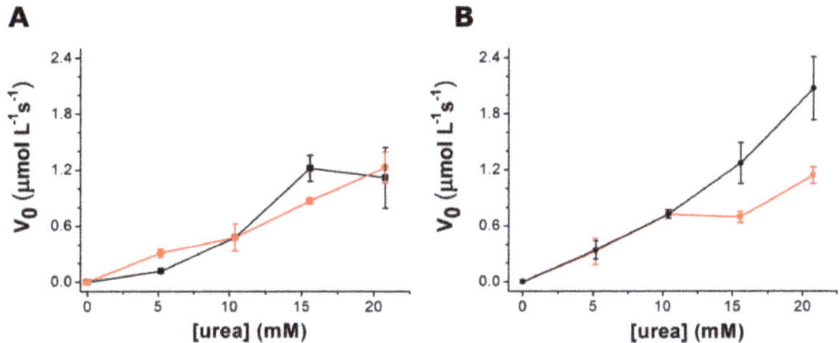

Figure 12. Initial rate of urea hydrolysis reaction versus urea concentration performed by $Cu^{II}L3$ (**A**) and $Cu^{II}L2$ (**B**). The black line and squares are relative to the increase in water content of the reaction from 0 to 40%, whereas the red line and circles are relative to the water content kept constant at 20%.

Since other aspects of the reaction could be affecting the reaction kinetics such as the activity of water and reaction pH, we decided to verify their significance in our systems. For instance, the composition of a solvent mixture containing water will change the activity of water (a_w) [64] and the increase in a_w could impact the reaction mechanism and kinetics of a hydrolytic reaction. For instance, the water activity of the methanol/water mixtures used in our systems ranged from 0.26 to 0.65 (Figure S76) (determined by the equation from Zhu et al. [65]), whereas the acetonitrile/water (10% volume) mixture had a water activity of 0.9. Therefore, it would be reasonable to ascribe the observed differences of reaction rates to the different water activity of these systems. However, the catalytic reaction performed in ACN containing 2% ($a_w \sim 0.4$) (Figure S74) still showed a high reaction rate of urea hydrolysis in this solvent mixture. This finding reinforces the proposed idea of solvent effect in the reaction kinetics, since acetonitrile aids in the equilibrium shift to monomeric species.

In addition to the water activity, the pH in our reactions was not controlled and the increase in pH during the reaction could affect the coordination sphere and the reaction rate. Taking into consideration this possibility, we performed a reaction including a pH indicator (phenol red) and we did not observe a significant effect in the pH due to the produced ammonia over the reaction time in the UV–Vis spectrum (Figure S73A). To remove any possible interference produced by the indicator, we performed a reaction control without the pH indicator and added it after the reaction reached a saturation level of ammonia. Again, no color change was observed (Figure S73B), indicating that in the methanol/water and acetonitrile/water mixtures (Figure S73C), the produced ammonia did not seem to severely affect the reaction pH. In fact, the pH of the buffered solutions in solvent mixtures has been previously analyzed by other groups, and they noticed that the NH_3/NH_4^+ buffer presented a lower pH in the solvent mixture than the observed one in pure water [66]. An in situ FTIR experiment (Figure S75) showed an increase of a band in the 3000 cm^{-1} region, indicative of the formation of ammonium ions. After 90 s, we observed that this band oscillated between a minimum and a maximum, suggestive of the equilibrium between ammonia and ammonium. Therefore, due to the fact that the pH of the reaction is not expressively altered during the reaction, we suspect that the main interaction of ammonium is with the hydroxide formed in the hydrolysis reaction of ammonia.

In order to evaluate the impact of the pH in our catalytic reactions, we decided to use buffered aqueous solutions of urea at three different pHs: 3, 6, and 8, and in this case, we observed a profile indicative of a catalyzed reaction at both basic and acidic pHs (Figure S73D). Interestingly, the reaction rates using buffered solutions were lower than the non-buffered one, which might indicate an inhibition of the hydrolysis by the buffers.

Since the in situ reversibility of the system was not observed, the best classification of the effect of water in the urea hydrolysis reaction is "upregulation" or "regulation" [12]. We believe that the strong water dependence of this system will be able to be explored in the future as water sensors.

4. Conclusions

A strong correlation between dimer/monomer equilibrium and catalysis was observed in the copper complexes synthesized in this work. These complexes were synthesized from Schiff bases from L-proline, exhibiting a square planar geometry. EPR analysis enabled us to verify the existence of a mixture of compounds in solution, especially in the methanol/water mixtures. We explored the hydrolytic capacity of these complexes in urea hydrolysis as a model reaction. As observed by EPR, in the acetonitrile/water mixture, the equilibrium shifted to a monomer and the hydrolysis of urea was too fast to detect the kinetics of ammonia formation by the Berthelot method. However, when the equilibrium monomer/dimer was present, as in the methanol/water mixture, the reaction proceeded slower and the ammonia formation kinetics could be detected in a saturation profile. This effect was shown to be due to the preferential solvation effect, by which hydrogen bonds formed between the secondary and tertiary coordination spheres stabilized the initial state of the aquation reaction. A strong influence of water concentration was observed in the methanol/water system, with special attention to $Cu^{II}L2$ complex. A comparison with the dimeric perchlorate complex enabled us to visualize the importance of the monomer in the reaction, since the dimer produced ammonia from urea very slowly. In conclusion, this work relates a key feature of the chlorido bridges in a supramolecular structure of Cu^{II} complexes for an allosteric (upregulation) behavior in catalysis.

Supplementary Materials: The following are available online at http://www.mdpi.com/2624-8549/2/2/32/s1 [67–71], Figures S1–S17: NMR spectra of ligands, Figures S18–S22: HRMS of ligands, Figure S23: Optimized crystalline structure and number assignment of HL1, HL2, and HL3, Figures S24–S28: Comparison of FTIR between ligands and complexes, Figure S29: FTIR spectra of $[Cu^{II}L2(CH_3OH)]ClO_4$ dispersed in KBr, Figures S30–S35: Comparison of UV–Vis between ligands and complexes, Figures S36–S53: EPR spectra of complexes, Figures S53–S55: Conductivity measurements of complexes in methanol/water, Figures S56–S62: HRMS of complexes, Figures S63–S68: Ammonia quantification produced by the complexes, Figure S69: Infrared spectra of reactions of the complexes (a) 30 s, (b) 600 s, and with (c) urea. Table S1: Crystal data and structure refinement of L1–L3, Table S2: Bond length for L1–L3, Table S3: Bond angles for L1–L3, Table S4: Comparison of the main infrared bands between ligands and complexes, Table S5: Comparison of transitions in the ultraviolet and visible region between ligands and complexes, Table S6: Comparison of oxidation and reduction potentials due to the cyclic voltammetry of ligands and complexes, Table S7: EPR parameters of aggregates and monomeric species of the Cu^{II} complexes of this work in dichloromethane at 77 K, Table S8: EPR parameters for the Cu^{II} complexes of this work in acetonitrile at 298 K, Table S9: EPR parameters for the Cu^{II} complexes of this work in acetonitrile/water (80/20) mixture at 298 K, Table S10: EPR parameters for the $Cu^{II}L2$ and $[Cu^{II}L2(CH_3OH)]ClO_4$ of this work in the methanol and methanol/water (80/20) mixture at 298 K, Table S11: Maximum amount of ammonia formed by the complexes of this work under the conditions of the acetonitrile/water and methanol/water mixture at 308 K. CCDC 1983221, 1983222 and 1983224 contain the supplementary crystallographic data for L1, L2, and L4, respectively.

Author Contributions: Conceptualization, C.B.C. and C.G.C.M.N.; Methodology, C.B.C.; Synthesis, C.B.C.; Characterization, C.B.C., O.R.N., and R.G.S.; Computer simulations, A.F.d.M. and F.M.C.; Formal analysis, C.B.C. and C.G.C.M.N.; Investigation, C.B.C. and C.G.C.M.N.; Resources, C.G.C.M.N., A.F.d.M., and O.R.N.; Data curation, C.B.C.; Writing—original draft preparation, C.B.C. and C.G.C.M.N.; Writing—review and editing, C.B.C. and C.G.C.M.N.; Supervision, C.G.C.M.N.; Project administration, C.G.C.M.N.; Funding acquisition, C.G.C.M.N. and A.F.d.M. All authors have read and agreed to the published version of the manuscript.

Funding: This research was funded by FAPESP (projects 2012/15147-4, 2013/07296-2 and 2016/01622-3), CNPq for a master fellowship (132842/2017-3), and CAPES (project 001). Otaciro Rangel Nascimento thanks CNPq (project 305668/2014-5). A.F.d.M. thanks CNPq for a Research Fellowship.

Acknowledgments: We would also like to thank Diego E. Sastre for editing the images and Elton Sitta for the aid in the in situ FTIR experiments. The authors acknowledge the National Laboratory for Scientific Computing (LNCC/MCTI, Brazil) and UFSCar for providing the high-performance computing resources of the SDumont supercomputer (http://sdumont.lncc.br) and of the Could@UFSCar, respectively, both of which have contributed to the results reported in this paper.

Conflicts of Interest: The authors declare no conflicts of interest.

References

1. Abrusán, G.; Marsh, J.A.; Wilke, C. Ligand-Binding-Site Structure Shapes Allosteric Signal Transduction and the Evolution of Allostery in Protein Complexes. *Mol. Biol. Evol.* **2019**, *36*, 1711–1727. [CrossRef] [PubMed]
2. Nussinov, R.; Tsai, C.J.; Liu, J. Principles of Allosteric Interactions in Cell Signaling. *J. Am. Chem. Soc.* **2014**, *136*, 17692–17701. [CrossRef] [PubMed]
3. Kiyonaka, S.; Kubota, R.; Michibata, Y.; Sakakura, M.; Takahashi, H.; Numata, T.; Inoue, R.; Yuzaki, M.; Hamachi, I. Allosteric Activation of Membrane-Bound Glutamate Receptors Using Coordination Chemistry within Living Cells. *Nat. Chem.* **2016**, *8*, 958–967. [CrossRef] [PubMed]
4. Ito, S.; Takeya, M.; Osanai, T. Substrate Specificity and Allosteric Regulation of a D-Lactate Dehydrogenase from a Unicellular Cyanobacterium Are Altered by an Amino Acid Substitution. *Sci. Rep.* **2017**, *7*, 1–9. [CrossRef] [PubMed]
5. Scarso, A.; Zaupa, G.; Houillon, F.B.; Prins, L.J.; Scrimin, P. Tripodal, Cooperative, and Allosteric Transphosphorylation Metallocatalysts. *J. Org. Chem.* **2007**, *72*, 376–385. [CrossRef] [PubMed]
6. Fritsky, I.O.; Ott, R.; Krämer, R. Allosteric Regulation of Artificial Phosphoesterase Activity by Metal Ions. *Angew. Chem. Int. Ed.* **2000**, *39*, 3255–3258. [CrossRef]
7. Kovbasyuk, L.; Pritzkow, H.; Krämer, R.; Fritsky, I.O. On/off Regulation of Catalysis by Allosteric Control of Metal Complex Nuclearity. *Chem. Commun.* **2004**, *4*, 880–881. [CrossRef]
8. Cheng, H.F.; D'Aquino, A.I.; Barroso-Flores, J.; Mirkin, C.A. A Redox-Switchable, Allosteric Coordination Complex. *J. Am. Chem. Soc.* **2018**, *140*, 14590–14594. [CrossRef]
9. Kuwabara, J.; Yoon, H.J.; Mirkin, C.A.; Dipasquale, A.G.; Rheingold, A.L. Pseudo-Allosteric Regulation of the Anion Binding Affinity of a Macrocyclic Coordination Complex. *Chem. Commun.* **2009**, *45*, 4557–4559. [CrossRef]
10. D'Aquino, A.I.; Cheng, H.F.; Barroso-Flores, J.; Kean, Z.S.; Mendez-Arroyo, J.; McGuirk, C.M.; Mirkin, C.A. An Allosterically Regulated, Four-State Macrocycle. *Inorg. Chem.* **2018**, *57*, 3568–3578. [CrossRef]
11. McGuirk, C.M.; Mendez-Arroyo, J.; Lifschitz, A.M.; Mirkin, C.A. Allosteric Regulation of Supramolecular Oligomerization and Catalytic Activity via Coordination-Based Control of Competitive Hydrogen-Bonding Events. *J. Am. Chem. Soc.* **2014**, *136*, 16594–16601. [CrossRef] [PubMed]
12. Wiester, M.J.; Ulmann, P.A.; Mirkin, C.A. Enzyme Mimics Based upon Supramolecular Coordination Chemistry. *Angew. Chem. Int. Ed.* **2011**, *50*, 114–137. [CrossRef] [PubMed]
13. Masar, M.S.; Gianneschi, N.C.; Oliveri, C.G.; Stern, C.L.; Nguyen, S.B.T.; Mirkin, C.A. Allosterically Regulated Supramolecular Catalysis of Acyl Transfer Reactions for Signal Amplification and Detection of Small Molecules. *J. Am. Chem. Soc.* **2007**, *129*, 10149–10158. [CrossRef] [PubMed]
14. Gianneschi, N.C.; Nguyen, S.B.T.; Mirkin, C.A. Signal Amplification and Detection via a Supramolecular Allosteric Catalyst. *J. Am. Chem. Soc.* **2005**, *127*, 1644–1645. [CrossRef]
15. Hyo, J.Y.; Mirkin, C.A. PCR-like Cascade Reactions in the Context of an Allosteric Enzyme Mimic. *J. Am. Chem. Soc.* **2008**, *130*, 11590–11591.
16. Kennedy, R.D.; MacHan, C.W.; McGuirk, C.M.; Rosen, M.S.; Stern, C.L.; Sarjeant, A.A.; Mirkin, C.A. General Strategy for the Synthesis of Rigid Weak-Link Approach Platinum(II) Complexes: Tweezers, Triple-Layer Complexes, and Macrocycles. *Inorg. Chem.* **2013**, *52*, 5876–5888. [CrossRef]
17. Jeon, Y.M.; Heo, J.; Brown, A.M.; Mirkin, C.A. Triple-Decker Complexes Formed via the Weak Link Approach. *Organometallics* **2006**, *25*, 2729–2732. [CrossRef]
18. Kuwabara, J.; Stern, C.L.; Mirkin, C.A. A Coordination Chemistry Approach to a Multieffector Enzyme Mimic. *J. Am. Chem. Soc.* **2007**, *129*, 10074–10075. [CrossRef]
19. Hyo, J.Y.; Heo, J.; Mirkin, C.A. Allosteric Regulation of Phosphate Diester Transesterification Based upon a Dinuclear Zinc Catalyst Assembled via the Weak-Link Approach. *J. Am. Chem. Soc.* **2007**, *129*, 14182–14183.
20. Yoon, H.J.; Kuwabara, J.; Kim, J.H.; Mirkin, C.A. Allosteric Supramolecular Triple-Layer Catalysts. *Science* **2010**, *330*, 66–69. [CrossRef]
21. Gianneschi, N.C.; Bertin, P.A.; Nguyen, S.B.T.; Mirkin, C.A.; Zakharov, L.N.; Rheingold, A.L. A Supramolecular Approach to an Allosteric Catalyst. *J. Am. Chem. Soc.* **2003**, *125*, 10508–10509. [CrossRef] [PubMed]

22. Pankhurst, J.R.; Paul, S.; Zhu, Y.; Williams, C.K.; Love, J.B. PolynuclearAlkoxy-Zinc Complexes of Bowl-Shaped Macrocycles and Their Use in the Copolymerisation of Cyclohexene Oxide and CO_2. *Dalton Trans.* **2019**, *48*, 4887–4893. [CrossRef] [PubMed]

23. Sutton, C.E.; Harding, L.P.; Hardie, M.; Riis-Johannessen, T.; Rice, C.R. Allosteric Effects in a Ditopic Ligand Containing Bipyridine and Tetra-Aza-Crown Donor Units. *Chem. A Eur. J.* **2012**, *18*, 3464–3467. [CrossRef] [PubMed]

24. Shinkai, S.; Ikeda, M.; Sugasaki, A.; Takeuchi, M. Positive Allosteric Systems Designed on Dynamic Supramolecular Scaffolds: Toward Switching and Amplification of Guest Affinity and Selectivity. *Acc. Chem. Res.* **2001**, *34*, 494–503. [CrossRef] [PubMed]

25. Strotmeyer, K.P.; Fritsky, I.O.; Ott, R.; Pritzkow, H.; Krämer, R. Evaluating the Conformational Role of an Allosteric CuII Ion in Anion Recognition and Catalysis by a Tricopper Complex. *Supramol. Chem.* **2003**, *15*, 529–547. [CrossRef]

26. Park, J.; Hong, S. Cooperative Bimetallic Catalysis in Asymmetric Transformations. *Chem. Soc. Rev.* **2012**, *41*, 6931–6943. [CrossRef]

27. Park, J.; Lang, K.; Abboud, K.A.; Hong, S. Self-Assembly Approach toward Chiral Bimetallic Catalysts: Bis-Urea-Functionalized (Salen)Cobalt Complexes for the Hydrolytic Kinetic Resolution of Epoxides. *Chem. A Eur. J.* **2011**, *17*, 2236–2245. [CrossRef]

28. Park, J.; Lang, K.; Abboud, K.A.; Hong, S. Self-Assembled Dinuclear Cobalt(II)-Salen Catalyst through Hydrogen-Bonding and Its Application to Enantioselective Nitro-Aldol (Henry) Reaction. *J. Am. Chem. Soc.* **2008**, *130*, 16484–16485. [CrossRef]

29. Shirvan, A.; Golchoubian, H.; Bouwman, E. Syntheses and Chromotropic Behavior of Two Halo Bridged DinuclearCopper(II) Complexes Containing Pyridine-Based Bidentate Ligand. *J. Mol. Struct.* **2019**, *1195*, 769–777. [CrossRef]

30. Albert, J.; Bosque, R.; D'Andrea, L.; Durán, J.A.; Granell, J.; Font-Bardia, M.; Calvet, T. Synthesis and Crystal Structure of the DinuclearCyclopalladated Compounds of Methyl (E)-4-(Benzylideneamino)Benzoate with Acetato and Chlorido Bridge Ligands: Study of Their Splitting Reactions with Pyridine. *J. Organomet. Chem.* **2016**, *815–816*, 44–52. [CrossRef]

31. Sangeetha, N.R.; Pal, S. Dimeric and Polymeric Square-Pyramidal Copper(II) Complexes Containing Equatorial-Apical Chloride or Acetate Bridges. *Polyhedron* **2000**, *19*, 1593–1600. [CrossRef]

32. Grosshauser, M.; Comba, P.; Kim, J.Y.; Ohto, K.; Thuéry, P.; Lee, Y.H.; Kim, Y.; Harrowfield, J. Ferro- and Antiferromagnetic Coupling in a Chlorido-Bridged, Tetranuclear Cu(II) Complex. *Dalton Trans.* **2014**, *43*, 5662–5666. [CrossRef] [PubMed]

33. Qi, C.M.; Sun, X.X.; Gao, S.; Ma, S.L.; Yuan, D.Q.; Fan, C.H.; Huang, H.B.; Zhu, W.X. Chlorido-Bridged MnII Schiff-Base Complex with Ferromagnetic Exchange Interactions. *Eur. J. Inorg. Chem.* **2007**, *2007*, 3663–3668. [CrossRef]

34. Luis Olivares-Romero, J.; Juaristi, E. Synthesis of Three Novel Chiral Diamines Derived from (S)-Proline and Their Evaluation as Precursors of Diazaborolidines for the Catalytic Borane-Mediated Enantioselective Reduction of Prochiral Ketones. *Tetrahedron* **2008**, *64*, 9992–9998. [CrossRef]

35. Reyes-Rangel, G.; Vargas-Caporali, J.; Juaristi, E. In Search of Diamine Analogs of the α,α-Diphenyl Prolinol Privileged Chiral Organocatalyst. Synthesis of Diamine Derivatives of α,α-Diphenyl-(S)-Prolinol and Their Application as Organocatalysts in the Asymmetric Michael and Mannich Reactions. *Tetrahedron* **2016**, *72*, 379–391. [CrossRef]

36. Sibi, M.P.; Zhang, R.; Manyem, S. A New Class of Modular Chiral Ligands with Fluxional Groups. *J. Am. Chem. Soc.* **2003**, *125*, 9306–9307. [CrossRef]

37. Pracht, P.; Caldeweyher, E.; Ehlert, S.; Grimme, S. A Robust Non-Self-Consistent Tight-Binding Quantum Chemistry Method for Large Molecules. *ChemRxiv* **2019**, 1–19.

38. Bannwarth, C.; Ehlert, S.; Grimme, S. GFN2-XTB—An Accurate and Broadly Parametrized Self-Consistent Tight-Binding Quantum Chemical Method with Multipole Electrostatics and Density-Dependent Dispersion Contributions. *J. Chem. Theory Comput.* **2019**, *15*, 1652–1671. [CrossRef]

39. Grimme, S.; Bannwarth, C.; Shushkov, P. A Robust and Accurate Tight-Binding Quantum Chemical Method for Structures, Vibrational Frequencies, and Noncovalent Interactions of Large Molecular Systems Parametrized for All Spd-Block Elements (Z = 1–86). *J. Chem. Theory Comput.* **2017**, *13*, 1989–2009. [CrossRef]

40. AtoholiSema, H.; Bez, G.; Karmakar, S. Asymmetric Henry Reaction Catalysed by L-Proline Derivatives in the Presence of Cu(OAc)2: Isolation and Characterization of an in Situ Formed Cu(II) Complex. *Appl. Organomet. Chem.* **2014**, *28*, 290–297. [CrossRef]

41. Kalita, M.; Bhattacharjee, T.; Gogoi, P.; Barman, P.; Kalita, R.D.; Sarma, B.; Karmakar, S. Synthesis, Characterization, Crystal Structure and Bioactivities of a New Potential Tridentate (ONS) Schiff Base Ligand N-[2-(Benzylthio) Phenyl] Salicylaldimine and Its Ni(II), Cu(II) and Co(II) Complexes. *Polyhedron* **2013**, *60*, 47–53. [CrossRef]

42. Balakrishnan, C.; Theetharappan, M.; Kowsalya, P.; Natarajan, S.; Neelakantan, M.A.; Mariappan, S.S. Biocatalysis, DNA–Protein Interactions, Cytotoxicity and Molecular Docking of Cu(II), Ni(II), Zn(II) and V(IV) Schiff Base Complexes. *Appl. Organomet. Chem.* **2017**, *31*, e3776. [CrossRef]

43. Valentová, J.; Varényi, S.; Herich, P.; Baran, P.; Bilková, A.; Kožíšek, J.; Habala, L. Synthesis, Structures and Biological Activity of Copper(II) and Zinc(II) Schiff Base Complexes Derived from Aminocyclohexane-1-Carboxylic Acid. New Type of Geometrical Isomerism in Polynuclear Complexes. *Inorg. Chim. Acta* **2018**, *480*, 16–26. [CrossRef]

44. Iftikhar, B.; Javed, K.; Khan, M.S.U.; Akhter, Z.; Mirza, B.; Mckee, V. Synthesis, Characterization and Biological Assay of Salicylaldehyde Schiff Base Cu(II) Complexes and Their Precursors. *J. Mol. Struct.* **2018**, *1155*, 337–348. [CrossRef]

45. Rosu, T.; Pahontu, E.; Maxim, C.; Georgescu, R.; Stanica, N.; Gulea, A. Some New Cu(II) Complexes Containing an on Donor Schiff Base: Synthesis, Characterization and Antibacterial Activity. *Polyhedron* **2011**, *30*, 154–162. [CrossRef]

46. Knoblauch, S.; Benedix, R.; Ecke, M.; Gelbrich, T.; Sieler, J.; Somoza, F.; Hennig, H. Synthesis, Crystal Structure, Spectroscopy, and Theoretical Investigations of Tetrahedrally Distorted Copper(II) Chelates with [CuN2S2] Coordination Sphere. *Eur. J. Inorg. Chem.* **1999**, *1999*, 1393–1403. [CrossRef]

47. Rauf, A.; Shah, A.; Khan, A.A.; Shah, A.H.; Abbasi, R.; Qureshi, I.Z.; Ali, S. Synthesis, PH Dependent Photometric and Electrochemical Investigation, Redox Mechanism and Biological Applications of Novel Schiff Base and Its Metallic Derivatives. *Spectrochim. Acta Part A Mol. Biomol. Spectrosc.* **2017**, *176*, 155–167. [CrossRef]

48. Butsch, K.; Günther, T.; Klein, A.; Stirnat, K.; Berkessel, A.; Neudörfl, J. Redox Chemistry of Copper Complexes with Various Salen Type Ligands. *Inorg. Chim. Acta* **2013**, *394*, 237–246. [CrossRef]

49. Hosseinzadeh, P.; Lu, Y. Design and Fine-Tuning Redox Potentials of Metalloproteins Involved in Electron Transfer in Bioenergetics. *Biochim. Biophys. Acta Bioenerg.* **2016**, *1857*, 557–581. [CrossRef]

50. Banerjee, S.; Mazumdar, S. Electrospray Ionization Mass Spectrometry: A Technique to Access the Information beyond the Molecular Weight of the Analyte. *Int. J. Anal. Chem.* **2012**, *2012*, 1–40. [CrossRef]

51. Geary, W.J. The Use of Conductivity Measurements in Organic Solvents for the Characterisation of Coordination Compounds. *Coord. Chem. Rev.* **1971**, *7*, 81–122. [CrossRef]

52. Fousiamol, M.M.; Sithambaresan, M.; Smolenski, V.A.; Jasinski, J.P.; Kurup, M.R.P. Halogen/Azide Bridged Box Dimer Copper(II) Complexes of 2-Benzoylpyridine-3-Methoxybenzhydrazone: Structural and Spectral Studies. *Polyhedron* **2018**, *141*, 60–68. [CrossRef]

53. Kunnath, R.J.; Sithambaresan, M.; Aravindakshan, A.A.; Natarajan, A.; Kurup, M.R.P. The Ligating Versatility of Pseudohalides like Thiocyanate and Cyanate in Copper(II) Complexes of 2-Benzoylpyridine Semicarbazone: Monomer, Dimer and Polymer. *Polyhedron* **2016**, *113*, 73–80. [CrossRef]

54. RecioDespaigne, A.A.; Da Costa, F.B.; Piro, O.E.; Castellano, E.E.; Louro, S.R.W.; Beraldo, H. Complexation of 2-Acetylpyridine- and 2-Benzoylpyridine-Derived Hydrazones to Copper(II) as an Effective Strategy for Antimicrobial Activity Improvement. *Polyhedron* **2012**, *38*, 285–290. [CrossRef]

55. Hricovíni, M.; Mazúr, M.; Sîrbu, A.; Palamarciuc, O.; Arion, V.B.; Brezová, V. Copper(II) Thiosemicarbazone Complexes and Their Proligands upon Uva Irradiation: An Epr and Spectrophotometric Steady-State Study. *Molecules* **2018**, *23*, 721. [CrossRef]

56. Hangan, A.C.; Borodi, G.; Stan, R.L.; Páll, E.; Cenariu, M.; Oprean, L.S.; Sevastre, B. Synthesis, Crystal Structure, DNA Cleavage and Antitumor Activity of Two Copper(II) Complexes with N-Sulfonamide Ligand. *Inorg. Chim. Acta* **2018**, *482*, 884–893. [CrossRef]

57. Stoll, S.; Schweiger, A. EasySpin, a Comprehensive Software Package for Spectral Simulation and Analysis in EPR. *J. Magn. Reson.* **2006**, *178*, 42–55. [CrossRef]

58. Szell, P.M.J.; Zablotny, S.; Bryce, D.L. Halogen Bonding as a Supramolecular Dynamics Catalyst. *Nat. Commun.* **2019**, *10*, 1–8. [CrossRef]

59. Selyutin, G.E.; Podrugina, T.A.; Gal'pern, M.G.; Luk, E.A. Aggregation Properties of Copper dibenzo-barrelenophthalocyanine. *Chem. Heterocycl. Compd.* **1995**, *31*, 146–149. [CrossRef]

60. Moliner-Martínez, Y.; Herráez-Hernández, R.; Campíns-Falcó, P. Improved Detection Limit for Ammonium/Ammonia Achieved by Berthelot's Reaction by Use of Solid-Phase Extraction Coupled to Diffuse Reflectance Spectroscopy. *Anal. Chim. Acta* **2005**, *534*, 327–334. [CrossRef]

61. Stehle, S.; Braeuer, A.S. Hydrogen Bond Networks in Binary Mixtures of Water and Organic Solvents. *J. Phys. Chem. B* **2019**, *123*, 4425–4433. [CrossRef] [PubMed]

62. Langford, C.H.; Tong, J.P.K. Preferential Solvation and the Role of Solvent in Kinetics. Examples from Ligand Substitution Reactions. *Acc. Chem. Res.* **1977**, *10*, 258–264. [CrossRef]

63. Brown, R.S. Bio-Inspired Approaches to Accelerating Metal Ion-Promoted Reactions: Enzyme-like Rates for Metal Ion Mediated Phosphoryl and Acyl Transfer Processes. *Pure Appl. Chem.* **2015**, *87*, 601–614. [CrossRef]

64. Blandamer, M.J.; Engberts, J.B.F.N.; Gleeson, P.T.; Reis, J.C.R. Activity of Water in Aqueous Systems; a Frequently Neglected Property. *Chem. Soc. Rev.* **2005**, *34*, 440–458. [CrossRef]

65. Zhu, H.; Yuen, C.; Grant, D.J.W. Influence of Water Activity in Organic Solvent + Water Mixtures on the Nature of the Crystallizing Drug Phase. 1. Theophylline. *Int. J. Pharm.* **1996**, *135*, 151–160. [CrossRef]

66. Subirats, X.; Rosés, M.; Bosch, E. On the Effect of Organic Solvent Composition on the PH of Buffered HPLC Mobile Phases and the PKa of Analytes—A Review. *Sep. Purif. Rev.* **2007**, *36*, 231–255. [CrossRef]

67. Colombari, F.M.; Bernardino, K.; Gomes, W.R.; Lozada-Blanco, A.; Moura, A.F. Themis: A Software to Assess Association Free Energies Via Direct Estimative of Partition Functions. Unpublished work.

68. Sun, M.; Xu, L.; Qu, A.; Zhao, P.; Hao, T.; Ma, W.; Hao, C.; Wen, X.; Colombari, F.M.; de Moura, A.F.; et al. Site-selective photoinduced cleavage and profiling of DNA by chiral semiconductor nanoparticles. *Nature Chem.* **2018**, *10*, 821–830. [CrossRef]

69. Bernardino, K.; Lima, T.A.; Ribeiro, M.C.C. Low-Temperature Phase Transitions of the Ionic Liquid 1-Ethyl-3-methylimidazolium Dicyanamide. *J. Phys. Chem. B* **2019**, *123*, 9418–9427. [CrossRef]

70. Bernardino, K.; Goloviznina, K.; Gomes, M.C.; Pádua, A.A.H.; Ribeiro, M.C.C. Ion pair free energy surface as a probe of ionic liquid structure. *J. Chem. Phys.* **2020**, *152*, 014103. [CrossRef]

71. Colombari, F.M.; da Silva, M.A.R.; Homsi, M.; Souza, B.R.L.; Araujo, M.; Francisco, J.L.; Silva, G.T.S.T.; Silva, I.F.; de Moura, A.F.; Teixeira, I.F. Graphitic Carbon Nitrides asPlatforms for Single-Atom Photocatalysis. *Faraday Discuss* **2020**. In Press. [CrossRef]

Article

Towards a Generalized Synthetic Strategy for Variable Sized Enantiopure M$_4$L$_4$ Helicates

Stephanie A. Boer [1,2], Winnie Cao [1], Bianca K. Glascott [1] and David R. Turner [1,*]

[1] School of Chemistry, Monash University, Clayton, VIC 3800, Australia; boers@ansto.gov.au (S.A.B.); wei.cao2@monash.edu (W.C.); biancaglascott13@gmail.com (B.K.G.)

[2] Australian Synchrotron, ANSTO, 800 Blackburn Road, Clayton, VIC 3168, Australia

[*] Correspondence: david.turner@monash.edu; Tel.: +61-3-9905-6293

Received: 31 May 2020; Accepted: 19 June 2020; Published: 28 June 2020

Abstract: The reliable and predictable synthesis of enantiopure coordination cages is an important step towards the realization of discrete cages capable of enantioselective discrimination. We have built upon our initial report of a lantern-type helical cage in attempts to expand the synthesis into a general approach. The use of a longer, flexible diacid ligand results in the anticipated cage [Cu$_4$(**L1**)$_4$(solvent)$_4$] with a similar helical pitch to that previously observed and a cavity approximately 30% larger. Using a shorter, more rigid ligand gave rise to a strained, conjoined cage-type complex when using DABCO as an internal bridging ligand, [{Co$_4$(**L2**)$_4$(DABCO)(OH$_2$)$_x$}$_2$ (DABCO)]. The expected paddlewheel motif only forms for one of the Co$_2$ units within each cage, with the other end adopting a "partial paddlewheel" with aqua ligands completing the coordination sphere of the externally facing metal ion. The generic approach of using chiral diacids to construct lantern-type cages is partially borne out, with it being apparent that flexibility in the core group is an essential structural feature.

Keywords: metal–organic cage; helicate; metallosupramolecular; chirality

---

1. Introduction

The formation of enclosed cages, capable of guest encapsulation, has been a mainstay of the supramolecular field since the early work on carcerands by Cram [1]. There has been particular attention given to the use of metal ions as structural agents in the formation of coordination cages, with their relatively predictable coordination geometries allowing for a good degree of control over the self-assembling synthesis [2–8]. The use of cages as agents for guest discrimination is prevalent in the literature, with the internal cavities providing excellent size and shape discrimination between analytes. Such discrimination should lend itself well to separations of racemic mixtures and there is growing literature regarding the synthesis of chiral self-assembled cages [9–12].

There have been a number of different approaches towards chiral coordination cages reported in the literature, although far fewer than reports of achiral (or racemic) systems. Tetrahedral metallo-supramolecular cages, with metal ions situated at the corners of the cage, can have tris-chelated octahedral metal centres that, by definition, have either Λ or Δ chirality with local C$_3$ symmetry [13–15]. In the absence of any templating effect there is nothing to direct the chirality to a specific handedness and a mixture of cages with multiple diastereoisomers may result. Whilst this can provide a route to chiral cages using achiral precursors, it is clearly not the most desirable synthetic strategy, as the resolution of these complexes can be challenging. Similar approaches can be employed to make cages with other geometries, such as cubes and octahedra, with similar issues associated with the chirality of tris-chelate metal centres [16–20]. Rather than local chirality around metal centres, it is desirable that the chirality is communicated to the overall cage, including the interior cavity. Strategies towards this include the use of chiral ligands to connect the metals together [21–23] or using supplementary non-bridging chiral ligands that cap the metal centres [24–26]. The major synthetic difficulty of these routes is the requirement for the ligands to be enantiomerically pure. There is also the issue that peripheral chiral groups, whilst reducing the symmetry of the cage as a whole, may not impact the interior space. One particularly relevant methodology has been to incorporate chiral functionalities at the ends of bis-chelate ligands, such that the molecular chirality dictates the Δ/Λ chirality around the resulting tris-chelated metal and therefore gives bulk enantiopurity in the cages [27–31]. Chiral coordination cages have recently been demonstrated to show enantioselective recognition [15,24], enantio- and chemo-selective catalysis [23,25], and recognition by fluorescence or luminescence methods [12]. Most pertinent to this current work are examples of using amino-acid based ligands to form chiral metallosupramolecular species containing copper paddlewheels, such as triangles and face-capped octahedra [32–34].

Recently we showed that it is possible to form charge-neutral, helical complexes of the form Cu_4L_4 containing two copper paddlewheel motifs and four dicarboxylate ligands containing amino acids, as shown in Figure 1 [35]. A similar, smaller cage complex was reported around the same time by Chen et al. [36], and many achiral analogues are known [37–42]. Subsequently we have explored how changing the nature of the dicarboxylate affects the type and size of cage that can be obtained. Using a more rigid naphthalenediimide-based ligand we, and others, isolated $Cu_{12}L_{12}$ octahedra [43,44] and, by using a more rotationally unrestricted biphenyl core, we obtained Cu_8L_8 "double-walled" squares [45]. It is also possible to isolate discrete catenanes using similar ligands [46], which are analogous to motifs observed in extended networks [47–50].

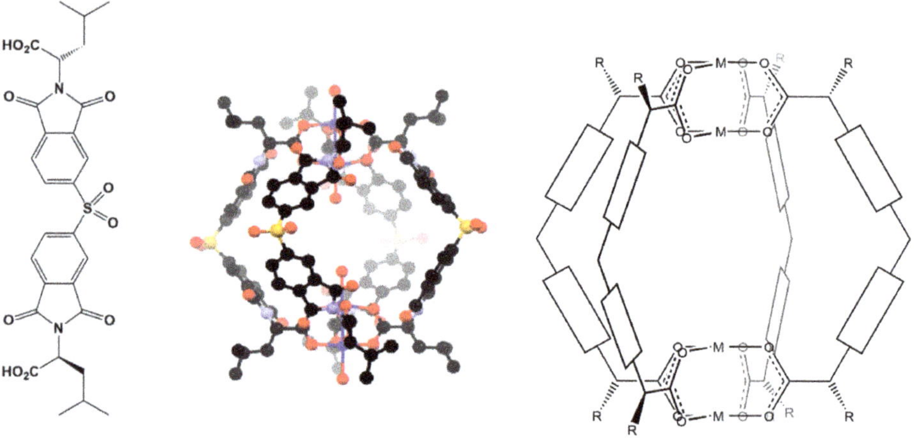

Figure 1. The previously reported chiral diacid (H₂LeuBPSD) and its Cu_4L_4 cage complex alongside a schematic of the generic formulation of these helical cages, in which the cores of the ligands can be altered in attempts to construct analogous cages.

Herein we report the synthesis and structure of two enantiopure cages constructed using dicarboxylate ligands with different core groups, shown in Scheme 1. It is found that a longer, more curved ligand (**L1²⁻**) is able to form an analogue of the original Cu_4L_4 cage, whereas a shorter and less flexible ligand (**L2²⁻**) clearly shows strain and is unable to form a fully closed cage complex.

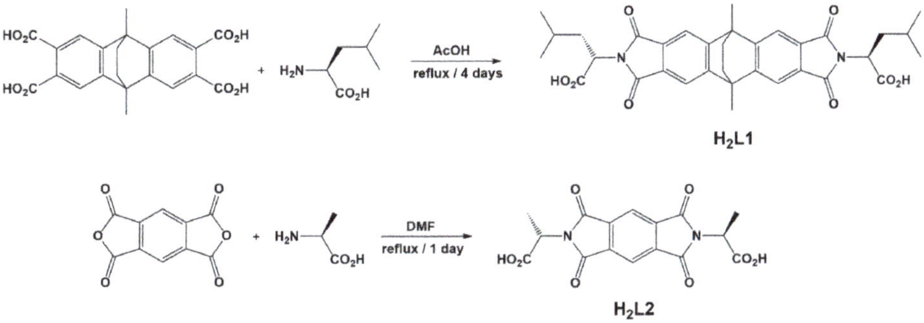

Scheme 1. Synthetic methodology for the two chiral diacids that have been used to synthesise cage complexes; the extended species **H$_2$L1** and the shorter **H$_2$L2**.

2. Materials and Methods

2.1. General Details

All reagents and solvents were purchased from standard commercial suppliers and used as received, except for 9,10-dimethyl-9,10-dihydro-2,3,6,7-tetracrboxyl-9,10-ethanoanthracene, which was synthesised according to a literature preparation [51]. NMR spectra were obtained using a Bruker Avance 400 spectrometer operating at 400 MHz (^{1}H) or 100 MHz (^{13}C). FT–IR spectra were collected using an Agilent Carey 630 diamond attenuated total reflectance spectrometer. Mass spectra were collected using a Micromass platform electrospray mass spectrometer. TGA data were collected using a Mettler TGA/DSC 1 instrument (5 K/min under N$_2$ atmosphere). Elemental analyses were conducted at the London Metropolitan University.

2.2. Synthesis and Characterisation

Synthesis of **H$_2$L1**: A suspension of L-leucine (150 mg, 1.15 mmol) and 9,10-dimethyl-9,10-dihydro-2,3,6,7-tetracarboxyl-9,10-ethanoanthracene (190 mg, 0.463 mmol) was added to a round-bottomed flask with acetic acid (20 mL) and heated at 120 °C for four nights, during which time all solids were taken into solution. After this time the light-yellow solution was poured over ice, forming a light-yellow precipitate, which was recovered by filtration, washed with water and dried in vacuo. Yield: 172.5 mg, 62%. m.p. 195–198 °C. Found: C, 63.56; H, 5.96; N, 3.99%; C$_{34}$H$_{36}$N$_2$O$_8$. Requires: C, 63.32; H, 6.15; N, 3.89%. δH (400 MHz, d_6–DMSO): 0.81 (m, 12H, CH$_2$CH*Me*$_2$); 1.36 (m, 2H, CH$_2$CHMe$_2$); 1.67 (s, 4H, CH$_2$CH$_2$); 1.8 (m, 2H, CH$_2$CHMe$_2$); 2.21 (s, 6H, Me$_{core}$); 2.16 (m, 2H, CH$_2$CHMe$_2$); 4.74 (dd, ^3J = 12, ^3J = 4.2 Hz, 2H, NCH); 7.78 (s, 4H, ArH). δC (100 MHz, d_6–DMSO): 21.1; 23.5; 26.1; 37.1; 43.4; 44.3; 50.6; 116.6; 129.7; 152.8; 168.0; 171.3. υ_{max}/cm^{-1}: 2958w; 2871w; 1841w; 17,700m; 1702s; 1617m; 1534w; 1465w; 1442w; 1377s; 1321w; 1250m; 1205m; 1155m; 1088w; 991w; 911w; 869w; 752m. *m/z* (ES−): 504.1 ([M-Leu + OH]$^-$, calculated for C$_{26}$H$_{26}$NO$_8{}^-$, 504.2), 100%; 599.2 ([M − H]$^-$, calculated for C$_{34}$H$_{35}$N$_2$O$_8{}^-$, 599.2) 93%.

Synthesis of **H$_2$L2**: The synthesis was based on a literature report with minor modifications [52]. Pyromellitic dianhydride (0.707 g, 3.24 mmol) and L-alanine (0.752 g, 8.44 mmol) were added to a round-bottomed flask with acetic acid (40 mL) and heated under reflux overnight, giving a colourless solution. Concentrated hydrochloric acid (20 mL) was added to the solution, which was allowed to concentrate by evaporation overnight to give a white powder. The product was collected by vacuum filtration, washed with cold water, added dropwise, and dried in air. Yield: 0.703 g, 60%. δH (400 MHz, d_6–DMSO): 1.59 (d, 6H, Me); 4.96 (q, 2H, NCH); 8.30 (s, 2H, ArH); 13.32 (s, 2H, CO$_2$H). δC (100 MHz, d_6–DMSO): 15.14; 48.02; 118.45; 137.26; 165.92; 171.14. υ_{max}/cm^{-1}: 3570w; 3146w; 3007w; 2953w; 1758w; 1702s; 1618w; 1561w; 1459m; 1383m; 1364m; 1304m; 1252w; 1208w; 1171w; 1154w; 1132w;

1104w; 1067m; 1018wm; 934m; 854m; 822w; 802w; 761w; 726m; 677w. *m/z* (ES−): 315.1 ([M-COO-H]⁻, calculated for $C_{15}H_{11}N_2O_6{}^-$, 315.1) 100%; 359.1 ([M − H]⁻, calculated for $C_{16}H_{11}N_2O_8{}^-$, 359.1) 20%.

Synthesis of [Cu₄(**L1**)₄(DMF)(OH₂)₃]·3.5DMF·9.5H₂O (compound 1): **H₂L1** (10 mg, 16.7 μmol) and Cu₂(OAc)₄ (6.0 mg, 33.3 μmol) were added to DMF (1 mL) and sonicated to dissolve. The solution was heated at 85 °C for two nights during which time light blue/green block-shaped crystals formed, which were recovered by vacuum filtration. Yield: 2.2 mg, 61%. Found: C, 56.04; H, 5.58, N, 5.59%; $C_{139}H_{149}N_9O_{36}Cu_4$·3.5DMF·9.5H₂O ([Cu₄(**L1**)₄ (DMF) (OH₂)₃]·3.5DMF·9.5H₂O). Requires: C, 56.14; H, 6.02; N, 5.31%. υ_{max}/cm^{-1}: 2955w; 2869w; 1771w; 1709 s; 1637m; 1465w; 1411m; 1377s; 1351m; 1259w; 1204w; 1156w; 1101w; 1062w; 992w; 913w; 870w; 805w; 771w; 755m; 729w; 660m. *m/z* (ES+): 2649.7 (cage + H)⁺, calculated for $C_{136}H_{136}Cu_4N_8O_{32}{}^+$, 2647.645; 2708.7 (cage + H₂O + MeCN + H)⁺, calculated for $C_{138}H_{141}Cu_4N_9O_{33}{}^+$, 2706.683. TGA: On-set, 70 °C, mass loss = 10% (calculated 11.8% for loss of two non-coordinated H₂O and 4.5 DMF molecules and all coordinated solvent); decomp, 350 °C.

Synthesis of [Co₈(**L2**)₈(DABCO)₃(OH₂)₈]·solvent (compound 2): **H₂L2** (10 mg, 28 μmol), DABCO (1.6 mg, 14 μmol), and Co(NO₃)₂·6H₂O (8.1 mg, 28 μmol) were added to DMF (2 mL) and sonicated to dissolve. The solution was heated at 100 °C for three nights during which time crystalline royal blue plates formed. Only one or two small single crystals formed, preventing bulk analysis. *m/z* (ES−): 955.9 ([Co₄(**L2**)₄(DABCO)(H₂O)₃(OH)]NO₃)²⁻, calculated for $C_{70}H_{59}Co_4N_{11}O_{39}{}^{2-}$, 956.0; 1841.9 ([Co₄(**L2**)₄(DABCO)]NO₃)-, calculated for $C_{70}H_{52}Co_4N_{11}O_{35}{}^-$, 1842.0.

2.3. X-ray Crystallography

Single crystal X-ray diffraction data were obtained using the MX1 beamline at the Australian Synchrotron, operating at 17.4 keV (λ = 0.7108 Å) [53]. The data were collected at 100 K using an open-flow N₂ cryostream. The data collections were controlled using the software BluIce [54]. The initial data indexing and reduction was conducted using the XDS software suite [55]. Structures were solved using SHELXT and refined against F² by standard least squares methods using SHELXL-2018 [56,57]. The programs X-Seed or Olex2 were used as a graphical interface to the SHELX program suite [58,59]. Non-hydrogen atoms were refined using an anisotropic model, except for a few exceptions, as shown in Appendix A. All hydrogen atoms were placed in calculated positions and refined using a riding model. The data were treated with the SQUEEZE routine of PLATON to account for regions of poorly ordered solvent, as shown in Appendix A [60]. Crystallographic data have been deposited in the Cambridge Structural Database (2006794–2006795) and can be obtained from www.ccdc.cam.ac.uk. Special refinement details are provided in both Appendix A and CIF files.

3. Results

3.1. Design and Synthesis of Ligands

Two dicarboxylate ligands, containing amino acid residues, were synthesized as analogues of the sulfone–diphthalimide species that we previously used to form helical [Cu₄(LeuBPSD)₄] cages [35]. The two core groups used represent a longer, bent ligand using a 9, 10-dimethylethanoanthracene derivative (**H₂L1**) and a shorter, more rigid ligand using a pyromellitic acid building block (**H₂L2**). Pyromelliticdiimides are well known in the literature. For example, isophthalate derivatives have been used in MOFs with high sorption capacities [60], dipyridyl derivatives have been used in a number of coordination polymers [61–64], as have thioether derivatives [65], and the diimide core has also been incorporated into a number of macrocycles and receptors [66–71]. Pyromelliticdiimides based on amino acid derivatives have been reported in hydrogen-bonding systems and coordination polymers [72–79]. The applicability of the dihydroanthracene core group is hinted at in the literature with metallocages reported containing shorter ligands using bicyclooctene as a core group [36]; the ethanodihydroanthracene moiety can be viewed as an analogous, extended core group with the same geometry. This core group, as a bis-catechol, has been used in the synthesis of a uranyl lantern-type cage, discrete silane assemblies, and a Mn_{12} metallacycle [80–82].

9,10-Dimethyl-9,10-ethanoanthracene-2,3,6,7-tetracarboxylic acid was synthesised in three steps by a literature procedure [51]. A low yielding cyclisation step proved rather prohibitive for the large scale synthesis of this material. The tetra-acid was reacted with leucine in acetic acid under reflux for four nights, resulting in **H₂L1** being isolated as a pale, yellow solid, as shown in Scheme 1. **H₂L2** was synthesized by the slight amendment of a known method [52].

3.2. Synthesis and Structure of Helicate-Type Complexes

The reaction of **H₂L1** with copper acetate at an elevated temperature in DMF yielded a blue/green crystalline material suitable for analysis by single crystal X-ray diffraction (compound **1**). The structure, solved and refined in the orthorhombic setting $C222_1$, reveals the anticipated lantern-type helical complex, $[Cu_4(L1)_4(solvent)_4]$. The asymmetric unit contains one half of a cage (comprising one copper paddlewheel and one and two half ligands). The overall cage has the two halves related by a two-fold proper rotation that bisects the mid points of two ethanoanthracene core groups. The external facing sites of the two paddlewheels are occupied by aqua ligands. The internal sites are occupied by one DMF and one aqua ligand, which are disordered within the crystal structure. The cage has an internal Cu···Cu distance of 9.225(3) Å, substantially longer than the prior sulfone-based analogue (ca. 7.2 Å). Taking the internal cavity, minus coordinated solvent, as an approximate prolate spheroid, the internal volume is ca. 380 Å^3. The persistence of the cage in solution was confirmed by mass spectrometry as a $H_2O/MeCN$ solvated complex. The cage does not possess a C_4 axis along the $Cu_2···Cu_2$ vector, with the paddlewheels themselves not aligned, although a pseudo four-fold rotation can be seen in Figure 2 on the left. The complex has a helical pitch that results in the ligands coordinating at sites 90° apart between the two paddlewheels, analogous to the structures observed in studies of the smaller sulfone-based ligand. It seems that the use of a longer ligand results in a complex that is somewhat less regular in shape, presumably due to the enhanced flexibility of the core group, yet the overall structural design is retained. The arrangement around the paddlewheels is not distorted, with imide N···N distances between adjacent ligands in the range 6.29–6.39 Å.

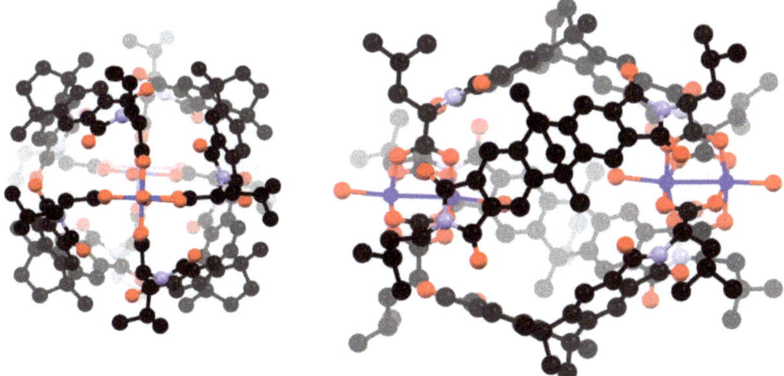

Figure 2. Two views of the cage complex $[Cu_4(L1)_4]$ showing only the coordinating oxygen atoms in the solvent positions. Hydrogen atoms are omitted for clarity.

The reaction of **H₂L2** with cobalt nitrate at an elevated temperature in DMF yielded a few royal blue crystals (compound **2**) that were suitable to obtain moderate resolution single crystal X-ray diffraction data using synchrotron radiation. Unfortunately, the yields were so low as to prohibit bulk analysis other than by mass spectrometry of the reaction solution, as described below. The crystal structure was solved and refined in the orthorhombic space group $P2_12_12$ and contains halves of two crystallographically unique complexes in the asymmetric unit. The nature of the complex is somewhat complicated to determine unambiguously due to crystallographic disorder, with several possible sites for (de)protonation. However, the gross formulation is clear as a "dumbbell-like" complex consisting of two pseudo-$[Co(L2)_4(DABCO)]$ units, bridged by a central DABCO ligand with an overall formulation of $[\{Co_4(L2)_4(DABCO)(H_2O)_x\}_2(DABCO)]$, as shown in Figure 3. The two crystallographically unique dumbbell complexes both contain an axis of rotation passing through the central DABCO ligand, meaning that the two ends are symmetry equivalent. The two complexes appear to be compositionally identical, although the crystallographic disorder is different and, for brevity, only the complex with the more minor disorder is discussed below—see Appendix A for further crystallographic discussion.

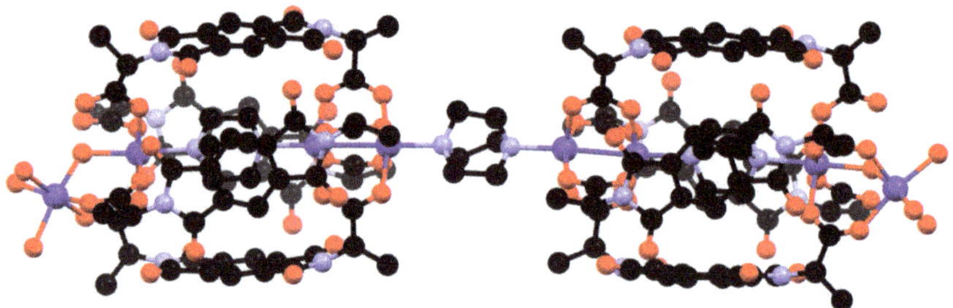

Figure 3. The "dumbbell-type" complex $[\{Co_4(L2)_4(DABCO)(H_2O)_4\}2(DABCO)]$. Hydrogen atoms are omitted for clarity.

In both complexes, there is one major component of the disorder. The ends of this dumbbell complex are a variation of the M_4L_4 cage motif, with the cage closed by the expected paddlewheel motif at one end (that closest to the centre of the complex) yet capped by an unusual coordination environment at the other end (the far ends of the complex, Figure 4). The two ends are bridged by a DABCO ligand within the cavity of the cage. The Co_2 unit at the end of the complex consists of two octahedral cobalt ions. The one nearest the centre of the cavity is coordinated by four carboxylate oxygen atoms and the DABCO ligand, as expected, but also by an aqua ligand. This aqua ligand forms a bridge to the second Co ion, which is coordinated by only two of the carboxylate groups and has three aqua ligands to complete its coordination sphere. The remaining two carboxylates of this M_4L_4 unit are non-coordinating. Ambiguity arises from the nature of the bridging ligand (H_2O, OH^-, or O^{2-}) and the protonation state of the carboxylate groups (i.e., carboxylate vs. carboxylic acid); bond lengths and crystal colour confirm the oxidation state of the metal ions as cobalt (II). The distances between the bridging oxygen atom and non-coordinated oxygen atoms of the carboxylate groups are ca. 2.7 Å and the angle subtended at the bridging oxygen atom is 104°. These measurements strongly suggest that this is a μ_2 bridging H_2O, acting as a hydrogen bond donor to the two carboxylates, and therefore has the formula $[\{Co_4(L2)_4(DABCO)(H_2O)_4\}_2(DABCO)]$. There are minor components in both instances for which solvation within the coordination sphere cannot be fully resolved due to their small occupancy and proximity to heavier elements. What can be found suggests tetrahedral or non-standard coordination geometries for these positions, supported by the observation of crystal colour (blue), which suggests some amount of tetrahedral Co^{II} is present.

The two crystallographically unique complexes are different in terms of the distortion of the cage, presumably a consequence of accommodating the disorder around the terminal Co_2 units. One useful way of measuring this distortion is by the distances between the imide nitrogen atoms of the adjacent **L2** ligands. The complex containing two disordered Co positions adopts a "pinched" conformation, with three N···N distances of 6.2 Å and one of 7.2 Å (measured at both ends of the cage). The measurements are the same at both ends of the cage (i.e., there is no enhanced distortion of the N···N distance at the non-paddlewheel node). The result is that there is a larger "window" on one face of the cage. The other complex, with three disordered cobalt positions, contains N···N distances in the range 6.19–6.54 Å, again with no significant difference between the distances at the two ends of the cage. There is some evidence of strain in the cages by examination of the Co-O bond lengths and the appreciable bowing of the **L2** ligands. The average of the Co-O bond lengths in the two paddlewheels is 2.03 Å, whereas those in the pseudo-paddlewheels (measured only at the bottoms of these motifs, which are analogous to a paddlewheel) average 2.06 Å. Whilst only a slight difference, this does suggest some strain at the extremities of the complexes. The **L2** ligands are significantly bowed away from planarity, which is normally observed for pyromelliticdiimides [60–79]. Calculating a mean plane through the C and N atoms of the pyromelliticdiimide core of **L2** highlights that there is significant deviation away from planarity, with the nitrogen atoms ca. 0.2 Å removed from the plane. The carbon atoms of the alanine groups (formerly the α-carbon of the amino acid) are even further removed from the plane by 0.55–0.90 Å, highlighting the convex nature of the ligands in this complex and the apparent strain.

ESI mass spectrometry of the reaction solution shows both singly- and doubly-charged signals corresponding to one end of the dumbbell complex, $[Co_4(\mathbf{L2})_4(DABCO)]$ with some degree of solvation, indicating that the cage is stable in solution, although not the dumbbell complex (not unexpected, given the rather curious nature of the complex). A signal for a 1- ion matches for $[Co_4(\mathbf{L2})_4(DABCO)]NO_3$, suggesting the that the complex can exist in a non-solvated form, presumably with the implication that the paddlewheels are fully closed. The 2- ion has a *m/z* value that corresponds to $[Co_4(\mathbf{L2})_4(DABCO)(H_2O)_3(OH)]NO_3$, which matches with the formulation seen in the crystal structure (with the deprotonation of one water). These results suggest that the cage complex is retained in the bulk, but not the dumbbell complex.

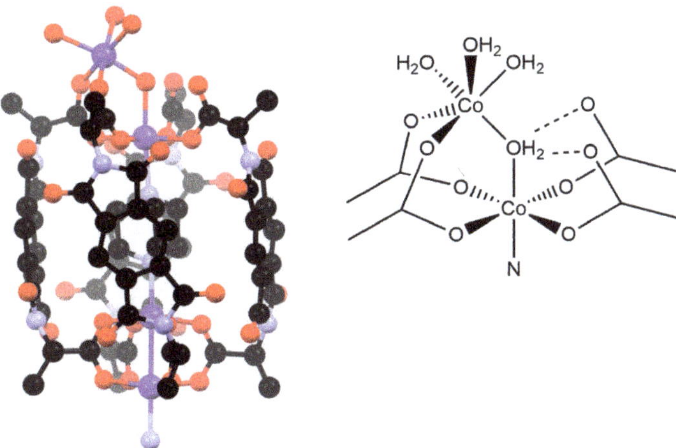

Figure 4. One end of the "dumbbell-type" complex shown in Figure 3, in which the coordination environment of the terminal end (top of picture) and the appreciable bowing of the **L2** ligands can be seen (hydrogen atoms are omitted for clarity) alongside a schematic of the unusual terminal coordination environment.

4. Discussion

It is well known that copper forms a paddlewheel motif when combined with carboxylates, as reported for a large number of cage complexes [32–42]. For this reason, it has been the mainstay of our efforts to form lantern-type cages. Unfortunately, we were only able to isolate an identifiable product using $\mathbf{H_2L1}$ in combination with copper, with the reactions using $\mathbf{H_2L2}$ and copper giving solutions that could not be properly characterized. NMR, in particular, proves challenging, due to the paramagnetism of the copper(II) ion. Perhaps it can be reasoned that a copper (II) analogue of compound **2** would not form, as the coordination environment is not ideally suited to the coordination preferences of the ion (presumably necessitating the N-donor DABCO ligand in a Jahn–Teller distorted axial position).

Similarly, a cobalt (II) analogue of compound **1** could not be prepared, again with characterization issues regarding paramagnetic nuclei. Cobalt–carboxylate paddlewheels are somewhat rarer in the literature than their copper counterparts, and perhaps the DABCO ligand assists in stabilizing the paddlewheel that we do observe. Whilst it is unfortunate that direct comparisons between complexes with identical metal ions can be made, there does appear to be some rationale for why this may be the case.

The cage-type complexes obtained using **H₂L1** and **H₂L2** demonstrate both the versatility of the synthetic approach and its limitations. It is clear that there is a drive towards the formation of lantern-type structures, a well-known M_4L_4 motif for achiral cages (in which rigid ligands connect the paddlewheels in a mesocate-type arrangement with the ligands running straight between the paddlewheels), but one which is less explored for helical complexes [35–42]. The $[Cu_4(\mathbf{L1})_4(solvent)_4]$ cage forms as anticipated. Whilst the conformation of the cage is distorted away from an idealized C_4 symmetric species in the solid-state structure, it is likely that there is some structural relaxation in solution (one must always be careful not to extrapolate solution behaviour from a static crystallographic model).

It seems that the anticipated M_4L_4-type motif in compound **2** is being disrupted due to the use of a shorter and more rigid ligand than **H₂L1** and those used in previous studies. Although the change is most pronounced at the terminal coordination sites, there are other features in the complex, namely, the extreme bowing of the ligands and subtle changes to Co-O bond lengths, which indicate that the system is under some strain. Clearly the rigidity of the ligand, combined with size, has a substantial impact. The earlier report by Chen et al. demonstrated that short diimides can be used to form cages, in this case using a bicyclooctene core [36]. Given the similarity in length between bicyclooctene and phenyl cores, it is therefore evident that the rigidity of the pyromelliticdiimide is the governing factor in limiting the formation of a complete lantern-type cage. Perhaps this is also borne out by our published use of larger, NDI-based ligands that formed octahedral cages, as the rigidity presumably disfavoured the formation of M_4L_4 complexes [43,44]. The internal bridging DABCO may help to stabilize the cages in $[\{Co_4(\mathbf{L2})_4(DABCO)(H_2O)_4\}2(DABCO)]$. Whilst it is far from definitive, we were unable to obtain any crystalline material using **H₂L2** in the absence of DABCO, other than analogues of previously reported coordination polymers, thus suggesting that the role of DABCO is essential [79].

5. Conclusions

In our efforts to expand the scope of a generalized approach to forming chiral lantern-type cages, complexes containing paddlewheel and "pseudo-paddlewheel" motifs were synthesized using the chiral diacids **H₂L1** and **H₂L2**. The $[Cu_4(\mathbf{L1})_4]$ cage adopts the anticipated lantern-type helical structure with a larger central cavity than our previously reported sulfone-based cage (ca. 380 Å³). The cages within the $[\{Co_4(\mathbf{L2})_4(DABCO)(H_2O)_x\}_2(DABCO)]$ complexes retain the overall general design but are clearly under some strain, highlighted by the broken paddlewheels at their extremities. These results show promising initial steps towards a generic approach for the formation of enantiopure helical compounds, with the condition/limitation that the core group of the ligand contains the correct angular preference and flexibility. These results are encouraging to our ongoing efforts to expand the structural types accessible from this general approach.

Author Contributions: Conceptualization, D.R.T.; methodology, S.A.B. and D.R.T.; investigation, S.A.B., W.C., B.K.G. and D.R.T.; writing—original draft preparation, D.R.T.; writing—review and editing, S.A.B., W.C., B.K.G. and D.R.T.; supervision, D.R.T.; project administration, D.R.T.; funding acquisition, D.R.T. All authors have read and agreed to the published version of the manuscript.

Funding: This research was partly funded by the Australian Research Council, grant number FT120100300.

Acknowledgments: Part of this work was conducted using the MX1 beamline at the Australian Synchrotron, part of ANSTO.

Conflicts of Interest: The authors declare no conflict of interest. The funders had no role in the design of the study; in the collection, analyses, or interpretation of data; in the writing of the manuscript, or in the decision to publish the results.

Appendix A

Crystallographic Data and Special Refinement Details

Crystal data for 1: $C_{139}H_{149}Cu_4N_9O_{36}$, M = 2775.82, clear green block, $0.02 \times 0.01 \times 0.01$ mm^3, orthorhombic, space group $C222_1$, a = 17.052(3) Å, b = 27.160(5) Å, c = 37.010(7) Å, V = 17,140(6) Å^3, Z = 4, D_c = 1.076 g/cm^3, F_{000} = 5800, ADSC Quantum 210r, synchrotron radiation (MX1, Australian Synchrotron), λ = 0.7108 Å, T = 100.0K, $2\theta_{max}$ = 52.9°, 45,853 reflections collected, 14,407 unique (R_{int} = 0.1457). Final GooF = 1.029, R1 = 0.0861, wR2 = 0.2282, R indices based on 7756 reflections with I > 2σ(I) (refinement on F^2), 884 parameters, 178 restraints. Lp and absorption corrections applied, μ = 0.554 mm^{-1}.

Data for compound **1** were treated using the SQUEEZE routine of PLATON, finding two large voids per unit cell and several smaller pockets. Density accounting for 292 e⁻ was located per cage complex. This tentatively accounts for 15 H_2O and 3.5 DMF molecules, more than is found by TGA, although solvent is likely lost in handling and transit. A number of restraints were required for bond lengths and displacement elipsoids, primarily those associated with the iosbutyl chains and disordered solvent sites, as shown in the information embedded in CIF file.

Crystal data for 2: $C_{146}H_{124}Co_8N_{22}O_{72}$, M = 3810.10, blue plate, $0.16 \times 0.10 \times 0.04$ mm^3, orthorhombic, space group $P2_12_12$, a = 35.489(7) Å, b = 21.851(4) Å, c = 25.061(5) Å, V = 19,434(7) Å^3, Z = 4, D_c = 1.302 g/cm^3, F_{000} = 7784, ADSC Quantum 210r, synchrotron radiation (MX1, Australian Synchrotron), λ = 0.7108 Å, T = 100.0K, $2\theta_{max}$ = 50.0°, 125,332 reflections collected, 34,118 unique (R_{int} = 0.0826). Final GooF = 1.023, R1 = 0.0859, wR2 = 0.2263, R indices based on 27,536 reflections with I > 2σ(I) (refinement on F^2), 2356 parameters, 54 restraints. Lp and absorption corrections applied, μ = 0.753 mm^{-1}.

The structural model contains significant disorder at the terminal positions of the complexes, and the data were treated using SQUEEZE, due to large spaces between the complexes where no model could be applied. However, no restraints of any kind (other than some restraints for disordered DABCO ligands) were applied to the refinement, and the structure is reported in an unadulterated form. All atoms, other than those associated with the minor components of the disorder, were located and freely refined with anisotropic models. No hydrogen atoms were modelled on the aqua ligands, but these are included in the molecular formula. Some large displacement parameters remain in the model, indicative of disorder in the ligands that could not be modelled. The final refinement has a non-zero Flack parameter, 0.088 (7), which is likely a result of slightly low quality data and large areas of disorder in the model.

The complex containing atoms Co1–Co4 contains three disordered Co positions at the ends, refined with 50:30:20 occupancies by means of equalizing the μ_{iso} averaged displacement parameters. There is some evidence of a very minor fourth position (Q peak of apparent Co site < 2 e⁻) for which modelling was not attempted. The aqua ligands were located for the major occupancy site and refined isotropically; those belonging to the minor occupancy positions could not be located. Both DABCO ligands contained disorder that was modelled with fixed 50:50 occupancies.

The complex containing atoms Co5–Co8 contains two disordered Co positions at the ends, refined with 70:30 occupancies by means of equalizing the μ_{iso} averaged displacement parameters. Aqua ligands to complete the octahedral coordination sphere were all located for the major component (O only, isotropic model). The central DABCO ligand of the dumbbell contains disorder with two C_2H_4 groups modelled with fixed 50:50 occupancies, and one C_2H_4 group is modelled with large displacement parameters, as a disordered model could not be satisfactorily refined. Two carboxylate oxygen atoms (O44 and O52) are modelled as disordered (30:70) in line with being either coordinated to the minor occupancy cobalt position or not.

References

1. Cram, D.J.; Cram, J.M. *Container Molecules and Their Guests*; Royal Society of Chemistry: Cambridge, UK, 1997.
2. Chakrabarty, R.; Mukherjee, P.S.; Stang, P.J. Supramolecular Coordination: Self-Assembly of Finite Two- and Three-Dimensional Ensembles. *Chem. Rev.* **2011**, *111*, 6810–6918. [CrossRef] [PubMed]
3. Zarra, S.; Wood, D.M.; Roberts, D.A.; Nitschke, J.R. Molecular containers in complex chemical systems. *Chem. Soc. Rev.* **2015**, *44*, 419–432. [CrossRef] [PubMed]
4. Friese, V.A.; Kurth, D.G. From coordination complexes to coordination polymers through self-assembly. *Curr. Opinon Colloid Interface Sci.* **2009**, *14*, 81–93. [CrossRef]
5. Vardhan, H.; Yusubov, M.; Verpoort, F. Self-assembled metal-organic polyhedra: An overview of various applications. *Coord. Chem. Rev.* **2016**, *306*, 171–194. [CrossRef]
6. Saha, M.L.; De, S.; Pramanik, S.; Schmittel, M. Orthogonality in discrete self-assembly—Survey of current concepts. *Chem. Soc. Rev.* **2013**, *42*, 6860–6909. [CrossRef]
7. Northrop, B.H.; Zheng, Y.-R.; Chi, K.-W.; Stang, P.J. Self-Organization in Coordination-Driven Self-Assembly. *Acc. Chem. Res.* **2009**, *42*, 1554–1563. [CrossRef]
8. Casini, A.; Woods, B.; Wenzel, M. The Promise of Self-Assembled 3D Supramolecular Coordination Complexes for Biomedical Applications. *Inorg. Chem.* **2017**, *56*, 14715–14729. [CrossRef]
9. Chen, L.-J.; Yang, H.-B.; Shionoya, M. Chiral metallosupramolecular architectures. *Chem. Soc. Rev.* **2017**, *46*, 2555–2576. [CrossRef]
10. Clegg, J.K.; McMurtrie, J.C. Chiral Metallosupramolecular Polyhedra. In *Chirality in Supramolecular Assemblies: Causes and Consequences*; Keene, F.R., Ed.; John Wiley & Sons: Chichester, UK, 2016; pp. 218–256.
11. Pan, M.; Wu, K.; Zhang, J.-H.; Su, C.-Y. Chiral metal–organic cages/containers (MOCs): From structural and stereochemical design to application. *Coord. Chem. Rev.* **2019**, *378*, 333–349. [CrossRef]
12. Zhang, J.-H.; Xie, S.-M.; Zi, M.; Yuan, L.-M. Recent advances of application of porous molecular cages for enantioselective recognition and separation. *J. Sep. Sci.* **2020**, *43*, 134–149. [CrossRef]
13. Saalfrank, R.W.; Glaser, H.; Demleitner, B.; Hampel, F.; Chowdhry, M.M.; Schünemann, V.; Trautwein, A.X.; Vaughan, G.B.M.; Yeh, R.; Davis, A.V.; et al. Self-Assembly of Tetrahedral and Trigonal Antiprismatic Clusters $[Fe_4(L^4)_4]$ and $[Fe_6(L^5)_6]$ on the Basis of Trigonal Tris-Bidentate Chelators. *Chem. Eur. J.* **2002**, *8*, 493–497. [CrossRef]
14. Zhao, C.; Toste, F.D.; Raymond, K.N.; Bergman, R.G. Nucleophilic Substitution Catalyzed by a Supramolecular Cavity Proceeds with Retention of Absolute Stereochemistry. *J. Am. Chem. Soc.* **2014**, *136*, 14409–14412. [CrossRef] [PubMed]
15. He, Y.-P.; Yuan, L.-B.; Song, J.-S.; Chen, G.-H.; Lin, Q.; Li, C.; Zhang, L.; Zhang, J. Optical Resolution of the Water-Soluble $Ti_4(embonate)_6$ Cages for Enantioselective Recognition of Chiral Drugs. *Chem. Mater.* **2018**, *30*, 7769–7775. [CrossRef]
16. Browne, C.; Brenet, S.; Clegg, J.K.; Nitschke, J.R. Solvent-Dependent Host–Guest Chemistry of an Fe_8L_{12} Cubic Capsule. *Angew. Chem. Int. Ed.* **2013**, *52*, 1944–1948. [CrossRef] [PubMed]
17. Yang, Y.; Wu, Y.; Jia, J.-H.; Zheng, X.-Y.; Zhang, Q.; Xiong, K.-C.; Zhang, Z.-M.; Wang, Q.-M. Enantiopure Magnetic Heterometallic Coordination Cubic Cages $[M^{II}_8Cu^{II}_6]$ (M = Ni, Co). *Cryst. Growth Des.* **2018**, *18*, 4555–4561. [CrossRef]
18. Guo, J.; Xu, Y.-W.; Li, K.; Xiao, L.-M.; Chen, S.; Wu, K.; Chen, X.-D.; Fan, Y.-Z.; Liu, J.-M.; Su, C.-Y. Regio- and Enantioselective Photodimerization within the Confined Space of a Homochiral Ruthenium/Palladium Heterometallic Coordination Cage. *Angew. Chem. Int. Ed.* **2017**, *56*, 3852–3856. [CrossRef]
19. Wang, X.; Peng, P.; Xuan, W.; Wang, Y.; Zhuang, Y.; Tian, Z.; Cao, X. Narcissistic chiral self-sorting of molecular face-rotating polyhedral. *Org. Biomol. Chem.* **2018**, *16*, 34–37. [CrossRef]
20. Schweiger, M.; Seidel, S.R.; Schmitz, M.; Stang, P.J. Rational Design of Chiral Nanoscale Adamantanoids. *Org. Lett.* **2000**, *2*, 1255–1257. [CrossRef]
21. Dong, J.; Zhou, Y.; Zhang, F.; Cui, Y. A Highly Fluorescent Metallosalen-Based Chiral Cage for Enantioselective Recognition and Sensing. *Chem. Eur. J.* **2014**, *20*, 6455–6461. [CrossRef]
22. Sun, B.; Nurttila, S.S.; Reek, J.N.H. Synthesis and Characterization of Self-Assembled Chiral $Fe^{II}_2L_3$ Cages. *Chem. Eur. J.* **2018**, *24*, 14693–14700. [CrossRef]

23. Tan, C.; Jiao, J.; Li, Z.; Liu, Y.; Han, X.; Cui, Y. Design and Assembly of a Chiral Metallosalen-Based Octahedral Coordination Cage for Supramolecular Asymmetric Catalysis. *Angew. Chem. Int. Ed.* **2018**, *57*, 2085–2090. [CrossRef] [PubMed]

24. Zhang, P.; Wang, X.; Xuan, W.; Peng, P.; Li, Z.; Lu, R.; Wu, S.; Tian, Z.; Cao, X. Chiral separation and characterization of triazatruxene-based face-rotating polyhedral: The role of non-covalent facial interactions. *Chem. Commun.* **2018**, *54*, 4685–4688. [CrossRef] [PubMed]

25. Bhat, I.A.; Devaraj, A.; Howlader, P.; Chi, K.-W.; Mukherjee, P.S. Preparation of a chiral Pt_{12} tetrahedral cage and its use in catalytic Michael addition reaction. *Chem. Commun.* **2018**, *54*, 4814–4817. [CrossRef] [PubMed]

26. Murase, T.; Peschard, S.; Horiuchi, S.; Nishioka, Y.; Fujita, M. Remote chiral transfer into [2 + 2] and [2 + 4] cycloadditions within self-assembled molecular flasks. *Supramol. Chem.* **2011**, *23*, 199–208. [CrossRef]

27. Argent, S.P.; Riis-Johannessen, T.; Jeffery, J.C.; Harding, L.P.; Ward, M.D. Diastereoselective formation and optical activity of an M_4L_6 cage complex. *Chem. Commun.* **2005**, 4647–4649. [CrossRef] [PubMed]

28. Albrecht, M.; Burk, S.; Weis, P. Chiral Confined Space: Induction of Stereochemistry in a M_4L_4 Metallosupramolecular Container. *Synthesis* **2008**, 2963–2967. [CrossRef]

29. Ren, D.-H.; Qiu, D.; Pang, C.-Y.; Li, Z.; Gu, Z.-G. Chiral tetrahedral iron(II) cages: Diastereoselective subcomponent self-assembly, structure interconversion and spin-crossover properties. *Chem. Commun.* **2015**, *51*, 788–791. [CrossRef]

30. Ousaka, N.; Clegg, J.K.; Nitschke, J.R. Nonlinear Enhancement of Chiroptical Response through Subcomponent Substitution in M_4L_6 Cages. *Angew. Chem. Int. Ed.* **2012**, *51*, 1464–1468. [CrossRef]

31. Yan, L.-L.; Tan, C.-H.; Zhang, G.-L.; Zhou, L.-P.; Bünzli, J.-C.; Sun, Q.-F. Stereocontrolled Self-Assembly and Self-Sorting of Luminescent Europium Tetrahedral Cages. *J. Am. Chem. Soc.* **2015**, *137*, 8550–8555. [CrossRef]

32. Wisser, B.; Chamayou, A.-C.; Miller, R.; Scherer, W.; Janiak, C. A chiral C_3-symmetric hexanuclear triangular-prismatic copper(ii) cluster derived from a highly modular dipeptidic N,N′-terephthaloyl-bis(S-aminocarboxylato) ligand. *CrystEngComm* **2008**, *10*, 461–464. [CrossRef]

33. Chen, Z.; Liu, X.; Wu, A.; Liang, Y.; Wang, X.; Liang, F. Synthesis, structure and properties of an octahedral dinuclear-based Cu_{12} nanocage of trimesoyltri(L-alanine). *RSC Adv.* **2016**, *6*, 9911–9915. [CrossRef]

34. Zhang, Z.-J.; Shi, W.; Niu, Z.; Li, H.-H.; Zhao, B.; Cheng, P.; Liao, D.-Z.; Yan, S.-P. A new type of polyhedron-based metal–organic frameworks with interpenetrating cationic and anionic nets demonstrating ion exchange, adsorption and luminescent properties. *Chem. Commun.* **2011**, *47*, 6425–6427. [CrossRef]

35. Boer, S.A.; Turner, D.R. Self-selecting homochiral quadruple-stranded helicates and control of supramolecular chirality. *Chem. Commun.* **2015**, *51*, 17375–17378. [CrossRef] [PubMed]

36. Chen, L.; Kang, J.; Cui, H.; Wang, Y.; Liu, L.; Zhang, L.; Su, C.-Y. Homochiral coordination cages assembled from dinuclear paddlewheel nodes and enantiopure ditopic ligands: Syntheses, structures and catalysis. *Dalton Trans.* **2015**, *44*, 12180–12188. [CrossRef] [PubMed]

37. Xie, Y.; Yang, H.; Wang, Z.U.; Liu, Y.; Zhou, H.-C.; Li, J.-R. Unusual preservation of polyhedral molecular building units in a metal–organic framework with evident desymmetrization in ligand design. *Chem. Commun.* **2014**, *50*, 563–565. [CrossRef]

38. Lu, W.; Yuan, D.; Yakovenko, A.; Zhou, H.-C. Surface functionalization of metal–organic polyhedron for homogeneous cyclopropanation catalysis. *Chem. Commun.* **2011**, *47*, 4968–4970. [CrossRef] [PubMed]

39. Li, J.-R.; Yu, J.; Lu, W.; Sun, L.-B.; Sculley, J.; Balbuena, P.B.; Zhou, H.-C. Porous materials with pre-designed single-molecule traps for CO_2 selective adsorption. *Nat. Commun.* **2013**, *4*, 1538. [CrossRef]

40. Prakash, M.J.; Oh, M.; Liu, X.; Han, K.N.; Seong, G.H.; Lah, M.S. Edge-directed $[(M_2)_2L_4]$ tetragonal metal–organic polyhedra decorated using a square paddle-wheel secondary building unit. *Chem. Commun.* **2010**, *46*, 2049–2051. [CrossRef]

41. Craig, G.A.; Larpent, P.; Kusaka, S.; Matsuda, R.; Kitagawa, S.; Furukawa, S. Switchable gate-opening effect in metal–organic polyhedra assemblies through solution processing. *Chem. Sci.* **2018**, *9*, 6463–6469. [CrossRef]

42. Brega, V.; Zeller, M.; He, Y.; Lu, H.P.; Klosterman, J.K. Multi-responsive metal–organic lantern cages in solution. *Chem. Commun.* **2015**, *51*, 5077–5080. [CrossRef]

43. Boer, S.A.; White, K.F.; Slater, B.; Emerson, A.J.; Knowles, G.P.; Donald, W.A.; Thornton, A.W.; Ladewig, B.P.; Bell, T.D.M.; Hill, M.R.; et al. A Multifunctional, Charge-Neutral, Chiral Octahedral $M_{12}L_{12}$ Cage. *Chem. Eur. J.* **2019**, *25*, 8489–8493. [CrossRef]

44. Mollick, S.; Mukherjee, S.; Kim, D.; Zhiwei, Q.; Desai, A.V.; Saha, R.; More, Y.D.; Jiang, J.; Lah, M.S.; Ghosh, S.K. Hydrophobic Shielding of Outer Surface: Enhancing the Chemical Stability of Metal–Organic Polyhedra. *Angew. Chem. Int. Ed.* **2019**, *58*, 1041–1045. [CrossRef] [PubMed]

45. Boer, S.A.; Turner, D.R. Metallosupramolecular Architectures of Ambivergent Bis(Amino Acid) Biphenyldiimides. *Chem. Asian J.* **2019**, *14*, 2853–2860. [CrossRef]

46. Boer, S.A.; Cox, R.P.; Beards, M.J.; Wang, H.; Donald, W.A.; Bell, T.D.M.; Turner, D.R. Elucidation of naphthalene diimide metallomacrocycles and catenanes by solvent dependent excimer and exciplex emission. *Chem. Commun.* **2019**, *55*, 663–666. [CrossRef] [PubMed]

47. Kyratzis, N.; Cao, W.; Izogordina, E.I.; Turner, D.R. Structural changes in coordination polymers in response to small changes in steric bulk (H vs. Me): An experimental and theoretical study. *CrystEngComm* **2018**, *20*, 4115–4126. [CrossRef]

48. Boer, S.A.; Turner, D.R. A robust metallomacrocyclic motif for the formation of interpenetrated coordination polymers. *CrystEngComm* **2017**, *19*, 2402–2412. [CrossRef]

49. Boer, S.A.; Turner, D.R. Interpenetration in π-rich mixed-ligand coordination polymers. *Cryst. Growth Des.* **2016**, *16*, 6294–6303. [CrossRef]

50. Boer, S.A.; Hawes, C.S.; Turner, D.R. Engineering entanglement: Controlling the formation of polycatenanes and polyrotaxanes using π interactions. *Chem. Commun.* **2014**, *50*, 1125–1127. [CrossRef]

51. Rogan, Y.; Malpass-Evans, R.; Carta, M.; Lee, M.; Jansen, J.C.; Bernardo, P.; Clarizia, G.; Tocci, E.; Friess, K.; Lanc, M.; et al. A highly permeable polyimide with enhanced selectivity for membrane gas separations. *J. Mater. Chem. A* **2014**, *2*, 4874–4877. [CrossRef]

52. Faghihi, K.; Moghanian, H. ynthesis and Characterization of New Optically Active Poly(amide-imide)s Based on N,N'-(Pyromellitoyl)-bis-L-Amino Acids and 1,3,4-Oxadiazole Moieties. *Des. Monomers Polym.* **2010**, *13*, 207–220. [CrossRef]

53. Cowieson, N.P.; Aragao, D.; Clift, M.; Ericsson, D.J.; Gee, C.; Harrop, S.J.; Mudie, N.; Panjikar, S.; Price, J.R.; Riboldi-Tunnicliffe, A.; et al. MX1: A bending-magnet crystallography beamline serving both chemical and macromolecular crystallography communities at the Australian Synchrotron. *J. Synchrotron Rad.* **2015**, *22*, 187–190. [CrossRef] [PubMed]

54. McPhillips, T.M.; McPhillips, S.E.; Chiu, H.J.; Cohen, A.M.; Deacon, A.M.; Ellis, P.J.; Garman, E.; Gonzalez, A.; Sauter, K.; Phizackerley, R.P.; et al. Blu-Ice and the Distributed Control System: Software for data acquisition and instrument control at macromolecular crystallography beamlines. *J. Synchrotron Rad.* **2002**, *9*, 401–406. [CrossRef] [PubMed]

55. Kabsch, W. XDS. *Acta Cryst. Sect. D* **2010**, *66*, 125–132. [CrossRef] [PubMed]

56. Sheldrick, G.M. SHELXT—Integrated space-group and crystal-structure determination. *Acta Crystallogr. Sect. A* **2015**, *71*, 3–8. [CrossRef]

57. Sheldrick, G.M. Crystal structure refinement with SHELXL. *Acta Cryst. Sect. C* **2015**, *71*, 3–8. [CrossRef]

58. Dolomanov, O.V.; Bourhis, L.J.; Gildea, R.J.; Howard, J.A.K.; Puschmann, H. OLEX2: A complete structure solution, refinement and analysis program. *J. Appl. Cryst.* **2009**, *42*, 339–341. [CrossRef]

59. Barbour, L.J. X-Seed—A software tool for supramolecular crystallography. *J. Supramol. Chem.* **2001**, *1*, 189–191. [CrossRef]

60. Spek, A.L. PLATON SQUEEZE: A tool for the calculation of the disordered solvent contribution to the calculated structure factors. *Acta Cryst. Sect. C* **2015**, *71*, 9–18. [CrossRef]

61. Prasad, T.K.; Hong, D.H.; Suh, M.P. High Gas Sorption and Metal-Ion Exchange of Microporous Metal–Organic Frameworks with Incorporated Imide Groups. *Chem. Eur. J.* **2010**, *16*, 14043–14050. [CrossRef]

62. Leong, C.F.; Faust, T.B.; Turner, P.; Usov, P.M.; Kepert, C.J.; Babarao, R.; Thornton, A.W.; D'Alessandro, D.M. Enhancing selective CO_2 adsorption via chemical reduction of a redox-active metal–organic framework. *Dalton Trans.* **2013**, *42*, 9831–9839. [CrossRef]

63. Li, G.-B.; Yang, Q.-Y.; Pan, R.-K.; Liu, S.-G. Diverse cobalt(ii) coordination polymers for water/ethanol separation and luminescence for water sensing applications. *CrystEngComm* **2018**, *20*, 3891–3897. [CrossRef]

64. Li, G.-B.; Liu, J.-M.; Cai, Y.-P.; Su, C.-Y. Structural Diversity of a Series of Mn(II), Cd(II), and Co(II) Complexes with Pyridine Donor Diimide Ligands. *Cryst. Growth Des.* **2011**, *11*, 2763–2772. [CrossRef]

65. Kang, G.; Jeon, Y.; Lee, K.Y.; Kim, J.; Kim, T.H. Reversible Luminescence Vapochromism and Crystal-to-Amorphous-to-Crystal Transformations of Pseudopolymorphic Cu(I) Coordination Polymers. *Cryst. Growth Des.* **2015**, *15*, 5183–5187. [CrossRef]

66. Nakagaki, T.; Harano, A.; Fuchigami, Y.; Tanaka, E.; Kidoaki, S.; Okuda, T.; Iwanaga, T.; Goto, K.; Shinmyozu, T. Formation of Nanoporous Fibers by the Self-Assembly of a Pyromellitic Diimide-Based Macrocycle. *Angew. Chem. Int. Ed.* **2010**, *49*, 9676–9679. [CrossRef]

67. Nalluri, S.K.M.; Zhou, J.; Cheng, T.; Liu, Z.; Nguyen, M.T.; Chen, T.; Patel, H.A.; Krzyaniak, M.D.; Goddard, W.A.; Wasielewski, M.R.; et al. Discrete Dimers of Redox-Active and Fluorescent Perylene Diimide-Based Rigid Isosceles Triangles in the Solid State. *J. Am. Chem. Soc.* **2018**, *141*, 1290–1303. [CrossRef]

68. Colquhoun, H.M.; Zhu, Z.; Williams, D.J.; Drew, M.G.B.; Cardin, C.J.; Gan, Y.; Crawford, A.G.; Marder, T.B. Induced-Fit Binding of π-Electron-Donor Substrates to Macrocyclic Aromatic Ether Imide Sulfones: A Versatile Approach to Molecular Assembly. *Chem. Eur. J.* **2010**, *16*, 907–918. [CrossRef]

69. Liu, Z.-M.; Liu, Y.; Zheng, S.-R.; Yu, Z.-Q.; Pan, M.; Su, C.-Y. Assembly of Trigonal and Tetragonal Prismatic Cages from Octahedral Metal Ions and a Flexible Molecular Clip. *Inorg. Chem.* **2007**, *46*, 5814–5816. [CrossRef]

70. Zhang, Q.; Hamilton, D.G.; Feeder, N.; Teat, S.J.; Goodman, J.M.; Sanders, J.K.M. Synthesis and post-assembly modification of some functionalised, neutral π-associated [2]catenanes. *New J. Chem.* **1999**, *23*, 897–903. [CrossRef]

71. Iwanaga, T.; Nakamoto, R.; Yasutake, M.; Takemura, H.; Sako, K.; Shinmyozu, T. Cyclophanes within Cyclophanes: The Synthesis of a Pyromellitic Diimide-Based Macrocycle as a Structural Unit in a Molecular Tube and Its Inclusion Phenomena. *Angew. Chem. Int. Ed.* **2006**, *45*, 3643–3647. [CrossRef]

72. Ge, C.; Li, X.; Zhang, X.; Zhao, Y.; Zhang, R. 2,2′-(1,3,5,7-Tetra oxo-1,2,3,5,6,7-hexa-hydro-pyrrolo[3,4-f]isoindole-2,6-di-yl)diacetic acid N,N-dimethyl formamide disolvate. *Acta. Crystallogr. Sect. E* **2009**, *65*, o2400. [CrossRef]

73. Barooah, N.; Sarma, R.J.; Baruah, J.B. Solid-state hydrogen bonded assembly of N,N′-bis(glycinyl)-pyromellitic diimide with aromatic guests. *CrystEngComm* **2006**, *8*, 608–615. [CrossRef]

74. Barooah, N.; Singh, W.M.; Baruah, J.B. Preferential deprotonation and conformational stability of dicarboxylic acids: A packing effect. *J. Mol. Struc.* **2008**, *875*, 329–338. [CrossRef]

75. Ge, C.-H.; Zhang, X.-D.; Zhang, H.-D.; Zhao, Y.; Li, X.-Q.; Zhang, R. Color Variety of Organic Salt of N,N′-Bis(Glycinyl)Pyromellitic Diimide and N-Containing Base. *Mol. Cryst. Liq. Cryst.* **2011**, *534*, 114–123. [CrossRef]

76. Barooah, N.; Baruah, J.B. Effect of phenyl group on the self assembly of N,N′-bis-(2-phenylglycinyl)pyromellitic diimide with aromatic hydrocarbons. *J. Mol. Struc.* **2008**, *872*, 205–211. [CrossRef]

77. Jana, P.; Maity, S.; Maity, S.K.; Ghorai, P.K.; Halder, D. Insights into H-aggregates and CH···O hydrogen bond mediated self-assembly of pyromellitic bisimide. *CrystEngComm* **2012**, *14*, 6586–6592. [CrossRef]

78. Barooah, N.; Sarma, R.J.; Batsanov, A.S.; Baruah, J.B. Structural aspects of adducts of N-phthaloylglycine and its derivatives. *J. Mol. Struc.* **2006**, *791*, 122–130. [CrossRef]

79. Joarder, B.; Mukherjee, S.; Chaudhari, A.K.; Desai, A.V.; Manna, B.; Ghosh, S.K. Guest-Responsive Function of a Dynamic Metal–Organic Framework with a π Lewis Acidic Pore Surface. *Chem. Eur. J.* **2014**, *20*, 15303–15308. [CrossRef]

80. Thuéry, P.; Masci, B. Self-assembly of an Octa-uranate Cage Complex with a Rigid bis-Catechol Ligand. *Supramol. Chem.* **2003**, *15*, 95–99. [CrossRef]

81. Abrahams, B.F.; FitzGerald, N.J.; Robson, R. A Doughnut-Like $(Mn^{III})_{12}$ Metallocycle Formed by a Rigid Angular Bis-Catecholate with a Nanometer-Sized Central Hole. *Inorg. Chem.* **2010**, *49*, 5953–5956. [CrossRef]

82. Kawakami, Y.; Ogishima, T.; Kawara, T.; Yamauchi, S.; Okamoto, K.; Nikaido, S.; Souma, D.; Jim, R.-H.; Kabe, Y. Silane catecholates: Versatile tools for self-assembled dynamic covalent bond chemistry. *Chem. Commun.* **2019**, *55*, 6066–6069. [CrossRef]

 *chemistry*

Article

Supramolecular Assemblies of Trinuclear Copper(II)-Pyrazolato Units: A Structural, Magnetic and EPR Study

Kaige Shi [1], Logesh Mathivathanan [1], Radovan Herchel [2,*], Athanassios K. Boudalis [3,*] and Raphael G. Raptis [1,*]

[1] Department of Chemistry and Biochemistry, Florida International University, 11200 SW 8th Street, Miami, FL 33199, USA; shikaige1011@163.com (K.S.); lmathiva@fiu.edu (L.M.)
[2] Department of Inorganic Chemistry, Faculty of Science, Palacký University, CZ-77146 Olomouc, Czech Republic
[3] Institut de Chimie de Strasbourg (UMR 7177, CNRS-Unistra), Université de Strasbourg, 4 rue Blaise Pascal, CS 90032, F-67081 Strasbourg, France
* Correspondence: radovan.herchel@upol.cz (R.H.); bountalis@unistra.fr (A.K.B.); rraptis@fiu.edu (R.G.R.)

Received: 27 May 2020; Accepted: 17 June 2020; Published: 1 July 2020

Abstract: Two anionic complexes, $\{[Cu_3(\mu_3\text{-OH})(\mu\text{-4-Ph-pz})_3Cl_3]_2[Cu(4\text{-Ph-pzH})_4](\mu\text{-Cl})_2\}^{2-}$ (**1**) and $[Cu_3(\mu_3\text{-OH})(\mu\text{-pz})_3(\mu_{1,1}\text{-N}_3)_2(N_3)]^-$ (**2**), crystallize as one-dimensional polymers, held together by weak Cu-$(\mu$-Cl) and Cu-$(\mu$-N$_3$) interactions, respectively. Variable temperature magnetic susceptibility analyses determined the dominant antiferromagnetic intra-Cu_3 exchange parameters in the solid state for both complexes, whereas the weak ferromagnetic inter-Cu_3 interactions manifested also in the solid state EPR spectra, are absent in the corresponding frozen solution spectra. DFT calculations were employed to support the results of the magnetic susceptibility analyses.

Keywords: copper(II) complexes; pyrazolato ligands; supramolecular assembly; X-ray crystallography; magnetic susceptibility; EPR spectroscopy; isotropic exchange; antisymmetric exchange; dipolar interaction; DFT calculations

---

1. Introduction

Herein, we present a structural, magnetic susceptibility and EPR study of two supramolecular assemblies of metallacyclic Cu^{II} pyrazolates with $Cu_3(\mu_3\text{-OH})$ cores. The focus of this work is the elucidation of weak intermolecular interactions manifested in the magnetic properties and EPR spectra of the supramolecular assemblies.

Supramolecular interactions, such as H-bonding, dipolar, metallophilic and π–π interactions, are important not only for the structural organization of molecules in 3D, but because they often play a crucial role in determining the physical and spectroscopic properties of the assemblies. A corollary of the latter statement is that, in the absence of structural data, the detection of such spectroscopic "signatures" reveals the presence of supramolecular interactions—e.g., in biological systems. Magnetic exchange is among the properties that can be modulated by supramolecular interactions, thereby introducing new functionality into a system [1]. For instance, in the layered $Ni(H_2O)_2[Ni(CN)_4]\cdot xH_2O$ solid, H-bonding in the interlayer regions is shown to mediate weak ferromagnetic interactions, but when the coordinated water molecules were replaced by 3-halopyridine ligands, removing H-bonding, anti-ferromagnetic interactions through π-clouds became dominant [2]. Weak intramolecular antiferromagnetic exchange is active within a molecule containing two isolated Cu^{II} centers separated by a K^+ ion, whereas relatively strong ferromagnetic interactions were found between adjacent units, along a supramolecular pathway [3]. In yet another dinuclear Cu^{II} complex with chelating 2-hydroxy-1,10-phenanthroline and bridging thiocyanate ligands, it was shown that an intermolecular ferromagnetic exchange was facilitated by π–π interactions between phenanthroline ligands [4–6].

The antiferromagnetic exchange among the Cu^{II} centers of the triangular $Cu_3(\mu_3\text{-O/OH})$ units has been studied extensively by us and others [7–12]. We have recently turned our attention to interactions between weakly-coupled $Cu_3(\mu_3\text{-O/OH})$ species and shown that the presence of weak dipolar interactions str evident in the magnetic susceptibility and EPR spectra of H-bonded $Cu_3(\mu_3\text{-OH})$ units [13]. Continuing along the same lines, we report here the structure and magnetochemical studies of a heptanuclear Cu^{II} assembly employing 4-phenyl-pyrazole ligands and of a polymeric structure containing Cu_3N_6 metallacycles with terminal and bridging azido ligands.

2. Materials and Methods

2.1. Materials

4-Phenyl-pyrazole (4-Ph-pzH) was prepared according to a procedure from the literature [14]. All other reagents were purchased from commercial sources and used as received. Solvents were purified using standard techniques [15]. $[PPN]_2[Cu_3(\mu_3\text{-Cl})_2(\mu\text{-pz})_3Cl_3]$ and $[PPN]_2[Cu_3(\mu_3\text{-O})(\mu\text{-pz})_3Cl_3]$ were prepared according to published procedures [12]; PPN^+ = bis(triphenylphosphine)iminium.

2.2. Instruments

2.2.1. X-Ray Crystallography

Single crystal X-ray diffraction data were collected on a Bruker D8 QUEST CMOS system equipped with a TRIUMPH curved-crystal monochromator and a fine-focus X-ray tube with graphite monochromated Mo-Kα radiation (λ = 0.71073 Å) at ambient or low temperature using the APEX3 or APEX2 suite [16]. Crystal data, data collection and structure refinement details are listed in Supplementary Material (Table S1). Frames were integrated with the Bruker SAINT software package using a narrow-frame algorithm. Absorption effects were corrected using the multi-scan method (SADABS) [17]. Structures were solved by intrinsic phasing methods with ShelXT [18] and refined with ShelXL [19] using full-matrix least-squares minimization using Olex2 [20]. All non-hydrogen atoms were refined anisotropically. H atoms were included in calculated positions riding on the C atoms to which they are bonded, with C–H = 0.93 Å and $Uiso(H) = 1.2\ Ueq(C)$. Electron densities of poorly ordered lattice solvent molecules could not be modeled satisfactorily, and they were removed by using the SQUEEZE routine in PLATON [21]. In (**1**), the C47 and C48 atoms were constrained to have equivalent atomic displacement parameters and the relatively large thermal ellipsoids of C atoms of one of the phenyl rings (C46–C51) were restrained with enhanced rigid bond restraint [22].

2.2.2. EPR Spectroscopy

X-band spectra were recorded on a Bruker ESP300 spectrometer using a 4102ST rectangular cavity operating in the TE_{102} mode. For variable-temperature experiments the cavity was fitted in an ESR900 dynamic continuous flow cryostat and the temperature was regulated with an Oxford ITC4 servocontrol. Q-band spectra were recorded on an EMXplus spectrometer fitted with an EMX premiumQ microwave bridge and an ER5106QTW microwave resonator operating in the TE_{012} mode and controlled by the Bruker Xenon software. For variable-temperature experiments the resonator was fitted in an Oxford CF935 dynamic continuous flow cryostat and the temperature was regulated with an Oxford ITC503 servocontrol. The magnetic field was applied by a Bruker BE25 electromagnet using a Bruker ER082(155/45)Z power supply allowing a field sweep between −5 to 16,000 G.

2.2.3. Magnetic Measurements

The temperature dependence of the magnetization at an applied field of $B = 1$ T was acquired for powder samples of (**1**) and (**2**) using PPMS Dynacool magnetometer (Quantum Design Inc., San Diego, CA, USA). The experimental data were corrected for the underlying diamagnetism and signal of the sample holder. The experimental data were fitted with program Polymagnet [23].

2.3. Synthesis of Compounds (**1**) and (**2**)

2.3.1. Synthesis of [PPN]$_2$\{[Cu$_3$(μ_3-OH)(μ-4-Ph-pz)$_3$Cl$_3$]$_2$[Cu (4-Ph-pzH)$_4$]\}Cl$_2$\} (**1**)

CuCl$_2$·2H$_2$O (0.6 mmol, 102.3 mg), 4-Ph-pzH (0.8 mmol, 115.3 mg), NaOH (1 mmol, 40 mg) and PPNCl (0.1 mmol, 57.4 mg) were added to 15 mL CH$_3$CN and the reaction mixture was stirred overnight. A small amount of a grey solid was filtered off and the solvent volume was reduced to 4 mL under reduced pressure. Suitable crystals for X-ray diffraction were grown by slow evaporation. Yield: 45% (126 mg, 0.039 mmol). Crystal data for (**1**): Triclinic, $P\bar{1}$, a = 14.161(7) Å, b = 17.814(8) Å, c = 18.173(9) Å, α = 80.66(1)°, β = 68.33(1)°, γ = 85.25(1)°, V = 4202(4) Å3, Z = 1, R$_1$ = 0.0661, GoF = 1.020, for 985 parameters and 17,116 observed reflections.

2.3.2. Synthesis of (PPN)[Cu$_3$(μ_3-OH)(μ-pz)$_3$($\mu_{1,1}$-N$_3$)$_2$(N$_3$)] (**2**)

Caution! *Azide complexes of metal ions in the presence of organic ligands are potentially explosive. Only small amounts should be prepared, and they should be handled with care.*

A solution of NaN$_3$ (0.375 mmol, 24.42 mg) in 5 mL of MeCN was added dropwise to a solution of [PPN]$_2$ [Cu$_3$(μ_3-O)(μ-pz)$_3$Cl$_3$] (0.0628 mmol, 100 mg) suspended in 10 mL MeCN. The mixture was stirred overnight at rt. Well-shaped crystals suitable for X-ray diffraction were obtained upon slow evaporation of the filtrate at room temperature over three weeks. The crystals were isolated, washed three times with methanol and ether and dried in the vacuum. Yield: 65% (44 mg, 0.041 mmol), based on Cu. Anal. calcd/found for C$_{45}$H$_{40}$Cu$_3$N$_{16}$OP$_2$:C, 50.36/50.01; H, 3.84/3.62; N, 20.89/20.48. Compound (**2**) was similarly synthesized in MeOH (instead of MeCN), albeit with a 40% yield. Crystal data for (**2**): monoclinic, $P2_1/c$, a = 8.6121(9) Å, b = 17.034(2) Å, c = 32.237(3) Å, β = 96.493(2)°, V = 4698.8(8) Å3, Z = 4, R$_1$ = 0.0662, GoF = 0.987, for 623 parameters and 5601 observed reflections.

2.4. Theoretical Calculations

The theoretical calculations based on density functional theory (DFT) were done with ORCA 4.1 package [24] using B3LYP hybrid functional [25–27] accounting also for relativistic effects with ZORA Hamiltonian and respective ZORA-def2-TZVP basis set for Cu, N, O, Cl atoms and ZORA-def2-SVP basis set for C and H atoms [28]. Additionally, the calculations utilized the chain-of-spheres (RIJCOSX) approximation to exact exchange as implemented in ORCA [29,30] and the auxiliary basis SARC/J [31]. Increased integration grids (Grid7 and Gridx7 in ORCA convention), increased radial grid (IntAcc = 8) for Cu atoms and tight Self Consistent Field (SCF) convergence criteria were used in all calculations. The molecular fragment used in the calculations was extracted from the experimental X-ray structure. The calculated spin density was visualized with VESTA 3 program [32].

3. Results and Discussion

3.1. Synthesis

The reaction of CuCl$_2$·2H$_2$O, 4-Ph-pzH, NaOH and PPNCl in 6:9:13:1 ratio and approximately 5 mL of various solvents yielded the trinuclear complex, PPN[Cu$_3$(μ_3-OH)(μ-4-Ph-pz)$_3$Cl$_3$] [33]. Compound (**1**) was prepared from the same reagents employing a 6:6:13:1 reagent ratio in a more dilute reaction mixture (15 mL MeCN). Pettinari et al. have obtained a similar heptanuclear complex, [\{Cu$_3$(μ_3-OH)(μ-pz)$_3$(Cl)$_2$(Hpz)$_2$(H$_2$O)\}$_2$\{CuCl$_2$(Hpz)$_2$\}], by acid digestion of the trinuclear [Cu$_3$(μ_3-OH)(μ-pz)$_3$(CH$_3$COO)$_2$(pzH)] complex [34].

The azide complex (**2**) was synthesized in a metathesis reaction from (PPN)$_2$[Cu$_3$(μ$_3$-Cl)$_2$(μ-pz)$_3$Cl$_3$] by exchanging the terminal chlorides for azide using a slight excess of NaN$_3$ dissolved in MeOH.

3.2. Crystal Structure Description of [PPN]$_2$[{Cu$_3$(μ$_3$-OH)(μ-4-Ph-pz)$_3$Cl$_3$}$_2${Cu(4-Ph-pzH)$_4$}]Cl$_2$ (**1**)

Complex (**1**) crystallizes in the triclinic space group $P\bar{1}$ with the asymmetric unit containing one complete trinuclear and one-half of the mononuclear complex. Its crystal structure (Figure 1) is formed by repeating heptanuclear assemblies consisting of two trinuclear [Cu$_3$(μ$_3$-OH)(μ-4-Ph-pz)$_3$Cl$_3$]$^-$ anionic metallacycles on either side of a neutral, mononuclear, square planar [Cu(4-Ph-pzH)$_4$]; the latter is located on the crystallographic inversion center. In the solid state, two Cl ions act as bridges between the central mononuclear complex and the two trinuclear ones [Cl(4)-Cu(1), 2.744(2) Å; Cl(4)-Cu(4), 2.792(2) Å], occupying axial sites and forming a weakly bonded heptanuclear assembly. The two trinuclear anions contain 4-coordinate distorted square planar Cu-centers and a pyramidal μ$_3$-OH (the O atom is 0.473(4) Å away from the Cu$_3$-plane), have their Cu$_3$-planes parallel to each other and are connected via two long Cu(2) ... Cl(1) contacts of 3.023(2) Å to the adjacent heptanuclear unit; the μ-Cl atoms occupy one equatorial and one axial position with a Cu(1)–Cl(2)–Cu(2) angle of 101.00(6)°. The one-dimensional chains thereby generated run parallel to the crystallographic *a*-axis (Figure 2) separated by the PPN$^+$ counterions. The long Cu(4) ... Cl(4) distances of 2.792(2) Å between the mononuclear [Cu(4-Ph-pzH)$_4$]$^{2+}$ unit and the μ-Cl atoms are considered as non-bonding here; however, even longer distances of 2.817–2.839 Å have been reported in the corresponding *trans*-[CuCl$_2$(pz*H)$_4$] complexes (pz*H = pzH [35], 3-*t*Bu-pzH [36] and 3-Ph-pzH [37]). A complete list of bond lengths and angles for (**1**) is provided as Supplementary Material, Table S2.

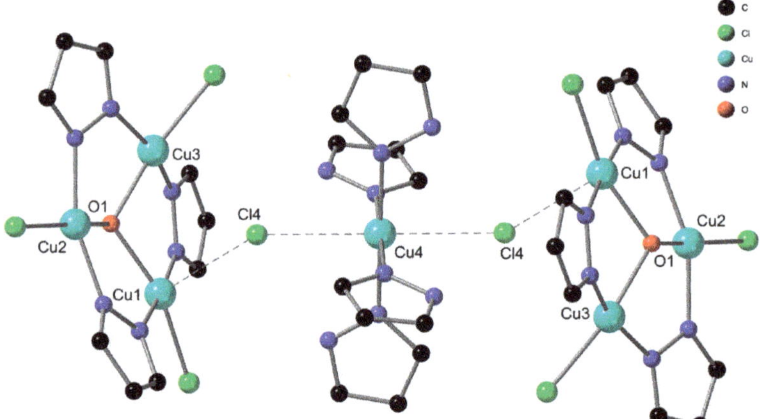

Figure 1. Crystal structure and partial atom labeling scheme of (**1**). Phenyl groups on the pyrazolate ligands, H atoms and PPN counterions are not shown for clarity. Selected interatomic distances (Å) and angles (°): Cu⋯Cu, 3.243(1), 3.417(1), 3.451(2), 5.228(3); Cu-O, 1.990(2)–2.012(3); Cu-N, 1.948(4)–1.957(3); Cu-Cl (terminal), 2.251(2)–2.307(2); Cu-Cl (bridging), 2.744(2); ∠CuOCu, 108.5(1)–118.3(2); ∠(μ$_3$-O)CuCl(terminal), 162.6(1)–169.3(1); ∠NCuN, 160.4(1), 163.3(2) and 175.7(2); ∠Cl(1)Cu(1)Cl(4), 110.10(4). For the mononuclear center, Cu-N, 2.017(4) and 2.017(4) Å; Cu-Cl (bridging), 2.792(2) Å; ∠NCuN, 91.6(1) and 180.

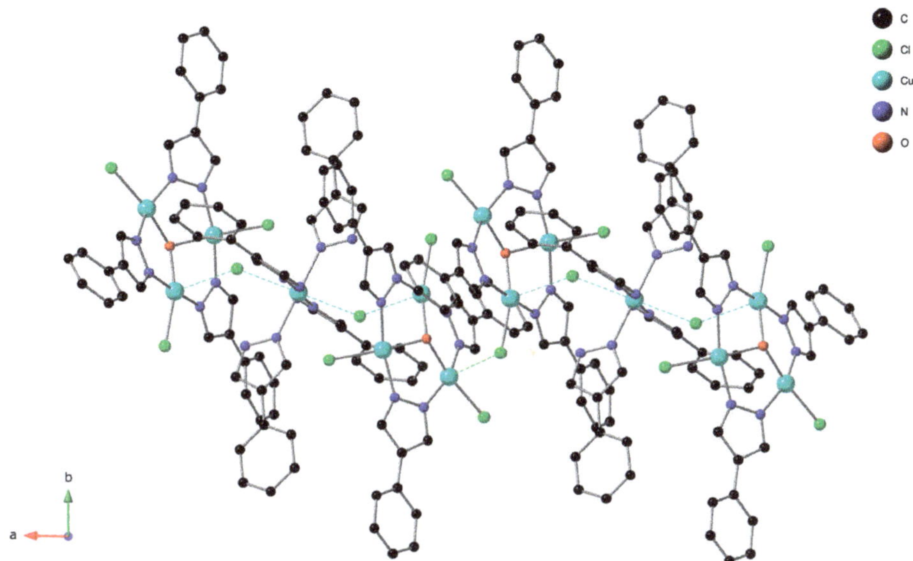

Figure 2. Molecular structure of (**1**) viewed parallel to the crystallographic *c* axis showing its polymeric character. Hydrogen atoms have been omitted for clarity.

3.3. Crystal Structure Description of (PPN)[Cu₃(μ₃-OH)(μ-pz)₃(μ,κ^{1,1}-N₃)₂(N₃)] (2)

The complex crystallizes in the monoclinic $P2_1/c$ space group with the whole molecule in the asymmetric unit. The structure consists of triangular μ₃-OH-capped metallacyclic units (O atoms at 0.345(2) Å from the Cu₃ plane, Figure 3) connected by end-on bridging azides, forming infinite chains along the crystallographic *b*-axis (Figure 4), separated by the PPN⁺ counterions. One of the three Cu centers is in square planar, whereas the other two are in distorted square pyramidal geometry. One of the three azide ligands is in a terminal monodentate coordination mode with Cu-N = 1.974(7) Å, and the other two are unsymmetrically bridging, in an end-on (μ,κ^{1,1}) fashion, between two Cu₃ units with Cu(1)–N(13) = 1.980(5), Cu(2')-N(13) = 2.421(5) and Cu(2)-N(3) = 2.001(5) Å, Cu(1')-N(3) = 2.322(5) Å at each bridgehead N, respectively. The bridging azides occupy one equatorial (shorter Cu-N bond) and one axial (longer Cu-N bond) at either side. The corresponding Cu–N_{azide}–Cu angles are 105.9(2) and 115.0(2)°, respectively, holding the Cu atoms at intermolecular distances of 3.386(1) Å and 3.470(1) Å. As expected, in the two tetragonal pyramidal Cu-centers, the axial Cu-N bonds are significantly longer than the equatorial ones. The azide ions are approximately linear with N–N–N angles of 176.7(7)° and 178.3(7)°; contain unequal N–N bond lengths, longer at the end involving the donor atoms, N3–N4 = 1.191(7) Å and longer at the dangling end, N4–N5 = 1.144(7) Å [38]. A complete list of bond lengths and angles for (**2**) is provided as Supplementary Material, Table S3.

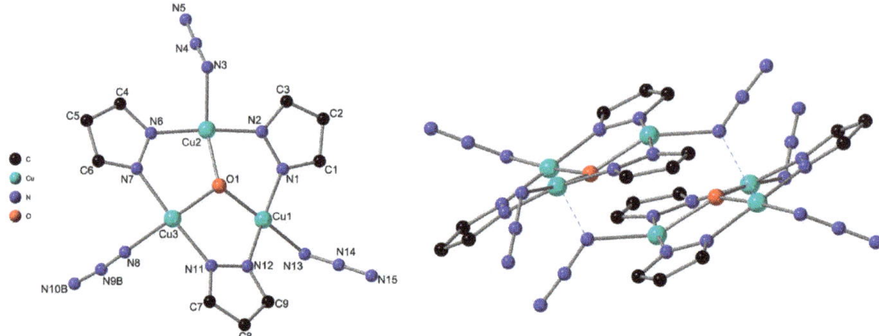

Figure 3. Asymmetric unit of (**2**) (**left**). Inter-Cu$_3$ contacts (**right**). H atoms, disordered azide ligand and PPN counterion are not shown for clarity. Selected interatomic distances (Å) and angles (°): Cu⋯Cu, 3.386(1), 3.389(1), 3.470(1); Cu–O, 1.976(4), 2.000(4), 2.023(4); Cu-N$_{pz}$, 1.935(5)–1.974(5); Cu-N$_{azide}$, 1.974(6), 1.980(5), 2.001(5); ∠CuOCu, 115.8(2), 116.8(2), 119.1(2).

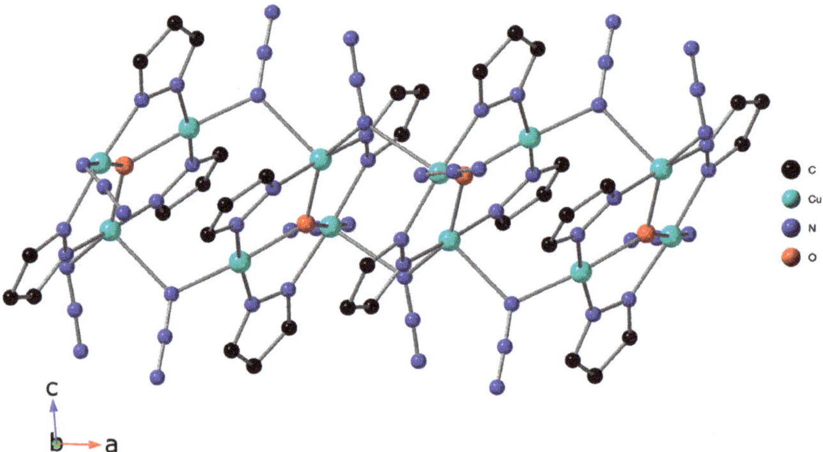

Figure 4. Packing diagram of compound (**2**) shown along the crystallographic *b*-axis.

3.4. Infrared Spectra

The coordination mode of azides to a transition metals is usually characterized by an intense IR band due to $\nu_{as}(N_3)$ at 2000–2055 cm^{-1} for a terminal and >2055 cm^{-1} for a bridging $N_3{}^-$; the larger values correspond to anions with unsymmetrical N–N–N bonding [39]. A broad trifurcated band with peaks at 2034, 2046 and 2060 cm^{-1} in the solid state spectrum of (**2**) (Figure S3) is attributed to the presence of both terminal and bridging azides.

3.5. Magnetic Susceptibility of (1)

The temperature dependence of the effective magnetic moment and the molar magnetization data for (**1**) are shown in Figure 5. The effective magnetic moment at room temperature is 4.3 μ_B and is rapidly decreasing, reaching a plateau of 3.6 μ_B at ca. 100 K, and then further decreasing below 30 K to 3.1 μ_B at 1.9 K. The theoretical spin-only value for seven non-interacting Cu^{II} ions with g = 2.0 is 4.58 μ_B, but usually the g-factor for this ion is much larger due to the angular orbital momentum contribution, so an even larger theoretical spin-only value is expected. The lower room temperature value of μ_{eff} together with its sharp decrease on subsequent cooling thus reflect strong antiferromagnetic exchange. Such strong antiferromagnetic exchange within each Cu$_3$(μ_3-OH) triangle leads to $S_{Cu1\text{-}2\text{-}3} = 1/2$ ground spin state. Therefore, the value of $\mu_{eff}/\mu_B \approx 3.6$ in the temperature interval 50–120 K can be explained by considering coexistence of two $S_{Cu1\text{-}2\text{-}3} = 1/2$ and one $S_{Cu4} = 1/2$ spin levels. A further decrease of μ_{eff} below 50 K is then ascribed to weak magnetic interactions between two $Cu_3(\mu_3\text{-}OH)$ triangles and eventually between $Cu_3(\mu_3\text{-}OH)$ triangles and the central $[Cu(4\text{-}Ph\text{-}pzH)_4]$ complex units (see Figure 2). Moreover, another important origin of decrease of μ_{eff} below 50 K can be attributed to the antisymmetric exchange interactions (ASE, also named Dzyaloshinsky–Moriya interactions) within two $Cu_3(\mu_3\text{-}OH)$ triangles, as this kind of interaction is typical of triangular molecular systems based on Kramers ions coupled with strong antiferromagnetic exchange [40]. Moreover, an ASE has been identified and quantified in similar coordination compounds with individual $Cu_3(\mu_3\text{-}OH)$ or $Cu_3(\mu_3\text{-}O)$ motifs, and its effects on magnetic and spectroscopic properties have been demonstrated [7,41]. Additionally, the presence of ASE in (**1**) was evidenced by low temperature EPR spectroscopy, as discussed in the following section. Therefore, the following spin Hamiltonian has been postulated in order to quantitatively analyze the experimental magnetic data

$$
\begin{aligned}
\hat{H} ={}& -J_{12}(\mathbf{S}_1 \cdot \mathbf{S}_2 + \mathbf{S}_{1'} \cdot \mathbf{S}_{2'}) - J_{13}(\mathbf{S}_1 \cdot \mathbf{S}_3 + \mathbf{S}_{1'} \cdot \mathbf{S}_{3'}) - J_{23}(\mathbf{S}_2 \cdot \mathbf{S}_3 + \mathbf{S}_{2'} \cdot \mathbf{S}_{3'}) - J_{12'}(\mathbf{S}_1 \cdot \mathbf{S}_2 + \mathbf{S}_{1'} \cdot \mathbf{S}_2) - J_{14}(\mathbf{S}_1 \cdot \mathbf{S}_4 + \mathbf{S}_{1'} \cdot \mathbf{S}_4) \\
&+ \mathbf{d}_{12} \cdot (\mathbf{S}_1 \times \mathbf{S}_2 + \mathbf{S}_{1'} \times \mathbf{S}_{2'}) + \mathbf{d}_{23} \cdot (\mathbf{S}_2 \times \mathbf{S}_3 + \mathbf{S}_{2'} \times \mathbf{S}_{3'}) + \mathbf{d}_{31} \cdot (\mathbf{S}_3 \times \mathbf{S}_1 + \mathbf{S}_{3'} \times \mathbf{S}_{1'}) + \mu_B \sum_{i=1}^{7} \mathbf{B} \cdot g \cdot \mathbf{S}_i
\end{aligned}
\tag{1}
$$

where the isotropic exchange, Zeeman terms and ASE expressed by $\mathbf{d}_{ij}$ vectors, $(d_x, d_y, d_z)_{ij}$, are included. The application of Moriya symmetry rules [42] for the $Cu_3(\mu_3\text{-}OH)$ triangles results in only one non-zero component: $\mathbf{d}_{ij} = (0, 0, d_z)_{ij}$ and it was assumed that $(d_z)_{ij}$ are equal for all pairs. Next, the molar magnetization in the direction of the magnetic field $\mathbf{B}_a = B \cdot (\sin\theta\cos\varphi, \sin\theta\sin\varphi, \cos\theta)$ was calculated as

$$
M_a = N_A kT \frac{\partial \ln Z}{\partial B_a}
\tag{2}
$$

and since the magnetic data were acquired on a polycrystalline sample, the powder average of the molar magnetization was then calculated as

$$
M_{mol} = 1/4\pi \int_0^{2\pi} \int_0^{\pi} M_a \sin\theta \, d\theta \, d\varphi
\tag{3}
$$

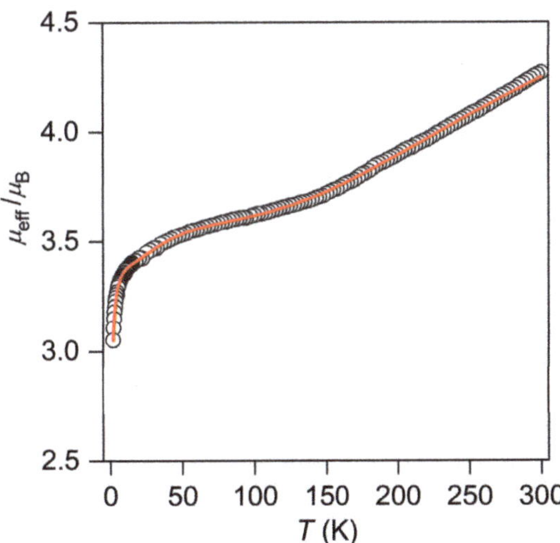

Figure 5. Temperature dependence of the effective magnetic moment for **1**. Empty circles—experimental data, full lines—calculated data with the spin Hamiltonian in Equation (1) and $J_{12} = J_{23} = -281$ cm^{-1}, $J_{13} = -226$ cm^{-1}, $J_{12'} = J_{14} = +19.7$ cm^{-1}, $|d_z| = 37.1$ cm^{-1}, $g = 2.35$.

In order to reduce the number of free parameters, DFT calculations were employed (*vide infra*) from which we may assume $J_{12} \approx J_{23}$, $|J_{12}|$, $|J_{23}| > |J_{13}|$ and $J_{12'} \approx J_{14}$, and $J_{12'}$, $J_{14} > 0$. Thus, magnetic data were fitted under the assumption that magnetic coupling through μ-Cl-ligands between two $Cu_3(\mu_3$-OH) triangles and between $Cu_3(\mu_3$-OH) triangles and the central [Cu(4-Ph-pzH)$_4$] complex unit is weakly ferromagnetic, whereas the strong antiferromagnetic exchange was expected within $Cu_3(\mu_3$-OH) triangles. Such analysis resulted in best-fit values of $J_{12} = J_{23} = -281$ cm^{-1}, $J_{13} = -226$ cm^{-1}, $J_{12'} = J_{14} = +19.7$ cm^{-1} and $|d_z| = 37.1$ cm^{-1} with an isotropic g-factor $g = 2.35$ (Figure 5). The temperature-independent paramagnetism was also accounted for by adding a constant term $\chi_{TIP} = 5.23 \times 10^{-9}$ m^3 mol^{-1} for seven copper atoms based on the generally accepted value for one Cu^{II} ion equal to 60×10^{-6} cm^3 mol^{-1} in cgs units [43], or 0.754×10^{-9} m^3 mol^{-1} in SI units. In summary, the strong antiferromagnetic exchange within the $Cu_3(\mu_3$-OH) triangles was confirmed together with the antisymmetric non-Heisenberg interaction, and the overall analysis was impossible without the introduction of a weak inter-triangle magnetic exchange.

The value of the magnetic exchange within the $Cu_3(\mu_3$-OH) triangles is comparable to those reported in the literature for similar Cu^{II}-pyrazolato/triazolato-bridged complexes containing μ_3-OH group. It seems that in compound (1) reported herein, the antiferromagnetic exchange is one of the strongest (Table 1) [44].

Table 1. Selected magnetostructural data for various $Cu_3(\mu_3$-O(H/R)) systems.

Compound [a]	Cu⋯Cu (Å)	$-J$, $-zJ'$/cm^{-1}	Ref.
[Cu^3(OH)(pz)3(Hpz)2(NO^3)^2]·H^{2}O	3.351	200, 0	[45]
[Cu^3(OH)(pz)3(py)2Cl^2]·py	3.112–3.321	140, 0	[46]
[Cu^3(OH)(aat)$_3$(CF^{3}SO^3)(H^{2}O)](CF^{3}SO^3)	3.355	197.7, 0	[47]
[Cu^3(OH)(aat)3(NO^3)(H^{2}O)^2](NO^3)·(H^{2}O)^2	3.341	190.9, 0	[47]
[Cu$_3$(OH)(aat)$_3$(ClO$_4$)(H$_2$O)$_2$](ClO$_4$)	3.371	198.2, 0	[47]
{[Cu$_3$(O)(triazolate)$_3$(OH)(H$_2$O)$_6$]}$_n$	3.388	112.6, 11.6	[48]
[Cu$_3$(triazolate)$_3$(OH)][Cu$_2$Br$_4$]	3.502	180, 68	[49]
[Cu$_3$Br(Hpz)$_2$(pz)$_3$(OCH$_3$)]Br	3.250–3.255	105, 0	[50]
[Cu$_3$(OH)(aaat)$_3$(H$_2$O)$_3$](NO$_3$)$_2$·H$_2$O	3.347–3.393	195, 0	[8]
{[Cu$_3$(OH)(aat)$_3$(SO$_4$)]·6H$_2$O}$_n$	3.337–3.364	185, 0	[8]
[Cu$_3$(admtrz)$_4$(SCN)$_3$(OH)(H$_2$O)](ClO$_4$)$_2$·H$_2$O	3.254–3.318	120, 53	[51]
[{Cu$_3$(OH)(pz)$_3$(Hpz)$_3$}$_2$SO$_4$](NO$_3$)$_2$·MeCN·MeOH·1.5H$_2$O	3.182–3.354	180, 12.7	[44]
[Ag(Hpz)$_2$]$_2$[{Ag$_2$(Hpz)$_2$(NO$_3$)$_2$}{Cu$_6$(OH)$_2$(pz)$_6$(Hpz)$_6$(SO$_4$)}$_2$](NO$_3$)$_6$·4H$_2$O	3.222–3.356	134, 10.5	[44]
[{Ag(H$_2$O)$_2$}{Cu$_3$(OH)(pz)$_3$(Hpz)$_3$(H$_2$O)(ClO$_4$)$_3$}]	3.302–3.372	158, 9.2	[44]
[Et$_3$NH][Cu$_3$(OH)(pz)$_3$(PhCOO)$_3$]	3.244–3.352	178, 57.5	[13]

[a] aat = 3-acetylamino-1,2,4-triazolate; Haaat = 3-acetylamino-5-amino-1,2,4-triazolate; admtrz = 4-amino-3,5-dimethyl-1,2,4-triazole.

3.6. EPR Spectroscopy of (1)

Solid state X-band EPR spectra of (1) at 4.2 K showed a complex derivative signal centered around 3000 G, with broad linewidths and a broad resonance around 3800 G ($g{\sim}1.8$), all characteristic of an exchange-coupled system (Figure 6). Upon heating, part of the signal decreased in intensity, leaving an axial signal, which persisted unchanged up to 290 K, and was assigned to the central [Cu(4-Ph-pzH)$_4$]$^{2+}$ complex, which appears not to be exchange-coupled to the two Cu$_3$ triangles. This is in agreement with the crystal structure, showing that the main coupling pathway between [Cu(4-Ph-pzH)$_4$] and the trinuclear units—Cl-counterions at axial sites on either side of the Cu$_3$-units—involves non-magnetic orbitals ($d_z{}^2$) and consistent with the analysis of magnetic susceptibility data, where J_{14} was shown to be the weakest interaction (*vide supra*, Section 3.5). Whereas the relaxation rate of the intradoublet signal of the Cu$_3$ accelerates rapidly with increasing temperature, the signal intensity of the mononuclear complex follows a Curie dependence, masking the contribution of the exchange-coupled system above 12 K. Attempts to increase the signal of the exchange-coupled system by exploiting the relaxation differences of the two components, in particular by increasing the microwave power at the low-temperature limit of the apparatus, failed; experiments at 4.0 K with microwave power of 20 mW did not selectively increase the signal intensity of that component.

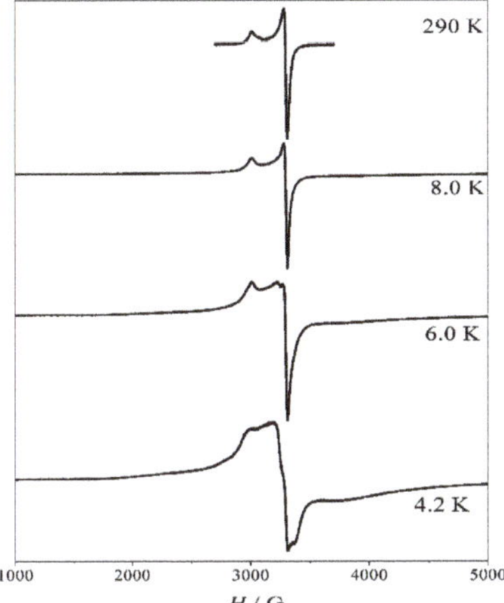

Figure 6. Solid-state X-band EPR spectra of (**1**) between 4.2 and 290 K. Experimental conditions: f_{EPR} = 9.43 GHz, MA = 2 G_{pp} (6–290 K) and 1 G_{pp} (4.2 K), P_{MW} = 2 mW (6–290 K) and 0.2 mW (4.2 K).

The above assignment is further corroborated by EPR studies in a frozen THF solution (Figure 7) showing significant differences between the signal attributed to the two components: one axial signal exhibited hyperfine features and the other exhibited a very broad $g_\perp$ < 2 feature. The latter feature is characteristic of half-integer trinuclear clusters and is due to the presence of magnetic asymmetries operating in tandem with antisymmetric exchange [52,53]. These characteristic features allowed simulations assuming two axial subcomponents. For the former, an s = 1/2 spin, described by the $\hat{H} = \beta \mathbf{H} \tilde{\mathbf{g}} \hat{\mathbf{S}} + \hat{\mathbf{I}} \tilde{\mathbf{A}} \hat{\mathbf{S}}$ Hamiltonian, and for the latter an effective s = 1/2 spin, described by a simple Zeeman Hamiltonian, were assumed. Simulations with parameters $g_{1\perp}$ = 2.145, $g_{1\|}$ = 2.334, $A_{1\|}$ = 527 (MHz), $g_{2\perp}$ = 1.82 (g-strain = 0.37 FWHM) and $g_{2\|}$ = 2.268 (relative intensities I_2:I_1 = 0.91:1) gave a satisfactory agreement to the experimental spectrum. Due to the large number of variables, the above parameter set is indicative, as far as line widths and g-strain parameters are concerned.

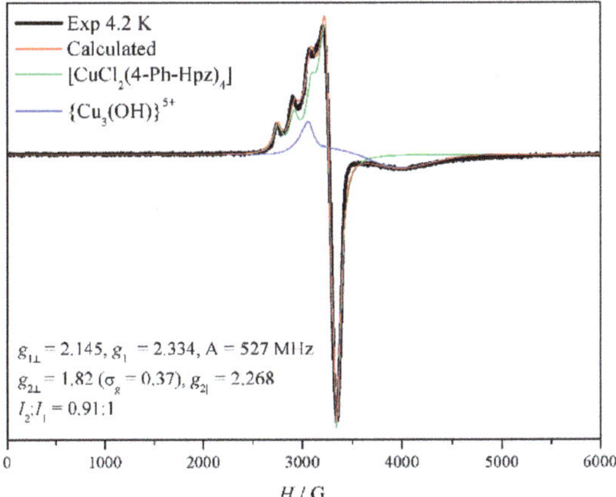

Figure 7. X-band EPR spectra of (**1**) in a frozen THF solution (black line) and calculated curve according to the discussion in the text (red line). The blue and green lines correspond to the two components. Experimental conditions: f_{EPR} = 9.42 GHz, MA = 2 G_{pp}, P_{MW} = 2 mW.

The presence of the hyperfine signals in solution and not in the solid state for the mononuclear component suggests the presence of dipolar interactions that are removed upon dissolution. To test this hypothesis and better understand the solid-state spectra of (1), complementary Q-band studies were carried out. The solid-state Q-band spectra (Figure S4) were by and large similar to the X-band ones, but they revealed additional features of the two subcomponents (Figure 8). In particular, the signal attributed to [Cu(4-Ph-pzH)$_4$] was found to be rhombic, with a small split in its x and y components, indiscernible in the X-band experiments. The overall behavior observed in the X-band spectra was confirmed, with a composite spectrum at 5 K.

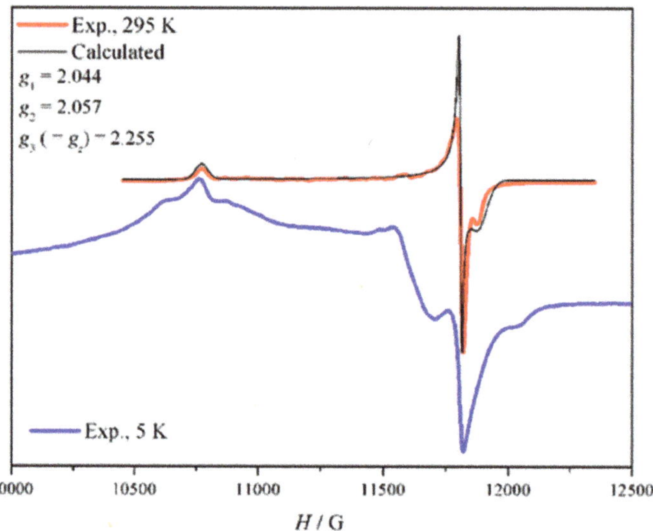

Figure 8. Q-band EPR spectra of (1) in a frozen THF solution (black line) and calculated curve according to the discussion in the text (red line). The blue and green lines correspond to the two components. Experimental conditions: f_{EPR} = 33.96 GHz, MA = 2 G$_{pp}$, P_{MW} = 0.29 mW (5 K) and 1.1 mW (295 K).

The solid-state structure is characterized by a network of possible dipolar interactions along the chains formed by the trinuclear and mononuclear complexes previously described (Figure S5). Depending on their magnetic symmetry, i.e., $|J| > |J'|$ versus $|J| < |J'|$, the spin densities of the triangles may be spread out over all three metal sites, or concentrated on one of them, respectively [13]. Our tentative conclusion from DFT calculation points toward the former case, which is also consistent with the magnetic susceptibility analysis, negating the applicability of the point-dipole approximation and hindering a straightforward analysis of dipolar interactions. Moreover, the non-trivial symmetry of the structure, combined with the extended nature of the system, seriously complicated any such analysis. Therefore, a detailed analysis of the dipole–dipole interactions was not pursued in this case.

3.7. Theoretical DFT Calculations of (1)

The complexity of magnetic interactions in (1) demands some theoretical insight guiding the analysis of the experimental magnetic data. Therefore, the isotropic exchange parameters J_i were calculated with the help of broken-symmetry calculations using several molecular fragments derived from experimental X-ray data (Figure 9). First, the triangular moiety was extracted and energies of high-spin state (HS) and broken-symmetry spin states (BS) were calculated with B3LYP to derive J-parameters for spin Hamiltonian

$$\hat{H} = -J_{12}(\mathbf{S}_1 \cdot \mathbf{S}_2 + \mathbf{S}_{1'} \cdot \mathbf{S}_{2'}) - J_{13}(\mathbf{S}_1 \cdot \mathbf{S}_3 + \mathbf{S}_{1'} \cdot \mathbf{S}_{3'}) - J_{23}(\mathbf{S}_2 \cdot \mathbf{S}_3 + \mathbf{S}_{2'} \cdot \mathbf{S}_{3'}) \tag{4}$$

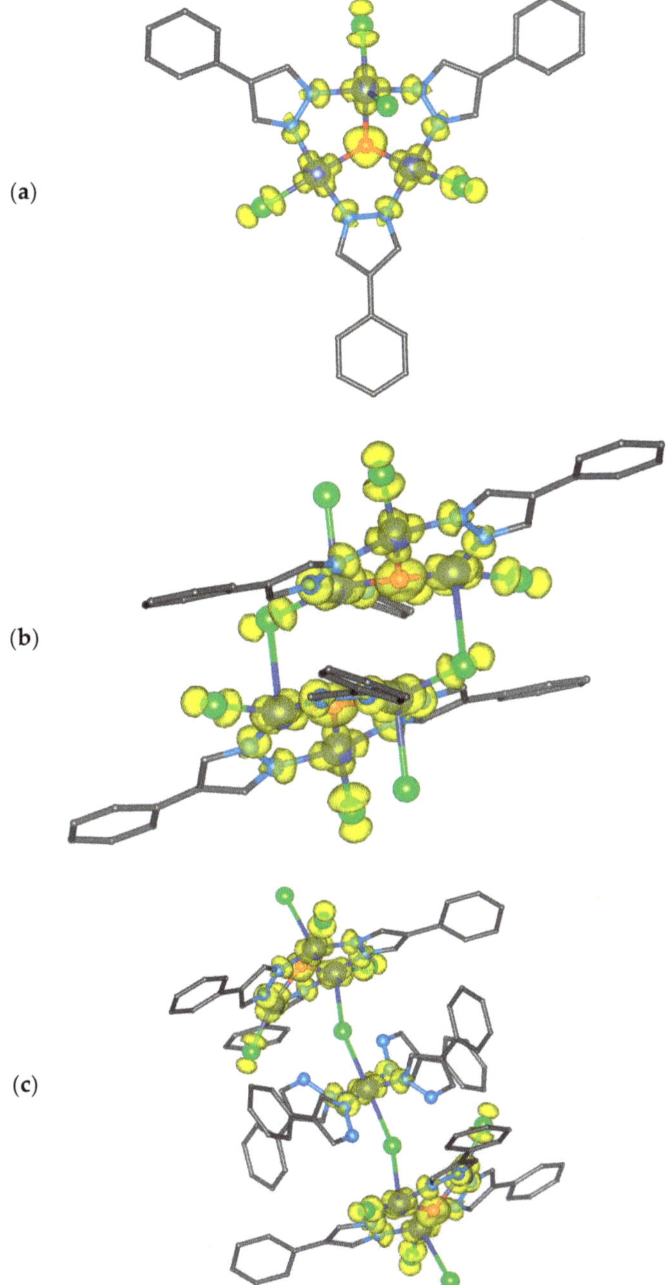

Figure 9. The calculated spin density distribution using B3LYP of (**1**) for the HS states of Cu_3 molecular fragment (**a**), Cu_6 fragment (**b**) and Cu_7 fragment (**c**). The spin density is represented by yellow surfaces. The isodensity surfaces are plotted with the cut-off value of $0.005\ ea_0{}^{-3}$. Hydrogen atoms are omitted for clarity.

As a result, the energies $\Delta_1 = -245.197$ cm^{-1}, $\Delta_2 = -323.346$ cm^{-1} and $\Delta_3 = -251.072$ cm^{-1} were calculated, where $\Delta_i = \varepsilon_{BS,i} - \varepsilon_{HS}$. From these energies, J-values were calculated by Ruiz's approach [54,55], resulting in $J_{12} = -159$ cm^{-1}, $J_{13} = -86.5$ cm^{-1} and $J_{23} = -165$ cm^{-1}. It must be noted that this approach is based on the so-called strong interaction limit, whereas the weak interaction limit treatment of Noodleman would have resulted in J-values generally twice larger [56]. Next, the hexanuclear molecular fragment was extracted in order to estimate the magnetic exchange mediated by chlorido-ligands between two trimeric units:

$$\hat{H} = -J_{12}(\mathbf{S}_1 \cdot \mathbf{S}_2 + \mathbf{S}_{1'} \cdot \mathbf{S}_{2'}) - J_{13}(\mathbf{S}_1 \cdot \mathbf{S}_3 + \mathbf{S}_{1'} \cdot \mathbf{S}_{3'}) - J_{23}(\mathbf{S}_2 \cdot \mathbf{S}_3 + \mathbf{S}_{2'} \cdot \mathbf{S}_{3'}) - J_{12'}(\mathbf{S}_1 \cdot \mathbf{S}_{2'} + \mathbf{S}_{1'} \cdot \mathbf{S}_2) \tag{5}$$

Therefore, energies of HS and BS123 states were calculated, leading to $\Delta_{123} = +5.808$ cm^{-1}, from which $J_{12'}$ equals $+2.90$ cm^{-1}. Finally, the heptanuclear molecular fragment was investigated using spin Hamiltonian

$$\hat{H} = -J_{12}(\mathbf{S}_1 \cdot \mathbf{S}_2 + \mathbf{S}_{1'} \cdot \mathbf{S}_{2'}) - J_{13}(\mathbf{S}_1 \cdot \mathbf{S}_3 + \mathbf{S}_{1'} \cdot \mathbf{S}_{3'}) - J_{23}(\mathbf{S}_2 \cdot \mathbf{S}_3 + \mathbf{S}_{2'} \cdot \mathbf{S}_{3'}) - J_{14}(\mathbf{S}_1 \cdot \mathbf{S}_4 + \mathbf{S}_{1'} \cdot \mathbf{S}_4) \tag{6}$$

and HS and BS4 spin states were calculated, resulting in $\Delta_{123} = +1.723$ cm^{-1}. Thus, also this interaction is weakly ferromagnetic, $J_{14} = +0.86$ cm^{-1}.

3.8. Magnetic Susceptibility Studies of (2)

The temperature dependence of effective magnetic moment data for compound (2) is depicted in Figure 10. The room temperature effective magnetic moment has value 3.1 μ_B, which is relatively close to the theoretical value 3.0 μ_B for three non-interacting spins $s_1 = s_2 = s_3 = 1/2$ with $g = 2.0$. Upon lowering the temperature, the effective magnetic moment continually decreases down to a value of 0.1 μ_B at 1.9 K, indicating the presence of strong antiferromagnetic exchange interactions. The observed magnetic behavior of (2) can be rationalized on a qualitative level by assuming dominant antiferromagnetic exchange within each triangle, which leads to $s_{eff} = 1/2$ ground spin state similarly to (1) and also supported by DFT (*vide infra*). These Cu$_3$-triangles are then coupled within the infinite chain by azido ligands, which also mediate a weak antiferromagnetic exchange. Established magnetostructural correlations [57,58] also suggest that the Cu–N–Cu angles of 115.0(2) and 105.9(2)° should mediate an antiferromagnetic exchange. Thus, from the magnetic point of view, the coordination polymer of (2) can be simplified to 1D chain of the antiferromagnetically coupled $s_{eff} = 1/2$ spins with this spin Hamiltonian:

$$\hat{H} = -J\sum_{i=1}^{\infty} \mathbf{S}_i \cdot \mathbf{S}_{i+1} + \mu_B \sum_{i=1}^{\infty} \mathbf{B} \cdot \mathbf{g}_i \cdot \mathbf{S}_i \tag{7}$$

where $s_i = s_{eff} = 1/2$. Fortunately, the analytical equation of the molar susceptibility for said system has already been derived by Johnston et al. [59] and the fitting procedure applied to the temperature dependence of the molar susceptibility of (2) resulted in $J = -53$ cm^{-1} with $g = 2.2$. The negative value of J confirms the antiferromagnetic exchange among the trinuclear building block within the coordination polymer in contrast to DFT calculations, which suggest ferromagnetic exchange mediated by azido ligands. Moreover, the deviation of calculated values of the effective magnetic moment at temperatures higher than ca 60 K is attributed to fact that at such high temperature, the proposed approximation of $s_{eff} = 1/2$ for each Cu$_3$-triangle loses its validity, because the excited $s_{eff} = 3/2$ state is also populated, explaining the higher values of the effective magnetic moment in comparison to the calculated ones. It must be noted that we have tried to employ spin Hamiltonian analogous to Equation1 also for (2), but the agreement with the experimental data was not achieved. Most probably, the case of both intra and inter Cu$_3$-triangle antiferromagnetic exchange would demand expanding the spin Hamiltonian to contain more spin centers to better simulate the polymeric character of the compound, which is unfortunately prohibited by large dimension of the respective Hilbert space.

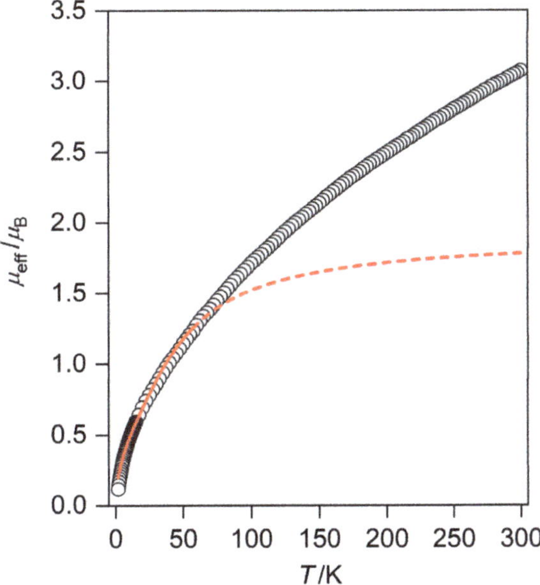

Figure 10. Temperature dependence of the effective magnetic moment for (**2**). The empty symbols—experimental data, the red line—calculated data with $J = -53$ cm^{-1} and $g = 2.2$ using spin Hamiltonian in Equation (2).

3.9. EPR Spectroscopy of (**2**)

The 4.2 K solid-state X-band EPR spectrum of (**2**) is characterized by a main derivative signal at $g = 2.05$, and a secondary half-field transition at $g = 4.06$; the latter transition was attributed to magnetic interactions with neighboring complexes of the polymeric structure (Figure 11). Due to the magnetic interaction pathway mediated by the bridging azides, which passes through non-magnetic orbitals of the copper(II) ions, this interaction may not be of exchange but of dipolar origins [60].

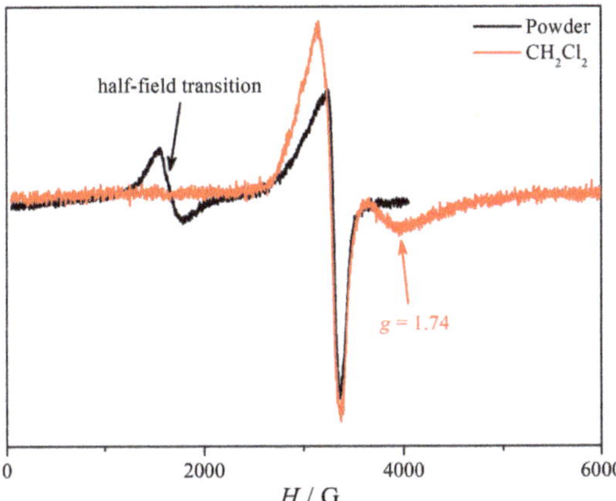

Figure 11. X-band EPR spectra of (**2**) in the solid state (black) and a frozen CH$_2$Cl$_2$ solution. Experimental parameters: $f_{EPR} = 9.429$ GHz, $P_{MW} = 2$ mW, MA $= 5$ G$_{pp}$ (powder); and $f_{EPR} = 9.425$ GHz, $P_{MW} = 2$ mW, MA $= 2$ G$_{pp}$ (solution).

In frozen CH_2Cl_2 solutions this half-field transition disappears, in line with a disruption of the polymeric network in solution. In turn, the solution spectrum exhibits a downfield shifted feature ($g = 1.74$), which is a characteristic signature of magnetic anisotropy induced by the combined operation of a moderate magnetic asymmetry ($J \neq J'$) and antisymmetric exchange interactions (Figure 10) [13,52,53].

3.10. Theoretical DFT Calculations of (2)

The magnetic interactions in (2) were also analyzed with the broken-symmetry calculations using two molecular fragments derived from experimental X-ray data (Figure 12). First, the triangular moiety was extracted and energies of high-spin state (HS) and broken-symmetry spin states (BS) were calculated with B3LYP to derive J-parameters for spin Hamiltonian

$$\hat{H} = -J_{12}(\mathbf{S}_1 \cdot \mathbf{S}_2 + \mathbf{S}_{1\prime} \cdot \mathbf{S}_{2\prime}) - J_{13}(\mathbf{S}_1 \cdot \mathbf{S}_3 + \mathbf{S}_{1\prime} \cdot \mathbf{S}_{3\prime}) - J_{23}(\mathbf{S}_2 \cdot \mathbf{S}_3 + \mathbf{S}_{2\prime} \cdot \mathbf{S}_{3\prime}) \tag{8}$$

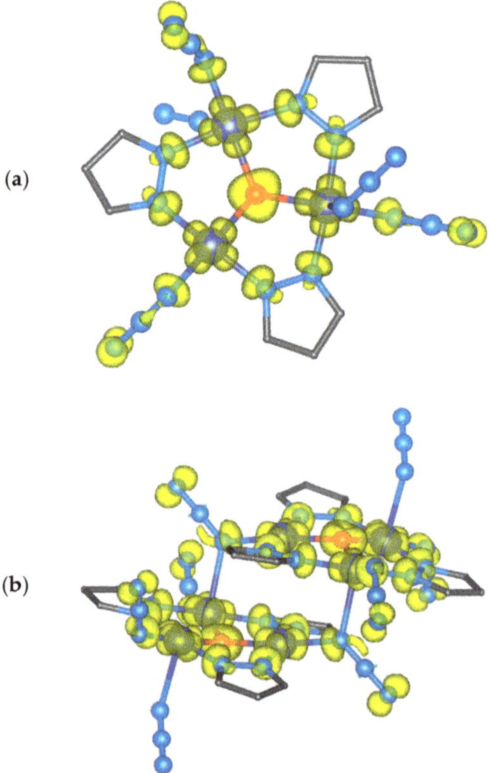

(a)

(b)

Figure 12. The calculated spin density distribution using B3LYP of (2) for the HS states of Cu_3 molecular fragment (a) and Cu_6 fragment (b). The spin density is represented by yellow surfaces. The isodensity surfaces are plotted with the cut-off value of $0.005\ ea_0^{-3}$. Hydrogen atoms are omitted for clarity.

As a result, the energies $\Delta_1 = -337.724$ cm^{-1}, $\Delta_2 = -362.944$ cm^{-1} and $\Delta_3 = -349.454$ cm^{-1} were computed. Next, J-values were calculated by Ruiz's approach, as for (1), resulting in $J_{12} = -176$ cm^{-1}, $J_{13} = -162$ cm^{-1} and $J_{23} = -187$ cm^{-1}. It should also be stressed that in weak interactions, limit J-values would have been two times larger. Afterwards, the hexamer molecular fragment was extracted in order to estimate the magnetic exchange mediated by azido-ligands between two trimeric units:

$$\hat{H} = -J_{12}(\mathbf{S}_1 \cdot \mathbf{S}_2 + \mathbf{S}_{1'} \cdot \mathbf{S}_{2'}) - J_{13}(\mathbf{S}_1 \cdot \mathbf{S}_3 + \mathbf{S}_{1'} \cdot \mathbf{S}_{3'}) - J_{23}(\mathbf{S}_2 \cdot \mathbf{S}_3 + \mathbf{S}_{2'} \cdot \mathbf{S}_{3'}) - J_{23'}(\mathbf{S}_2 \cdot \mathbf{S}_{3'} + \mathbf{S}_{2'} \cdot \mathbf{S}_3) \tag{9}$$

Thus, energies of HS and BS123 states were calculated, leading to $\Delta_{123} = +10.644$ cm^{-1}, from which $J_{23'}$ equals $+5.32$ cm^{-1}.

4. Conclusions

The one-dimensional solid state structures of (**1**) and (**2**) are held together by weak interactions via bridging anions—chlorides and azides, respectively—which are disrupted in solution. Because of the polymeric nature and low symmetry of these materials, the analysis of the magnetic properties has been supported by DFT calculations, suggesting strong antiferromagnetic exchange within Cu$_3$ units and weak ferromagnetic interactions among these units. The predominant antiferromagnetic exchange within the Cu$_3$ units has been unequivocally determined in both cases by the analysis of magnetic susceptibility characteristics. In addition, the operation of an antisymmetric exchange in (**1**) was evident by both magnetometry and EPR spectroscopy. The much weaker inter-Cu$_3$ exchange, ferromagnetic in (**1**) and antiferromagnetic in (**2**), is attributed to the fact that non-magnetic Cu-orbitals are involved at one or both ends. EPR spectroscopy determined the magnitude of dipolar interaction in solid state (**1**), while the possible dipolar interaction between Cu$_3$-units of (**2**) cannot be determined with certainty. The solution EPR spectra of both compounds are clearly distinguished from those in the solid state and are straightforwardly attributed to individual Cu$_3$(μ_3-OH) species of (**2**), and to the presence of isolated mononuclear and trinuclear species in the case of (**1**). Two new reports, describing the magnetic susceptibilities and (in one case) the EPR spectrum of polymeric supramolecular assemblies of copper(II) complexes, appeared in the literature recently [61,62].

Supplementary Materials: The following are available online at http://www.mdpi.com/2624-8549/2/3/39/s1. Table S1: Crystallographic and refinement data for (**1**) and (**2**). Table S2: Selected interatomic distances (Å) and angles (°) for (**1**). Table S3: Selected interatomic distances (Å) and angles (°) for (**2**) Figure S1: ORTEP diagram of (**1**). Figure S2: ORTEP diagram of (**2**). Figure S3: Infrared spectrum of (**2**). Figure S4: Solid-state X-band EPR spectra of (**1**) between 5 and 295 K. Figure S5: Dipolar exchange scheme indicating the symmetry codes of the spins Si (i = 1–4) and the main intermolecular distances of (**1**). The crystallographic data for this paper (CCDC 2005300 and 2005301) can also be obtained free of charge via www.ccdc.cam.ac.uk/data_request/cif, by emailing data_request@ccdc.cam.ac.uk, or by contacting The Cambridge Crystallographic Data Centre, 12 Union Road, Cambridge CB2 1EZ, UK; Fax: +44 1223 336033.

Author Contributions: Conceptualization, R.G.R.; formal analysis, L.M., R.H. and A.K.B.; investigation, K.S., L.M., R.H. and A.K.B.; writing—original draft preparation, K.S., L.M., R.H. and A.K.B.; writing—review and editing, L.M., R.H., A.K.B. and R.G.R. All authors have read and agreed to the published version of the manuscript.

Funding: Research at FIU was funded partially by the National Science Foundation, grant number CHE-1213683. R.H. acknowledges the financial support from institutional sources of the Department of Inorganic Chemistry, Palacky University in Olomouc, Czech Republic, and from the National Programme of Sustainability I (LO1305) of the Ministry of Education, Youth and Sports of the Czech Republic.

Acknowledgments: We thank Philippe Turek for support throughout this study.

Conflicts of Interest: The authors declare no conflict of interest.

References

1. Pascual-Colino, J.; Beobide, G.; Castillo, O.; Lodewyckx, P.; Luque, A.; Pérez-Yáñez, S.; Román, P.; Velasco, L.F. Adenine nucleobase directed supramolecular architectures based on ferrimagnetic heptanuclear copper(II) entities and benzenecarboxylate anions. *J. Inorg. Biochem.* **2020**, *202*, 110865. [CrossRef]
2. González, M.M.; Osiry, H.; Martínez, M.; Rodríguez-Hernández, J.; Lemus-Santana, A.A.; Reguera, E. Magnetic interaction in a 2D solid through hydrogen bonds and π-π stacking. *J. Magn. Magn. Mater.* **2019**, *471*, 70–76. [CrossRef]

3. Pham, C.T.; Nguyen, T.H.; Matsumoto, K.; Nguyen, H.H. CuI/CuII Complexes with Dipicolinoylbis(N,N-diethylthiourea): Structures, Magnetism, and Guest Ion Exchange. *Eur. J. Inorg. Chem.* **2019**, *2019*, 4142–4146. [CrossRef]

4. Chi, Y.-H.; Yu, L.; Shi, J.-M.; Zhang, Y.-Q.; Hu, T.-Q.; Zhang, G.-Q.; Shi, W.; Cheng, P. π–π Stacking and ferromagnetic coupling mechanism on a binuclear Cu(ii) complex. *Dalton Trans.* **2011**, *40*, 1453. [CrossRef] [PubMed]

5. Ehlert, M.K.; Rettig, S.J.; Storr, A.; Thompson, R.C.; Trotter, J. Metal pyrazolate polymers. Part 1. Synthesis, structure, and magnetic properties of the [Cu(pz)2]x polymer. *Can. J. Chem.* **1989**, *67*, 1970–1974. [CrossRef]

6. Ehlert, M.K.; Rettig, S.J.; Storr, A.; Thompson, R.C.; Trotter, J. Metal pyrazolate polymers. Part 2. Synthesis, structure, and magnetic properties of [Cu (4-Xpz) 2] x polymers (where X = Cl, Br, Me, H; pz = pyrazolate). *Can. J. Chem.* **1991**, *69*, 432–439. [CrossRef]

7. Zueva, E.M.; Petrova, M.M.; Herchel, R.; Trávníček, Z.; Raptis, R.G.; Mathivathanan, L.; McGrady, J.E. Electronic structure and magnetic properties of a trigonal prismatic CuII6 cluster. *Dalton Trans.* **2009**, *30*, 5924–5932. [CrossRef] [PubMed]

8. Ferrer, S.; Lloret, F.; Bertomeu, I.; Alzuet, G.; Borrás, J.; García-Granda, S.; Liu-González, M.; Haasnoot, J.G. Cyclic Trinuclear and Chain of Cyclic Trinuclear Copper(II) Complexes Containing a Pyramidal Cu3O(H) Core. Crystal Structures and Magnetic Properties of [Cu$_3$(μ_3-OH)(aaat)$_3$(H2O)$_3$](NO$_3$)$_2$·H$_2$O [aaat = 3-Acetylamino-5-amino-1,2,4-triazolate] and {[Cu$_3$(μ_3-OH)(aat)$_3$(μ_3-SO$_4$)]·6H$_2$O}$_n$ [aat = 3-Acetylamino-1,2,4-triazolate]: New Cases of Spin-Frustrated Systems. *Inorg. Chem.* **2002**, *41*, 5821–5830. [CrossRef]

9. Ferrer, S.; Lloret, F.; Pardo, E.; Clemente-Juan, J.M.; Liu-González, M.; García-Granda, S. Antisymmetric Exchange in Triangular Tricopper(II) Complexes: Correlation among Structural, Magnetic, and Electron Paramagnetic Resonance Parameters. *Inorg. Chem.* **2012**, *51*, 985–1001. [CrossRef]

10. Belinsky, M.I. Hyperfine Splittings in Spin-Frustrated Trinuclear Cu$_3$ Clusters. *Inorg. Chem.* **2004**, *43*, 739–746. [CrossRef]

11. Boča, R.; Dlháň, L.; Mezei, G.; Ortiz-Pérez, T.; Raptis, R.G.; Telser, J. Triangular, Ferromagnetically-Coupled CuII3–Pyrazolato Complexes as Possible Models of Particulate Methane Monooxygenase (pMMO). *Inorg. Chem.* **2003**, *42*, 5801–5803. [CrossRef] [PubMed]

12. Angaridis, P.A.; Baran, P.; Boča, R.; Cervantes-Lee, F.; Haase, W.; Mezei, G.; Raptis, R.G.; Werner, R. Synthesis and Structural Characterization of Trinuclear CuII–Pyrazolato Complexes Containing μ_3-OH, μ_3-O, and μ_3-Cl Ligands. Magnetic Susceptibility Study of [PPN]$_2$[(μ_3-O)Cu$_3$(μ-pz)$_3$Cl$_3$]. *Inorg. Chem.* **2002**, *41*, 2219–2228. [CrossRef] [PubMed]

13. Mathivathanan, L.; Boudalis, A.K.; Turek, P.; Pissas, M.; Sanakis, Y.; Raptis, R.G. Interactions between H-bonded [CuII$_3$(μ_3-OH)] triangles; a combined magnetic susceptibility and EPR study. *Phys. Chem. Chem. Phys.* **2018**, *20*, 17234–17244. [CrossRef] [PubMed]

14. Olguín, J.; Brooker, S. Synthesis of 3- and 5-formyl-4-phenyl-1H-pyrazoles: Promising head units for the generation of asymmetric imine ligands and mixed metal polynuclear complexes. *New J. Chem.* **2011**, *35*, 1242–1253. [CrossRef]

15. Armarego, W.L.F.; Chai, C.L.L. *Purification of Laboratory Chemicals*, 7th ed.; Elsevier/Butterworth-Heinemann: Amsterdam, The Netherlands; London, UK, 2013; ISBN 978-0-12-382161-4.

16. *APEX3*; Bruker AXS Inc.: Madisson, WI, USA, 2017.

17. *SADABS*; Bruker AXS Inc.: Madisson, WI, USA, 2001.

18. Sheldrick, G.M. SHELXT—Integrated space-group and crystal-structure determination. *Acta Crystallogr. Sect. A Found. Adv.* **2015**, *71*, 3–8. [CrossRef]

19. Sheldrick, G.M. Crystal Structure refinement with SHELXL. *Acta Crystallogr. Sect. C Struct. Chem.* **2015**, *71*, 3–8. [CrossRef] [PubMed]

20. Dolomanov, O.V.; Bourhis, L.J.; Gildea, R.J.; Howard, J.A.K.; Puschmann, H. OLEX2: A complete structure solution, refinement and analysis program. *J. Appl. Crystallogr.* **2009**, *42*, 339–341. [CrossRef]

21. Spek, A.L. Structure validation in chemical crystallography. *Acta Crystallogr. D Biol. Crystallogr.* **2009**, *65*, 148–155. [CrossRef]

22. Thorn, A.; Dittrich, B.; Sheldrick, G.M. Enhanced rigid-bond restraints. *Acta Crystallogr. A* **2012**, *68*, 448–451. [CrossRef]

23. Herchel, R.; Boča, R. *Program Polymagnet*; Slovak Technical University: Bratislava, Slovakia, 2006–2020.

24. Neese, F. The ORCA program system. *WIREs Comput. Mol. Sci.* **2012**, *2*, 73–78. [CrossRef]

25. Becke, A.D. Density-functional exchange-energy approximation with correct asymptotic behavior. *Phys. Rev. A* **1988**, *38*, 3098–3100. [CrossRef] [PubMed]

26. Lee, C.; Yang, W.; Parr, R.G. Development of the Colle-Salvetti correlation-energy formula into a functional of the electron density. *Phys. Rev. B* **1988**, *37*, 785–789. [CrossRef] [PubMed]

27. Stephens, P.J.; Devlin, F.J.; Chabalowski, C.F.; Frisch, M.J. Ab Initio Calculation of Vibrational Absorption and Circular Dichroism Spectra Using Density Functional Force Fields. *J. Phys. Chem.* **1994**, *98*, 11623–11627. [CrossRef]

28. Weigend, F.; Ahlrichs, R. Balanced basis sets of split valence, triple zeta valence and quadruple zeta valence quality for H to Rn: Design and assessment of accuracy. *Phys. Chem. Chem. Phys.* **2005**, *7*, 3297–3305. [CrossRef]

29. Neese, F.; Wennmohs, F.; Hansen, A.; Becker, U. Efficient, approximate and parallel Hartree–Fock and hybrid DFT calculations. A 'chain-of-spheres' algorithm for the Hartree–Fock exchange. *Chem. Phys.* **2009**, *356*, 98–109. [CrossRef]

30. Izsák, R.; Neese, F. An overlap fitted chain of spheres exchange method. *J. Chem. Phys.* **2011**, *135*, 144105. [CrossRef] [PubMed]

31. Weigend, F. Accurate Coulomb-fitting basis sets for H to Rn. *Phys. Chem. Chem. Phys.* **2006**, *8*, 1057–1065. [CrossRef]

32. Momma, K.; Izumi, F. VESTA 3 for three-dimensional visualization of crystal, volumetric and morphology data. *J. Appl. Crystallogr.* **2011**, *44*, 1272–1276. [CrossRef]

33. Shi, K.; Mathivathanan, L.; Boudalis, A.K.; Turek, P.; Chakraborty, I.; Raptis, R.G. Nitrite Reduction by Trinuclear Copper Pyrazolate Complexes: An Example of a Catalytic, Synthetic Polynuclear NO Releasing System. *Inorg. Chem.* **2019**. [CrossRef]

34. Casarin, M.; Cingolani, A.; Di Nicola, C.; Falcomer, D.; Monari, M.; Pandolfo, L.; Pettinari, C. The Different Supramolecular Arrangements of the Triangular [Cu$_3$ (μ_3-OH)(μ-pz)$_3$]$^{2+}$ (pz = Pyrazolate) Secondary Building Units. Synthesis of a Coordination Polymer with Permanent Hexagonal Channels. *Cryst. Growth Des.* **2007**, *7*, 676–685. [CrossRef]

35. Direm, A.; Tursun, M.; Parlak, C.; Benali-Cherif, N. Trans-dichlorotetrakis(1H-pyrazole-κN$_2$)copper(II): Synthesis, crystal structure, hydrogen bonding graph-sets, vibrational and DFT studies. *J. Mol. Struct.* **2015**, *1093*, 208–218. [CrossRef]

36. Sun, Y.-J.; Cheng, P.; Yan, S.-P.; Liao, D.-Z.; Jiang, Z.-H.; Shen, P.-W. Synthesis, crystal structure and properties of copper(II) complexes with different axial ligands and substituted pyrazoles. *J. Mol. Struct.* **2001**, *597*, 191–198. [CrossRef]

37. Małecka, M.; Chęcińska, L. Di chloro tetrakis(3-phenyl pyrazole-κN$_2$)copper(II). *Acta Crystallogr. C* **2003**, *59*, m115–m117. [CrossRef] [PubMed]

38. Adak, P.; Das, C.; Ghosh, B.; Mondal, S.; Pakhira, B.; Sinn, E.; Blake, A.J.; O'Connor, A.E.; Chattopadhyay, S.K. Two pseudohalide-bridged Cu(II) complexes bearing the anthracene moiety: Synthesis, crystal structures and catecholase-like activity. *Polyhedron* **2016**, *119*, 39–48. [CrossRef]

39. Ray, M.S.; Ghosh, A.; Bhattacharya, R.; Mukhopadhyay, G.; Drew, M.G.B.; Ribas, J. Different supramolecular hydrogen bond structures and significant changes in magnetic properties in dinuclear μ_2-1,1-N$_3$ copper(II) complexes with very similar tridentate Schiff base blocking ligands. *Dalton Trans.* **2004**. [CrossRef] [PubMed]

40. Boča, R.; Herchel, R. Antisymmetric exchange in polynuclear metal complexes. *Coord. Chem. Rev.* **2010**, *254*, 2973–3025. [CrossRef]

41. Mathivathanan, L.; Al-Ameed, K.; Lazarou, K.; Trávníček, Z.; Sanakis, Y.; Herchel, R.; McGrady, J.E.; Raptis, R.G. A trigonal prismatic Cu6-pyrazolato complex containing a μ_6-F ligand. *Dalton Trans.* **2015**, *44*, 20685–20691. [CrossRef]

42. Moriya, T. Anisotropic Superexchange Interaction and Weak Ferromagnetism. *Phys. Rev.* **1960**, *120*, 91–98. [CrossRef]

43. Kahn, O. *Molecular Magnetism*; VCH: New York, NY, USA, 1993; ISBN 978-1-56081-566-2.

44. Zheng, L.-L.; Leng, J.-D.; Zheng, S.-L.; Zhaxi, Y.-C.; Zhang, W.-X.; Tong, M.-L. Engineering delocalizing π electronic [CuII3(μ_3-OH)(μ-pz)$_3$]$^{2+}$ species into organometallic frameworks by Ag-π coordination. *CrystEngComm* **2008**, *10*, 1467. [CrossRef]

45. Hulsbergen, F.B.; ten Hoedt, R.W.M.; Verschoor, G.C.; Reedijk, J.; Spek, A.L. Synthesis, magnetic properties, and X-ray structure of catena-μ_3-nitrato-O,O',O''-[μ_3-hydroxo-1-nitrato-1,2;1,3;2,3-tris(μ-pyrazolato-N,N')-2,3-bis(pyrazole-N$_2$)tricopper(II) monohydrate]. An unusual chain of trinuclear copper clusters. *J. Chem. Soc. Dalton Trans.* **1983**. [CrossRef]

46. Angaroni, M.; Ardizzoia, G.A.; Beringhelli, T.; Monica, G.L.; Gatteschi, D.; Masciocchi, N.; Moret, M. Oxidation reaction of [{Cu(Hpz)$_2$Cl}$_2$](Hpz = pyrazole): Synthesis of the trinuclear copper(II) hydroxo complexes [Cu$_3$(OH)(pz)$_3$(Hpz)$_2$Cl$_2$]·solv (solv = H$_2$O or tetrahydrofuran). Formation, magnetic properties, and X-ray crystal structure of [Cu$_3$(OH)(pz)$_3$(py)$_2$Cl$_2$]·py (py = pyridine). *J. Chem. Soc. Dalton Trans.* **1990**. [CrossRef]

47. Ferrer, S.; Haasnoot, J.G.; Reedijk, J.; Müller, E.; Biagini Cingi, M.; Lanfranchi, M.; Manotti Lanfredi, A.M.; Ribas, J. Trinuclear N,N-Bridged Copper(II) Complexes Involving a Cu$_3$OH Core: [Cu$_3$(μ_3-OH)L$_3$A(H$_2$O)$_2$]A(H$_2$O)$_x$\{L = 3-Acetylamino-1,2,4-triazolate; A = CF$_3$SO$_3$, NO$_3$, ClO$_4$; x = 0, 2\} Synthesis, X-ray Structures, Spectroscopy, and Magnetic Properties. *Inorg. Chem.* **2000**, *39*, 1859–1867. [CrossRef] [PubMed]

48. Ding, B.; Yi, L.; Cheng, P.; Liao, D.-Z.; Yan, S.-P. Synthesis and Characterization of a 3D Coordination Polymer Based on Trinuclear Triangular CuII as Secondary Building Units. *Inorg. Chem.* **2006**, *45*, 5799–5803. [CrossRef] [PubMed]

49. Ouellette, W.; Prosvirin, A.V.; Chieffo, V.; Dunbar, K.R.; Hudson, B.; Zubieta, J. Solid-State Coordination Chemistry of the Cu/Triazolate/X System (X = F$^-$, Cl$^-$, Br$^-$, I$^-$, OH$^-$, and SO$_4{}^{2-}$). *Inorg. Chem.* **2006**, *45*, 9346–9366. [CrossRef] [PubMed]

50. Liu, X.; de Miranda, M.P.; McInnes, E.J.L.; Kilner, C.A.; Halcrow, M.A. Antisymmetric exchange in two tricopper(II) complexes containing a [Cu$_3$(μ_3-OMe)]$^{5+}$ core. *Dalton Trans.* **2004**. [CrossRef]

51. Liu, J.-C.; Guo, G.-C.; Huang, J.-S.; You, X.-Z. Different Oxidation States of Copper(I, I/II, II) Thiocyanate Complexes Containing 1,2,4-Triazole as a Bridging Ligand: Syntheses, Crystal Structures, and Magnetic Properties of 2-D Polymer CuI(admtrz)SCN, Linear Trinuclear [CuI$_2$CuII(admtrz)$_6$ (SCN)$_2$](ClO$_4$)$_2$, and Triangular Trinuclear [CuII$_3$(admtrz)$_4$(SCN)$_3$(μ_3-OH)(H$_2$O)](ClO$_4$)$_2$·H$_2$O (admtrz = 4-Amino-3,5-dimethyl-1,2,4-triazole). *Inorg. Chem.* **2003**, *42*, 235–243. [CrossRef] [PubMed]

52. Georgopoulou, A.N.; Margiolaki, I.; Psycharis, V.; Boudalis, A.K. Dynamic versus Static Character of the Magnetic Jahn–Teller Effect: Magnetostructural Studies of [Fe$_3$O(O$_2$CPh)$_6$(py)$_3$]ClO$_4$ py. *Inorg. Chem.* **2017**, *56*, 762–772. [CrossRef] [PubMed]

53. Boudalis, A.K.; Rogez, G.; Turek, P. Determination of the Distributions of the Spin-Hamiltonian Parameters in Spin Triangles: A Combined Magnetic Susceptometry and Electron Paramagnetic Resonance Spectroscopic Study of the Highly Symmetric [Cr$_3$O(PhCOO)$_6$(py)$_3$](ClO$_4$)·0.5py. *Inorg. Chem.* **2018**, *57*, 13259–13269. [CrossRef]

54. Ruiz, E.; Cano, J.; Alvarez, S.; Alemany, P. Broken symmetry approach to calculation of exchange coupling constants for homobinuclear and heterobinuclear transition metal complexes. *J. Comput. Chem.* **1999**, *20*, 1391–1400. [CrossRef]

55. Ruiz, E.; Rodríguez-Fortea, A.; Cano, J.; Alvarez, S.; Alemany, P. About the calculation of exchange coupling constants in polynuclear transition metal complexes. *J. Comput. Chem.* **2003**, *24*, 982–989. [CrossRef]

56. Neese, F. Prediction of molecular properties and molecular spectroscopy with density functional theory: From fundamental theory to exchange-coupling. *Coord. Chem. Rev.* **2009**, *253*, 526–563. [CrossRef]

57. Hay, P.J.; Thibeault, J.C.; Hoffmann, R. Orbital interactions in metal dimer complexes. *J. Am. Chem. Soc.* **1975**, *97*, 4884–4899. [CrossRef]

58. Ruiz, E.; Alemany, P.; Alvarez, S.; Cano, J. Toward the Prediction of Magnetic Coupling in Molecular Systems: Hydroxo- and Alkoxo-Bridged Cu(II) Binuclear Complexes. *J. Am. Chem. Soc.* **1997**, *119*, 1297–1303. [CrossRef]

59. Johnston, D.C.; Kremer, R.K.; Troyer, M.; Wang, X.; Klümper, A.; Bud'ko, S.L.; Panchula, A.F.; Canfield, P.C. Thermodynamics of spin $S = 1/2$ antiferromagnetic uniform and alternating-exchange Heisenberg chains. *Phys. Rev. B* **2000**, *61*, 9558–9606. [CrossRef]

60. Eaton, S.S.; More, K.M.; Sawant, B.M.; Eaton, G.R. Use of the ESR half-field transition to determine the interspin distance and the orientation of the interspin vector in systems with two unpaired electrons. *J. Am. Chem. Soc.* **1983**, *105*, 6560–6567. [CrossRef]

61. Monroe, J.C.; Carvajal, M.A.; Deumal, M.; Landee, C.P.; Redemeyer, M.; Turnbull, M.M. Revisiting the Role of Hydrogen Bonding in the Strong Dimer Superexchange of a 2D Copper(II) Halide Honeycomb-Like Lattice: Structural and Magnetic Study. *Inorg. Chem.* **2020**, *59*, 6319–6331. [CrossRef] [PubMed]
62. Matos, C.R.M.; Junior, H.C.S.; D'Amato, D.L.; de Souza, A.C.; Pinheiro, S.; Guedes, G.P.; Ferreira, G.B.; Alves, O.C.; de Almeida, F.B.; Garcia, F.; et al. Spin-frustration with two quasi-degenerated spin states of a copper(II) heptanuclear complex obtained from an amino acid ligand. *Dalton Trans.* **2020**. [CrossRef]

chemistry
MDPI

Article

Hydrogen and Halogen Bond Mediated Coordination Polymers of Chloro-Substituted Pyrazin-2-Amine Copper(I) Bromide Complexes

Aaron Mailman [1,*], Rakesh Puttreddy [1,2], Manu Lahtinen [1], Noora Svahn [1] and Kari Rissanen [1,*]

[1]   Department of Chemistry, University of Jyväskylä, P.O. BOX 35, FI-40014 Jyväskylä, Finland;
      rakesh.puttreddy@tuni.fi (R.P.); manu.k.lahtinen@jyu.fi (M.L.); svahnnoora@gmail.com (N.S.)
[2]   Faculty of Engineering and Natural Sciences, Tampere University, P. O. Box 541, FI-33101 Tampere, Finland
*   Correspondence: aaron.m.mailman@jyu.fi (A.M.); kari.t.rissanen@jyu.fi (K.R.)

Received: 7 May 2020; Accepted: 23 July 2020; Published: 5 August 2020

Abstract: A new class of six mono- (**1**; 3-Cl-, **2**; 5-Cl-, **3**; 6-Cl-) and di-(**4**; 3,6-Cl, **5**; 5,6-Cl-, **6**; 3,5-Cl-) chloro-substituted pyrazin-2-amine ligands (**1–6**) form complexes with copper (I) bromide, to give 1D and 2D coordination polymers through a combination of halogen and hydrogen bonding that were characterized by X-ray diffraction analysis. These Cu(I) complexes were prepared indirectly from the ligands and $CuBr_2$ via an in situ redox process in moderate to high yields. Four of the pyrazine ligands, **1**, **4–6** were found to favor a monodentate mode of coordination to one Cu^I ion. The absence of a C6-chloro substituent in ligands **1**, **2** and **6** supported N1–Cu coordination over the alternative N4–Cu coordination mode evidenced for ligands **4** and **5**. These monodentate systems afforded predominantly hydrogen bond (HB) networks containing a catenated (μ_3-bromo)-Cu^I 'staircase' motif, with a network of 'cooperative' halogen bonds (XB), leading to infinite polymeric structures. Alternatively, ligands **2** and **3** preferred a μ_2-N,N' bridging mode leading to three different polymeric structures. These adopt the (μ_3-bromo)-Cu^I 'staircase' motif observed in the monodentate ligands, a unique single (μ_2-bromo)-Cu^I chain, or a discrete Cu_2Br_2 rhomboid (μ_2-bromo)-Cu^I dimer. Two main HB patterns afforded by self-complimentary dimerization of the amino pyrazines described by the graph set notation $R_2^2(8)$ and non-cyclic intermolecular N–H···N′ or N–H···Br–Cu leading to infinite polymeric structures are discussed. The cooperative halogen bonding between C–Cl···Cl–C and the C–Cl···Br–Cu XB contacts are less than the sum of the van der Waals radii of participating atoms, with the latter ranging from 3.4178(14) to 3.582(15) Å. In all cases, the mode of coordination and pyrazine ring substituents affect the pattern of HBs and XBs in these supramolecular structures.

Keywords: hydrogen bond; halogen bond; pyrazine; chloropyrazine; chloropyrazin-2-amine; copper halide

1. Introduction

Construction of supramolecular structures from small molecules that self-assemble using hydrogen bonds (HBs) and other non-covalent interactions is the ultimate goal of the crystal engineering discipline [1]. Hydrogen bonding, due to the smaller size and easy polarizable nature of H-atom, has become a reliable tool to fabricate highly symmetric and exotic solid-state networks with tunable properties for applications in biology [2], and materials sciences [3]. The H-bonding knowledge gleaned from organic co-crystals have been applied to self-assemble metal-organic/coordination networks [4–7]. In fact, under the "crystal engineering umbrella", the design and synthesis of coordination compounds have received wide attention, due to their intriguing structural topologies [8–10]. Unlike organic co-crystals, engineering inorganic compounds is dependent on two principals; primary coordination sphere (metal-ligand interactions) and secondary coordination sphere (non-covalent interactions) [11].

Despite several factors (e.g., pH, temperature) [12,13] could influence these two "parameters" for structurally diverse outcomes, modulation of networks based on the metal-ions geometry, organic ligands and their functional groups stand out in coordination chemistry research [14–21]. This fact is due to reproducible outcomes, that reflect the strong metal-ligand coordination bonds, and can be achieved with the judicious choice of organic ligands and metal-ions [14–21]. In this context, for example, O- and N-atoms are typical donors for the coordination bond formation; the former are derived from functional groups such as –COOH, –SO$_3$H and phosphonates [22–24], and the latter primarily from N-heterocycles [25]. The combined use of these two groups render ligands a strong coordinating ability, and are well-known for the preparation of homometallic and heterometallic coordination compounds [25].

In this regard, aminopyrazine carboxylic acids (L), of the forms L, and L$^+$, for the construction of hydrogen-bonded organic co-crystals [26–31], and L$^-$ functioning as a polydentate ligand in the preparation of coordination compounds [32–39] are reported in the literature. The presence of aromatic N-atoms and carboxylic acid/carboxylate groups within the same ligand may enhance the N–M bond strengths, often assisted by the polydentate bonding nature of O-atoms [26]. However, if the –COOH group is replaced by an aprotic donor substituent such as chlorine, and in combination with copper halides, what will happen to N-atom coordination nature? Will the chlorine substituents and metal-bound halides establish halogen bonds (XBs) [40–44], and halogen···halogen interactions [45–48]? This knowledge of XBs in metal complexes is derived from our previous experiences in halopyridine-Cu(I)/Cu(II) compounds [49–52]. When the bulky chlorines are installed close to an N-atom, can this affect the N–M coordination? How do the substituents mediate a hybrid topology containing –HN–H···N$_{pz}$ (pz = pyrazine) hydrogen bonds and C–Cl···Br halogen bonds? To test our hypothesis, we synthesized six chloro-substituted pyrazin-2-amine ligands (**1–6**) using procedures reported in the literature [53–57], and each ligand was combined with CuBr$_2$ in a 1:2 metal:ligand ratio. The CuBr complexes of these ligands were obtained by exploiting the known redox activity of Cu(II) halides in the presence of organic carbonyl compounds [31]. This work represents the first systematic study of the metal-ligand, HB, and XB interactions of chloro-substituted pyrazin-2-amines with copper halides and recent results will be discussed.

2. Materials and Methods

All reagents and solvents, including **2**, were obtained from commercial suppliers and were of at least reagent grade, and used as received, unless otherwise stated. The ligands **1** and **3–6** were prepared by modified literature procedures [53–58]. Full experimental, FT-IR spectroscopic details and the single-crystal [59–66] and powder X-ray experimental [67–69], and structure refinements for **1**·CuBr–**6**·CuBr (CCDC numbers: 2001484–2001491) are given in the Supporting Information.

3. Results and Discussion

This new class of chloro-substituted pyrazin-2-amine ligands given in Chart 1 were made to react with cupric bromide (CuBr$_2$), either by refluxing in a 1:1:1 (*v/v/v*) mixture of acetone (Ace), ethanol (EtOH) and acetonitrile (MeCN) or 1:5 (*v/v*) mixture of Ace:EtOH to afford the desired Cu(I)Br complexes, via an in situ redox process, according to Scheme 1. An excess of pyrazine ligand was used to act as ligand and auxiliary base for the HBr liberated during the reduction of CuBr$_2$ to CuBr by acetone [70]. The reduction was accompanied by a color change from a deep blue-green solution characteristic of Cu(II) ions, to a clear, yellow solution upon refluxing for several minutes. In total, eight different Cu(I)Br coordination complexes obtained from the combination of these six ligands will be discussed.

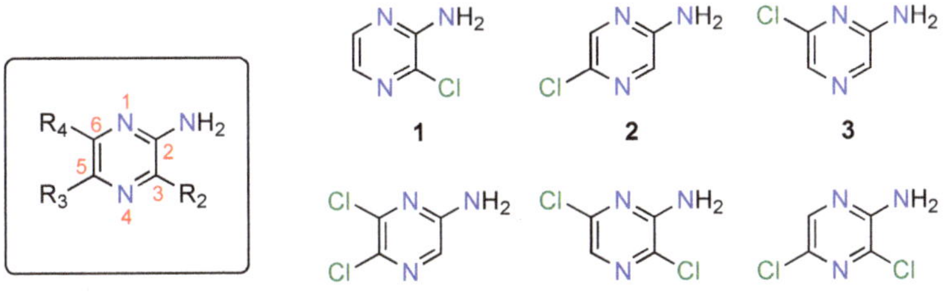

Chart 1. List of chloro-substituted pyrazin-2-amines and numbering scheme used throughout: 2-amino-3-chloropyrazine (**1**), 2-amino-5-chloropyrazine (**2**), 2-amino-6-chloropyrazine (**3**), 2-amino-5,6-dichloropyrazine (**4**), 2-amino-3,6-dichloropyrazine (**5**), and 2-amino-3,5-dichloropyrazine (**6**).

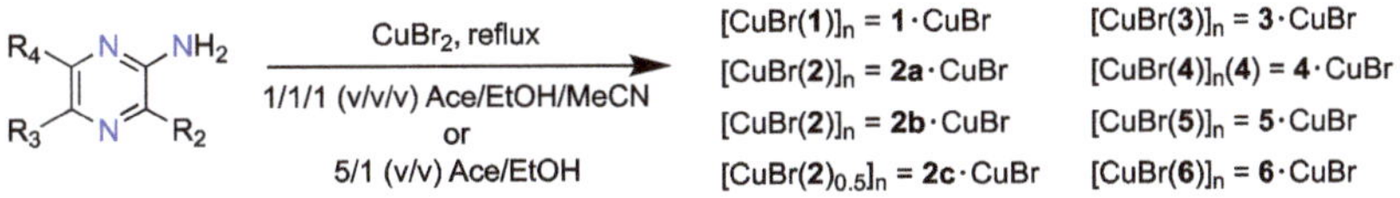

$[CuBr(\mathbf{1})]_n = \mathbf{1}\cdot CuBr \qquad [CuBr(\mathbf{3})]_n = \mathbf{3}\cdot CuBr$

$[CuBr(\mathbf{2})]_n = \mathbf{2a}\cdot CuBr \qquad [CuBr(\mathbf{4})]_n(\mathbf{4}) = \mathbf{4}\cdot CuBr$

$[CuBr(\mathbf{2})]_n = \mathbf{2b}\cdot CuBr \qquad [CuBr(\mathbf{5})]_n = \mathbf{5}\cdot CuBr$

$[CuBr(\mathbf{2})_{0.5}]_n = \mathbf{2c}\cdot CuBr \qquad [CuBr(\mathbf{6})]_n = \mathbf{6}\cdot CuBr$

Scheme 1. General synthetic route to Cu(I)-complexes of ligands **1–6** via an in situ redox process from $CuBr_2$, and the nomenclature used.

X-ray quality crystals of the complexes were obtained by slowly concentrating the reaction mixtures by controlled evaporation. Interestingly, **2c**·CuBr was only obtained from a 1:5 (*v/v*) Ace and EtOH reaction mixture, together with **2b**·CuBr. The concomitant polymorphism did not allow the isolation of **2c**·CuBr as a phase pure material; however, **2b**·CuBr could be prepared independently by alternative methodologies in moderate yield. Moreover, attempts to recrystallize the mixture of **2c**·CuBr and **2b**·CuBr from MeCN afforded only complexes **2a**·CuBr and **2b**·CuBr. This suggests the **2c**·CuBr is only somewhat stable when prepared from a 1:5 (*v/v*) Ace:EtOH mixture and readily dissociates in the polar aprotic solvent MeCN. Optical microscopy could be routinely used to distinguish the different color and crystal habits of the three different polymorphs of (**2a–c**) CuBr (See Figure S1). Powder X-ray diffraction methods were used to demonstrate the solid-state structures and estimate the phase purity of the bulk material by using Pawley full pattern fittings. The FT-IR spectra of the isolated complexes were compared to the respective ligands, and display two specific regions from 3500–2800 cm^{-1} and 1200–400 cm^{-1} for the hydrogen-bonded N–H stretching [71] and fingerprint regions, respectively, which are distinctive for each ligand and complex (see Figures S10–S16).

In general, reactions between donor ligands (L) and CuX (X = Cl, Br, I) yield complexes with formula $Cu_nX_nL_m$, that display diverse structural topologies, containing rhomboid dimer, zigzag polymer, staircase polymer, closed cubane, and hexagon clusters, have been reported in the literature [72]. The structural diversification stems from the very nature of the ligand coordination modes and the tendency of copper halides to form clusters via μ_2- and μ_3-halide bridges [73]. In our eight complexes, we obtained three topologies exclusively, namely staircase polymer (**1**·CuBr, **2a**·CuBr, **2c**·CuBr, **4**·CuBr, **5**·CuBr, and **6**·CuBr), rhomboid dimer (**2b**·CuBr), and zigzag polymer (**3**·CuBr), as depicted in Figure 1. The staircase and rhomboid structures feature well-known Cu···Cu distances (Cuprophilic interactions) [74], ranging from 2.7547(13) to 3.086(4) Å, but will not be discussed further in the text.

The complexes **1**·CuBr and **2a**·CuBr both crystallize as clear, pale-yellow or colorless needles in an orthorhombic space group, $Pna2_1$ and $P2_12_12_1$, respectively. Both contain discrete 1D polymeric chains of the catenated (μ_3-bromo)-Cu^I 'staircase' that run parallel to the *c*-axis and *a*-axis in **1**·CuBr and **2a**·CuBr, respectively. In these chains, the Cu(I) ions have tetrahedral geometry, coordinated by three bromines and the pyrazine N1-atom. Since the pyrazine N4-atom is not involved in the N–Cu bond formation, it plays a central role in the HB formation involving a C2-amino group of an adjacent chain. The orthogonal arrangement of the discrete 1-D polymeric chains leads to a herringbone packing structure in both **1·CuBr** and **2a·CuBr**, as illustrated in Figure 2. The N4···H–N HBs in **1·CuBr** of 2.245(11) Å, [∠N4···H–N = 155.6(9)°] are shorter than in **2a·CuBr** [2.349(7) Å, ∠N4···H–N = 133.4(5)°]. The XBs in **1·CuBr** and **2a·CuBr** complexes are, however, more comparable at 3.465(3) Å [∠Br···Cl–C = 165.4(3)°] and 3.509(3) Å [∠Br···Cl–C = 159.5(3)°], respectively. This network of HBs and XBs does lead to a complex molecular packing that extends in three dimensions.

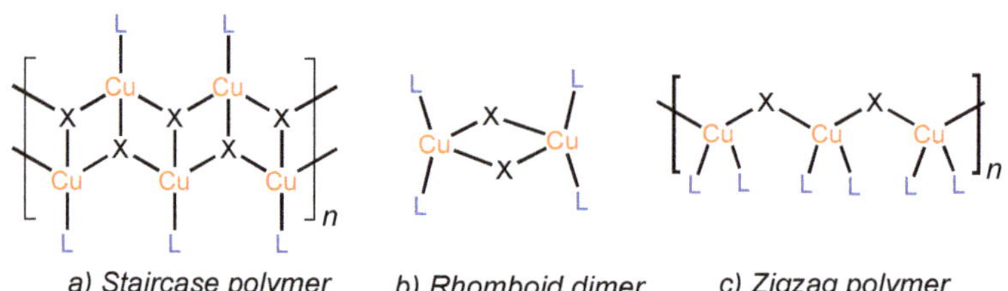

Figure 1. The three structural topologies realized in the Cu(I)-complexes of **1–6**, (**a**) catenated (μ₃-bromo)-Cuᴵ 'staircase' polymer, (**b**) Cu₂Br₂ rhomboid (μ₂-bromo)-Cuᴵ dimers, and (**c**) zigzag polymer (μ₂-bromo)-Cuᴵ chains.

The complex **6·CuBr** crystallizes in a monoclinic *P*2₁/*c* space group, and is composed of 1D polymeric chains of the catenated (μ₃-bromo)-Cuᴵ 'staircase' that run along the shortest unit cell *a*-axis. The asymmetric unit contains one bromide anion and a one Cu(I) cation coordinated by the sterically less hindered N1-atom of **6**. Not surprisingly, the C3 and C5 chloro-substituents vicinal to the N4-atom render this ring nitrogen Cu-coordination passive. The orthogonal arrangement of the discrete 1D polymeric chains leads to a herringbone packing structure that is similar to **1·CuBr** and **2a·CuBr** (for **6·CuBr**, see Figure S2). The 1D chains are aligned by a more extensive network of XBs, afforded by the C3- and C5-Cl substituents (C3–Cl···Br–Cu [3.4794(14) Å, ∠Br···Cl–C = 162.44(19)°], C5–Cl···Br–Cu [3.4178(14) Å, ∠Br···Cl–C = 175.34(19)°]) and longer N4···H–N of 2.511(4) Å, [∠N4···H–N = 135.8(3)°] HBs, compared to **1·CuBr** and **2a·CuBr**.

The N1–Cu mode of coordination realized in **1·CuBr**, **2a·CuBr**, and **6·CuBr** allows the vicinal C2-amino group hydrogen and a Cu(I) bound bromide to form cyclic N–H···Br–Cu HBs, with distances varying from ca. 2.609(7)–2.774(2) Å [∠Br···H–N = 146(1)°–168.9(4)°], within the discrete polymeric (μ₃-bromo)-Cuᴵ 'staircase' chain (See Figure S3). These cyclic intra-chain N–H···Br HBs afford a pseudo-helical type arrangement of the pyrazine ligands along the polymeric (μ₃-bromo)-Cuᴵ 'staircase' backbone. The HBs, in combination with the C–Cl···Br–Cu XBs from neighboring chains, aids the stabilization of the catenated (μ₃-bromo)-Cuᴵ 'staircase' polymeric chains in these complexes.

The complexes **4·CuBr** and **5·CuBr** crystallize in a triclinic *P*-1 and monoclinic *P*2₁/n space groups, respectively. In both complexes, the catenated (μ₃-bromo)-Cuᴵ 'staircase' motif is preserved, however, the tetrahedral Cu(I) centers are coordinated by three bromides and a pyrazine N4-atom. In the case of **4·CuBr**, it crystallizes with an additional pyrazine molecule in the asymmetric unit that is not involved in coordination with the Cu(I) centers. The N4–Cu coordination mode found in both complexes allows the C2-amino group to participate in hydrogen-bonding with a neighboring pyrazine in a self-complimentary N–H···N_{pz} pyrazine dimer. The resultant hydrogen-bonded dimers in **4·CuBr** and **5·CuBr** are described by the graph set notation R₂²(8), with HB parameters of ca. 2.308(12) Å, [166.4(7)°] and 2.138(5) Å [160.1(3)°], respectively (Figure 3). These Watson–Crick-like base pairing structures are characteristic of aminopyrazine derivatives [75,76]. The R₂²(8) hydrogen bonding in **4·CuBr** and **5·CuBr** leads to essentially linear chains that pack into laminar 2D polymeric sheets held together by the polymeric (μ₃-bromo)-Cuᴵ 'staircase' motif. The non-coordinating pyrazine in **4·CuBr** forms an additional set of R₂²(8) hydrogen-bonded dimers that cross-link the 2D sheets via non-cyclic

N–H⋯N-ring hydrogen bonds [2.219(11) Å ∠N–H⋯N4, 177.7(8)°], into a 3D structure shown in Figure 3a. Although **4**·CuBr and **5**·CuBr lack the intra-chain cyclic N–H⋯Br–Cu HBs (see Figure S4) found in **1**·CuBr, **2a**·CuBr, and **6**·CuBr, a network of non-cyclic inter-chain N–H⋯Br–Cu and C–H⋯Br–Cu HBs are established between neighboring chains in **5**·CuBr or the non-complexed pyrazine in **4**·CuBr. Overall, the N–H⋯Br–Cu and C–H⋯Br–Cu HBs together with the C–Cl⋯Br–Cu, and C–Cl⋯Cl–C XBs stabilize the polymeric μ₃-bromo 'staircase', and further cross-link the laminar 2D polymeric sheets into complex 3D structures. Notwithstanding the structural similarities between **4**·CuBr and **5**·CuBr, as determined by single-crystal X-ray diffraction, the solid-state structure of **4**·CuBr could not be demonstrated in the bulk material, as determined by powder X-ray diffraction. Full pattern Pawley analysis of **4**·CuBr suggests that the bulk material is effectively isomorphous with **6**·CuBr (See Table S6 and Figure S8). This suggests that additional polymorphs may exist where the Cu(I) center is coordinated by the sterically more congested N1 in **4**. This possibility has also been observed in **3**·CuBr, and polymorphism has been demonstrated in **2(a–c)**·CuBr, as described below. Despite repeated attempts, under different conditions, alternative single-crystal structures of Cu(I) complexes of **4** have still not been realized.

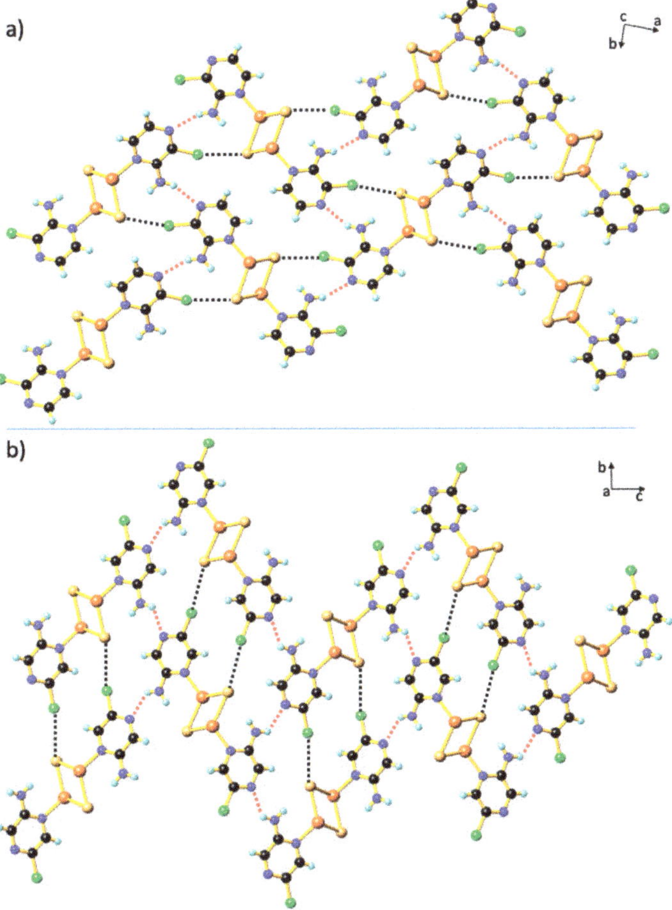

Figure 2. The herringbone packing arrangement of the discrete 1D polymeric chains of (**a**) **1**·CuBr, (**b**) and **2a**·CuBr. The red- and black- dotted lines represent hydrogen and halogen bonds, respectively.

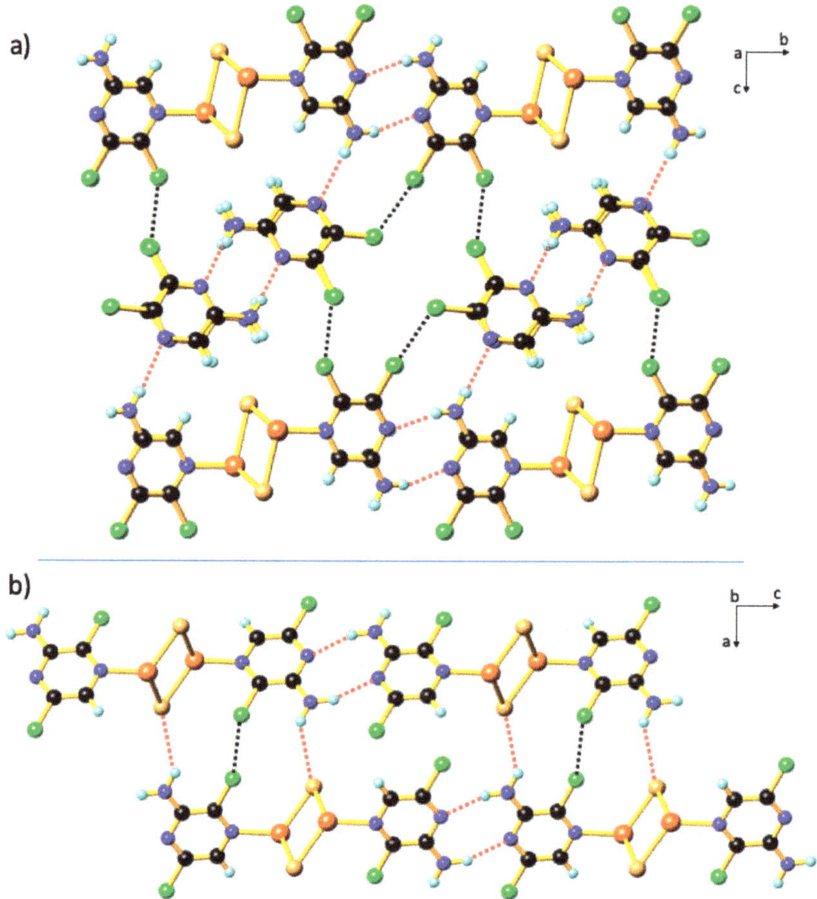

Figure 3. The $R_2^2(8)$ hydrogen-bonded ring motif realized in the crystal packing of (**a**) **4**·CuBr, and (**b**) **5**·CuBr. Red dotted lines are N–H···N $R_2^2(8)$ and N–H···Br–Cu hydrogen bonds, and black are C–Cl···Cl–Cu halogen bonds.

The three polymeric structures **2b**, **2c**·and **3**·CuBr all contain bridging μ_2-*N,N′* pyrazine ligands, and display the three (CuBr)$_n$ structural motifs outlined in Figure 1. Complex **2b**·CuBr crystallizes in the orthorhombic space group *Pbcn* as amber, prismatic blocks (See Figure S1a). In this complex, the pyrazine ligands are μ_2-*N,N′* bridging between Cu$_2$Br$_2$ rhomboid dimer units to afford a distorted tetrahedral coordination geometry at the Cu(I) ion formed by two symmetry-equivalent bromines, and the N1- and N4-atoms of two different pyrazine ligands. This leads to a 2D dimensional honey-comb sheet that propagates in the *ab* plane, as illustrated in Figure 4a. The honey comb structure is similar to the [CuCl(μ-2,5-dimethylpyrazine-*N,N′*)]$_n$ structure reported in the literature [77]. Adjacent layers are held together by C5–Cl···Br–Cu [3.5105(7) Å, ∠Br···Cl–C = 161.83(8)°] XBs and N–H···Br–Cu [2.6402(2) Å, ∠Br···H–N = 159.28(14)°] HBs, as given in Figure 4b. Moreover, these HBs and XBs, together with steric effects, prevent the formation of a catenated μ-bromo-Cu^I motif, as in the staircase polymers previously mentioned.

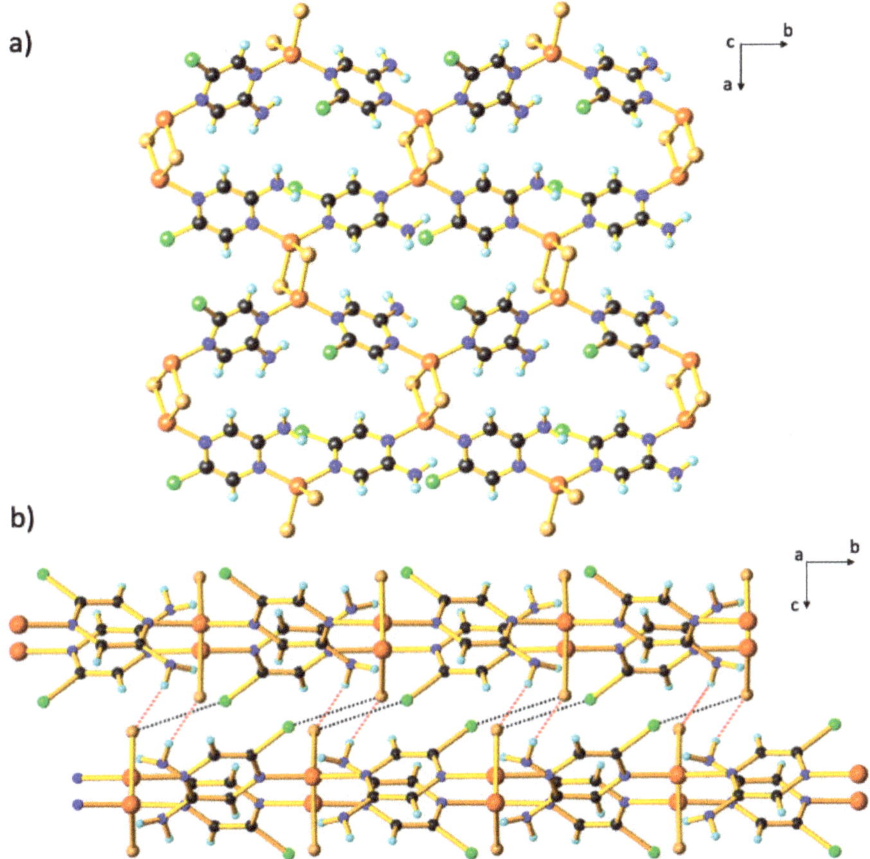

Figure 4. (**a**) 2-D Honey comb motif of **2b**·CuBr, and (**b**) 3-D packing unit displaying hydrogen (red dotted lines) and halogen bonds (black dotted lines) between 2-D sheets.

The bright yellow, needles of **2c**·CuBr crystallize in the monoclinic space group $P2_1/n$. The asymmetric unit consists of a Cu(I) ion; one bromide and one half of a pyrazine ligand. The C5-chlorine and C2-amine substituents exhibit a positional disorder with 50:50 occupancies. The distorted tetrahedral coordination environment of Cu(I) consists of one pyrazine nitrogen and three symmetry equivalent bromide anions. Ligand **2** is μ_2-*N,N′* bridging to yield a 2D polymeric sheet containing the catenated (μ_3-bromo)-Cu^{I} 'staircase' chain structure, as shown in Figure 5. The polymeric structure of **2c**·CuBr is similar to that found in the 2D coordination polymer of 2-aminopyrazine with CuI (i.e., [Cu$_2$I$_2$(2-aminopyrazine)]$_n$) [78] and [2(μ-2,5-dimethylpyrazine-*N,N′*)Cu$_2$]$_n$ where X = Br, I [77]. The adjacent sheets are linked by N–H···Br–Cu [2.6054(17) Å, ∠Br···H–N = 175(4)°] interactions between the C2-amino group and a Cu(I)-bound bromide. As a result of positional disorder, the C–Cl bond length associated with the 50% occupancy is close to the standard C–Cl bond distance, and results in a weak C5–Cl···Br–Cu [3.582(15) Å, ∠Cl···Br–Cu = 103.9(9)°] 'contact'. These weak interactions are likely triggered by the N–H···Br–Cu interactions between the neighboring molecules in this laminar structure. Due to the disorder, there are possible HB and XB interactions that could alternate along a plane that passes through the amino- and chloro-substituents of the pyrazines of adjacent sheets, thus forming moderately short Cl···Cl halogen contacts with R_{XB} = 0.88. These HBs and XB interactions further supports the catenated (μ_3-bromo)-Cu^I 'staircase' motif as a common structural feature in these complexes.

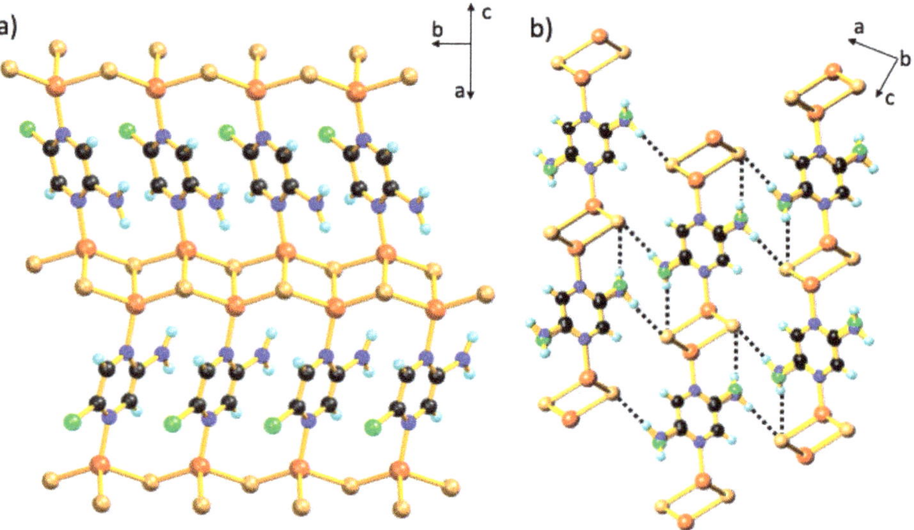

Figure 5. (**a**) Here, 2D polymeric sheet structure of **2c**·CuBr, and (**b**) the hydrogen and halogen bonds (black dotted lines) between sheets in the laminar structure. Note: the disordered atoms were not omitted to reflect the discussion.

The complex **3**·CuBr, crystallized in the monoclinic space group $P2_1/n$. The distorted tetrahedral coordination environment of the Cu(I) is composed of two symmetry-related bromides and two crystallographically different N-atoms from two distinct pyrazine ligands. The Cu(I) ions are bridged by μ_2-*N,N'* pyrazine ligands, forming 1D linear chains that run along the *c*-axis as illustrated in Figure 6a. These chains are further linked into a 2D sheet structure by polymeric (μ_2-bromo)-Cu^I-Br chains along the *a*-axis. The 2D layers are not flat, but instead adopt a slightly corrugated pattern because of the $\angle$N–Cu–N = 136.6(4)° at the distorted tetrahedron at the Cu(I) ion, as shown in Figure 6b. Two types of interactions assemble the layers into a 3D structure. First is the hydrogen bonding between a C2-amino group and Cu(I)-bound bromide from an adjacent layer [2.3615(10) Å, $\angle$Br···H–N = 164.2 (7)°; 2.4846(12) Å, $\angle$Br···H–N = 158.9(6)°], to afford a $R_2^2(8)$ ring motif from a set of two N–H···Br HBs. The second interaction is C6–Cl···Br–Cu halogen bonds [3.527(3) Å, $\angle$Br···Cl–C = 158.8(4)°] between a chlorine on the pyrazine ring and bromide coordinated to Cu(I) on an adjacent layers, as shown in Figure 6b. Similar to the (μ_3-bromo)-Cu^I 'staircase' motif, the μ_2-bromo polymeric chains found in **3**·CuBr is fully stabilized by the network of HB and XBs, and is likely adopted because of the steric demands of the N1 coordination, compared to the more accessible N4-coordination mode in ligand **3**.

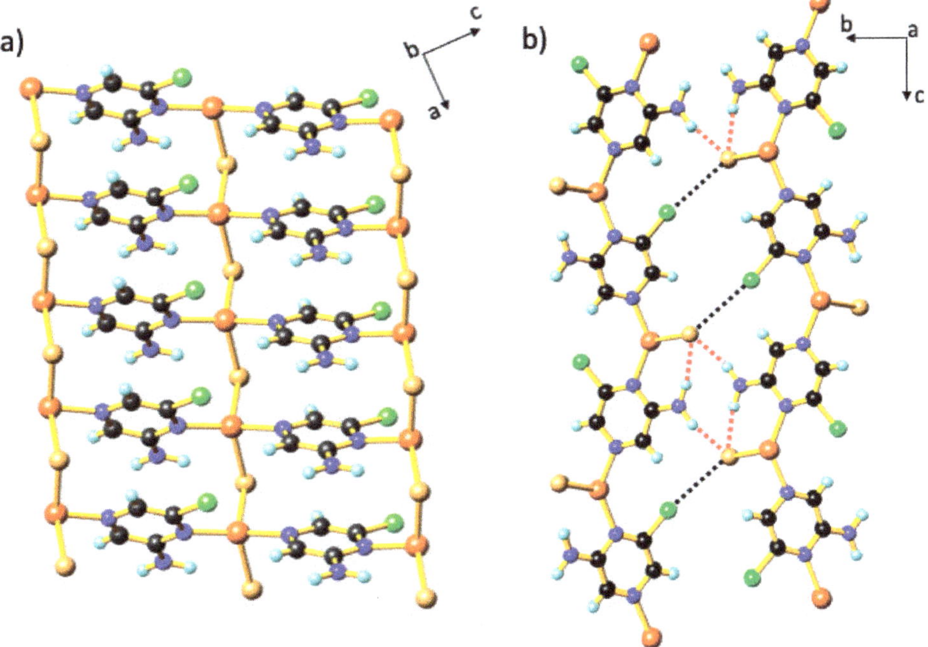

Figure 6. (**a**) Here, 2D Layered structure of **3**·CuBr, and (**b**) 2-D layers stack viewed along the a-axis. Red dotted lines are N–H···Br–Cu hydrogen bonds, and black are C6–Cl···Br–Cu halogen bonds.

4. Conclusions

In summary, eight Cu(I)Br coordination polymers have been synthesized, and their single-crystal structures characterized by using X-ray diffraction analysis. Three different types of (CuBr)$_n$ arrangements with n ≥ 2 were realized. Six of the eight complexes form characteristic catenated μ_2-bromo staircase polymers that are reminiscent of those reported in related copper(I) halide complexes. Our findings indicate that a combination of cyclic and non-cyclic hydrogen bonds (HBs), between the C2-amino group and Cu(I) bound bromides (N–H···Br–Cu) and halogen bonding (C–Cl···Br–Cu) interactions, underpins the formation of catenated μ_2- and μ_3-bromo (CuBr)$_n$ polymeric structures.

It was found that the mode of coordination by the dichloro-substituted aminopyrazines **4–6** to the Cu(I) could be readily predicted based on simple electronic and steric arguments. In the mono chloro-substituted pyrazines **1–3**, these effects were less obvious, as evidenced by the realization of both monodentate (**1**·CuBr and **2a**·CuBr), and polymeric structures (**2b**·CuBr, **2c**·CuBr and **3**·CuBr). The polymeric **3**·CuBr suggests that there is little steric hindrance afforded by the combination of a C2-amino and C6-chloro substituents vicinal to the N1–Cu coordination site, and may indicate that this is not a major factor in determining N1-Cu coordination passiveness in **4**·CuBr and **5**·CuBr. This could also account for the unique polymeric (μ_2-bromo)-Cu^I chain in **3**·CuBr, which is able to accommodate the more sterically demanding N1–Cu coordination, and allows stabilizing cyclic N–H···Br HBs.

Three polymorphs were isolated and structurally characterized based on the simple 2-amino-5-chloropyrazine ligand, **2**. We demonstrate that concomitant polymorphism leads to both 1D and 2D polymeric structures based on the μ_3-bromo staircase polymer motif in **2a**·CuBr and **2c**·CuBr, respectively. The third polymorph, **2b**·CuBr, displays a unique honeycomb network composed of discrete Cu$_2$Br$_2$ rhomboid (μ_2-bromo)-Cu^I dimers and μ_2-N,N′ bridging ligands. These three polymorphs demonstrate the role of the C2-amino group in forming three types of hydrogen bonding patterns recognized in all eight of these complexes. The self-complementary, symmetry-related N–H···N$_{pz}$ pyrazine dimers described by the graph set notation $R_2^2(8)$ with distances varying from ca. 2.138(5)–2.308(12) Å [$\angle$Br$\cdots$H–N = 160.1(3)°–177.7(8)°] were found in N4–Cu coordination structures **4**·CuBr and **5**·CuBr, and resulted in the formation of laminar 2D sheet structures. Based on the competitive binding sites in the simple mono-substituted aminopyrazine ligands, and the role of the amino-group to influence structural features through hydrogen bonding, we are currently investigating the formation of dihalopyrazine-copper complexes.

The Cu(I) halide complexes have attracted considerable attention because of their physical properties and functional applications in future technologies, such as solar energy conversion, light emitting devices, and possible sensing applications. As a result, there has been considerable focus on exploring and developing structural relationships that will provide insights for future materials design and synthesis. The interesting 2D-honeycomb network found in **2b**·CuBr is a particularly appealing architecture that could have applications in gas storage or separation, if the porosity can be tuned through ligand design. While systematic structural studies on Cu(I) halide and pyrazine derivatives have not received much attention to date, this work demonstrates a level of structural predictability not typically observed in Cu(I) halide complexes. We believe future investigations of halo-substituted aminopyrazines and related pyrazine derivatives will provide the basis for an empirical structural database that will aid future material designs.

Supplementary Materials: The following are available online at http://www.mdpi.com/2624-8549/2/3/45/s1, Full experimental, the single-crystal experimental and CCDC number details for **1**·CuBr–**6**·CuBr (Tables S1–S3), structure refinements for **1**·CuBr–**6**·CuBr, crystal habits of **2(a,b,c)**·CuBr (Figure S1), powder X-ray diffraction analysis (Tables S4–S6, Figures S5–S9), FT-IR spectroscopic details (Figures S10–S16), and X-ray crystal figures (Figures S2–S4).

Author Contributions: The manuscript was written and edited through the contributions of all authors. R.P. and A.M. were responsible for the original concept, methodology development and SCXRD analysis. The ligand synthesis was carried out by N.S., while the PXRD data collection, analysis and Pawley fittings were performed by M.L. K.R. and A.M. were responsible for proof reading and the final manuscript version. All authors have read and agreed to the published version of the manuscript.

Funding: This work was financially supported by the Academy of Finland (projects 298817 and 289172), the University of Jyväskylä, and in part by the European Union's H2020 program, under the Marie Skłodowska-Curie grant agreement 659123 for A.M.

Acknowledgments: The authors gratefully acknowledge the University of Jyväskylä for providing laboratory and SCXRD resources.

Conflicts of Interest: The authors declare no conflict of interest.

Reference and Note

1. Desiraju, G.R.; Vittal, J.J.; Ramanan, A. *Crystal Engineering: A Textbook*; World Scientific: Singapore, 2011.
2. Jeffrey, G.A.; Saenger, W. *Hydrogen Bonding in Biological Structures*; Springer: Berlin/Heidelberg, Germany, 1991.
3. Amabilino, D.B.; Smith, D.K.; Steed, J.W. Supramolecular materials. *Chem. Soc. Rev.* **2017**, *46*, 2404–2420. [CrossRef] [PubMed]
4. Biradha, K.; Desiraju, G.R.; Braga, D.; Grepioni, F. Hydrogen Bonding in Organometallic Crystals. 3.1Transition-Metal Complexes Containing Amido Groups. *Organometallics* **1996**, *15*, 1284–1295. [CrossRef]
5. Brammer, L.; Rivas, J.C.M.; Atencio, R.; Fang, S.; Pigge, F.C. Combining hydrogen bonds with coordination chemistry or organometallic π-arene chemistry: Strategies for inorganic crystal engineering. *J. Chem. Soc. Dalton Trans.* **2000**, 3855–3867. [CrossRef]
6. Beatty, A.M. Hydrogen bonded networks of coordination complexes. *CrystEngComm* **2001**, *3*, 243–255. [CrossRef]
7. Epstein, L.M.; Shubina, E.S. New types of hydrogen bonding in organometallic chemistry. *Coord. Chem. Rev.* **2002**, *231*, 165–181. [CrossRef]
8. Nair, K.P.; Pollino, J.M.; Weck, M. Noncovalently Functionalized Block Copolymers Possessing Both Hydrogen Bonding and Metal Coordination Centers. *Macromolecules* **2006**, *39*, 931–940. [CrossRef]

9. Chandrasekhar, P.; Mukhopadhyay, A.; Savitha, G.; Moorthy, J.N. Orthogonal self-assembly of a trigonal triptycene triacid: Signaling of exfoliation of porous 2D metal–organic layers by fluorescence and selective CO2 capture by the hydrogen-bonded MOF. *J. Mater. Chem. A* **2017**, *5*, 5402–5412. [CrossRef]

10. Aakeroy, C.B.; Beatty, A.M. Solid State, Crystal Engineering and Hydrogen Bonds. *ChemInform* **2004**, *35*. [CrossRef]

11. Cook, S.A.; Borovik, A.S. Molecular Designs for Controlling the Local Environments around Metal Ions. *Acc. Chem. Res.* **2015**, *48*, 2407–2414. [CrossRef]

12. Friese, V.A.; Kurth, D.G. Soluble dynamic coordination polymers as a paradigm for materials science. *Coord. Chem. Rev.* **2008**, *252*, 199–211. [CrossRef]

13. Seetharaj, R.; Vandana, P.V.; Arya, P.; Mathew, S. Dependence of solvents, pH, molar ratio and temperature in tuning metal organic framework architecture. *Arab. J. Chem.* **2019**, *12*, 295–315. [CrossRef]

14. Zhu, R.; Lübben, J.; Dittrich, B.; Clever, G.H. Stepwise Halide-Triggered Double and Triple Catenation of Self-Assembled Coordination Cages. *Angew. Chem. Int. Ed.* **2015**, *54*, 2796–2800. [CrossRef]

15. Sun, Q.-F.; Iwasa, J.; Ogawa, D.; Ishido, Y.; Sato, S.; Ozeki, T.; Sei, Y.; Yamaguchi, K.; Fujita, M. Self-Assembled M24L48 Polyhedra and Their Sharp Structural Switch upon Subtle Ligand Variation. *Science* **2010**, *328*, 1144–1147. [CrossRef]

16. Schultz, A.; Li, X.; Barkakaty, B.; Moorefield, C.N.; Wesdemiotis, C.; Newkome, G.R. Stoichiometric Self-Assembly of Isomeric, Shape-Persistent, Supramacromolecular Bowtie and Butterfly Structures. *J. Am. Chem. Soc.* **2012**, *134*, 7672–7675. [CrossRef]

17. Li, S.; Moorefield, C.N.; Wang, P.; Shreiner, C.D.; Newkome, G.R. Self-Assembly of Shape-Persistent Hexagonal Macrocycles with Trimeric Bis(terpyridine)–FeII Connectivity. *Eur. J. Org. Chem.* **2008**, *2008*, 3328–3334. [CrossRef]

18. Tominaga, M.; Suzuki, K.; Kawano, M.; Kusukawa, T.; Ozeki, T.; Sakamoto, S.; Yamaguchi, K.; Fujita, M. Finite, Spherical Coordination Networks that Self-Organize from 36 Small Components. *Angew. Chem. Int. Ed.* **2004**, *43*, 5621–5625. [CrossRef]

19. Ronson, T.K.; Fisher, J.; Harding, L.P.; Rizkallah, P.J.; Warren, J.E.; Hardie, M.J. Stellated polyhedral assembly of a topologically complicated Pd4L4 'Solomon cube'. *Nat. Chem.* **2009**, *1*, 212–216. [CrossRef]

20. Olenyuk, B.; Whiteford, J.A.; Fechtenkötter, A.; Stang, P.J. Self-assembly of nanoscale cuboctahedra by coordination chemistry. *Nature* **1999**, *398*, 796–799. [CrossRef]

21. Song, B.; Kandapal, S.; Gu, J.; Zhang, K.; Reese, A.; Ying, Y.; Wang, L.; Wang, H.; Li, Y.; Wang, M.; et al. Self-assembly of polycyclic supramolecules using linear metal-organic ligands. *Nat. Commun.* **2018**, *9*, 4575. [CrossRef]

22. Zhang, J.-W.; Kan, X.-M.; Li, X.-L.; Luan, J.; Wang, X.-L. Transition metal carboxylate coordination polymers with amide-bridged polypyridine co-ligands: Assemblies and properties. *CrystEngComm* **2015**, *17*, 3887–3907. [CrossRef]

23. Ali Akbar Razavi, S.; Morsali, A. Linker functionalized metal-organic frameworks. *Coord. Chem. Rev.* **2019**, *399*, 213023. [CrossRef]

24. Goura, J.; Chandrasekhar, V. Molecular Metal Phosphonates. *Chem. Rev.* **2015**, *115*, 6854–6965. [CrossRef]

25. Lawrance, G.A. Mixed Donor Ligands Based in part on the article Mixed Donor Ligands by Nikolay N. Gerasimchuk & Kristin Bowman-James which appeared in the Encyclopedia of Inorganic Chemistry, First Edition. Available online: https://doi.org/10.1002/9781119951438.eibc0132 (accessed on 15 December 2011).

26. Taghipour, F.; Mirzaei, M. A survey of interactions in crystal structures of pyrazine-based compounds. *Acta Crystallogr. Sect. C* **2019**, *75*, 231–247. [CrossRef] [PubMed]

27. Wang, Y.; Stoeckli-Evans, H. The inner-salt zwitterion, the dihydrochloride dihydrate and the dimethyl sulfoxide disolvate of 3,6-bis(pyridin-2-yl)pyrazine-2,5-dicarboxylic acid. *Acta Crystallogr. Sect. C* **2012**, *68*, o431–o435. [CrossRef]

28. Dobson, A.J.; Gerkin, R.E. 3-Aminopyrazine-2-carboxylic Acid. *Acta Crystallogr. Sect. C* **1996**, *52*, 1512–1514. [CrossRef] [PubMed]

29. Berrah, F.; Bouacida, S.; Roisnel, T. 2-Amino-3-carboxypyrazin-1-ium dihydrogen phosphate. *Acta Crystallogr. Sect. E* **2011**, *67*, o1409–o1410. [CrossRef] [PubMed]

30. Berrah, F.; Ouakkaf, A.; Bouacida, S.; Roisnel, T. Bis(2-amino-3-carboxypyrazin-1-ium) sulfate dihydrate. *Acta Crystallogr. Sect. E* **2011**, *67*, o677–o678. [CrossRef] [PubMed]

31. Berrah, F.; Bouacida, S.; Bouhraoua, A.; Roisnel, T. 2-Amino-3-carboxypyrazin-1-ium perchlorate bis(2-aminopyrazin-1-ium-3-carboxylate) monohydrate. *Acta Crystallogr. Sect. E* **2012**, *68*, o1714–o1715. [CrossRef]

32. Barszcz, B.; Masternak, J.; Hodorowicz, M.; Jabłońska-Wawrzycka, A. Cadmium(II) and calcium(II) complexes with N,O-bidentate ligands derived from pyrazinecarboxylic acid. *J. Therm. Anal. Calorim.* **2012**, *108*, 971–978. [CrossRef]

33. Goher, M.A.S.; Al-Salem, N.A.; Mautner, F.A.; Klepp, K.O. A copper(II) azide compound of pyrazinic acid containing a new dinuclear complex anion [$Cu_2(N_3)_6]_2^-$. Synthesis, spectral and study of KCu_2(pyrazinato) $(N_3)_4$. *Polyhedron* **1997**, *16*, 825–831. [CrossRef]

34. Zheng, X.; Chen, Y.; Ran, J.; Li, L. Synthesis, crystal structure, photoluminescence and catalytic properties of a novel cuprous complex with 2,3-pyrazinedicarboxylic acid ligands. *Sci. Rep.* **2020**, *10*, 6273. [CrossRef] [PubMed]

35. Li, Y.-W.; Tao, Y.; Hu, T.-L. Synthesis, structure, and photoluminescence of ZnII and CdII coordination complexes constructed by structurally related 5,6-substituted pyrazine-2,3-dicarboxylate ligands. *Solid State Sci.* **2012**, *14*, 1117–1125. [CrossRef]

36. Bouchene, R.; Khadri, A.; Bouacida, S.; Berrah, F.; Merazig, H. Bis(3-amino-pyrazine-2-carboxyl-ato-κ2 N(1) O)di-aqua-nickel(II) dihydrate. *Acta Crystallogr. Sect. E* **2013**, *69*, m309–m310. [CrossRef] [PubMed]

37. Koleša-Dobravc, T.; Maejima, K.; Yoshikawa, Y.; Meden, A.; Yasui, H.; Perdih, F. Vanadium and zinc complexes of 5-cyanopicolinate and pyrazine derivatives: Synthesis, structural elucidation and in vitro insulino-mimetic activity study. *New J. Chem.* **2017**, *41*, 735–746. [CrossRef]

38. Benhamada, N.; Bouchene, R.; Bouacida, S.; Zouchoune, B. Molecular structure, bonding analysis and redox properties of transition metal–Hapca [bis(3-aminopyrazine-2-carboxylic acid)] complexes: A theoretical study. *Polyhedron* **2015**, *91*, 59–67. [CrossRef]

39. Dehghanpour, S.; Jahani, K.; Mahmoudi, A.; Babakhodaverdi, M.; Notash, B. In situ hydrothermal synthesis of 2D mercury(I)–organic framework from 3-aminopyrazine-2-carboxylic acid and mercury(II) acetate. *Inorg. Chem. Commun.* **2012**, *25*, 79–82. [CrossRef]

40. Desiraju, G.R.; Ho, P.S.; Kloo, L.; Legon, A.C.; Marquardt, R.; Metrangolo, P.; Politzer, P.; Resnati, G.; Rissanen, K. Definition of the halogen bond (IUPAC Recommendations 2013). *Pure Appl. Chem.* **2013**, *85*, 1711–1713. [CrossRef]

41. Cavallo, G.; Metrangolo, P.; Milani, R.; Pilati, T.; Priimagi, A.; Resnati, G.; Terraneo, G. The Halogen Bond. *Chem. Rev.* **2016**, *116*, 2478–2601. [CrossRef]

42. Gilday, L.C.; Robinson, S.W.; Barendt, T.A.; Langton, M.J.; Mullaney, B.R.; Beer, P.D. Halogen Bonding in Supramolecular Chemistry. *Chem. Rev.* **2015**, *115*, 7118–7195. [CrossRef]

43. Troff, R.W.; Mäkelä, T.; Topić, F.; Valkonen, A.; Raatikainen, K.; Rissanen, K. Alternative Motifs for Halogen Bonding. *Eur. J. Org. Chem.* **2013**, *2013*, 1617–1637. [CrossRef]

44. Rissanen, K. Halogen bonded supramolecular complexes and networks. *CrystEngComm* **2008**, *10*, 1107–1113. [CrossRef]

45. Bui, T.T.T.; Dahaoui, S.; Lecomte, C.; Desiraju, G.R.; Espinosa, E. The Nature of Halogen⋯Halogen Interactions: A Model Derived from Experimental Charge-Density Analysis. *Angew. Chem. Int. Ed.* **2009**, *48*, 3838–3841. [CrossRef] [PubMed]

46. Metrangolo, P.; Resnati, G. Type II halogen⋯halogen contacts are halogen bonds. *IUCrJ* **2014**, *1*, 5–7. [CrossRef] [PubMed]

47. Awwadi, F.F.; Willett, R.D.; Peterson, K.A.; Twamley, B. The Nature of Halogen⋯Halogen Synthons: Crystallographic and Theoretical Studies. *Chem. A Eur. J.* **2006**, *12*, 8952–8960. [CrossRef] [PubMed]

48. Brammer, L.; Minguez Espallargas, G.; Adams, H. Involving metals in halogen-halogen interactions: Second-sphere Lewis acid ligands for perhalometallate ions (M-X⋯X′-C). *CrystEngComm* **2003**, *5*, 343–345. [CrossRef]

49. Von Essen, C.; Rissanen, K.; Puttreddy, R. Halogen Bonds in 2,5-Dihalopyridine-Copper(I) Halide Coordination Polymers. *Materials* **2019**, *12*, 3305. [CrossRef]

50. Puttreddy, R.; von Essen, C.; Peuronen, A.; Lahtinen, M.; Rissanen, K. Halogen bonds in 2,5-dihalopyridine-copper(ii) chloride complexes. *CrystEngComm* **2018**, *20*, 1954–1959. [CrossRef]

51. Puttreddy, R.; von Essen, C.; Rissanen, K. Halogen Bonds in Square Planar 2,5-Dihalopyridine–Copper(II) Bromide Complexes. *Eur. J. Inorg. Chem.* **2018**, *2018*, 2393–2398. [CrossRef]

52. Puttreddy, R.; Peuronen, A.; Lahtinen, M.; Rissanen, K. Metal-Bound Nitrate Anion as an Acceptor for Halogen Bonds in Mono-Halopyridine-Copper(II) Nitrate Complexes. *Cryst. Growth Des.* **2019**, *19*, 3815–3824. [CrossRef]

53. Kwak, S.H.; Lee, G.-H.; Gong, Y.-D. Synthesis of N-Substituted-2-Aminothiazolo[4,5-b]pyrazines by Tandem Reaction of o-Aminohalopyrazines with Isothiocyanates. *Bull. Korean Chem. Soc.* **2012**, *33*. [CrossRef]

54. Caldwell, J.J.; Veillard, N.; Collins, I. Design and synthesis of 2(1H)-pyrazinones as inhibitors of protein kinases. *Tetrahedron* **2012**, *68*, 9713–9728. [CrossRef]

55. Bartolomé-Nebreda, J.M.; Delgado, F.; Martín-Martín, M.L.; Martínez-Viturro, C.M.; Pastor, J.; Tong, H.M.; Iturrino, L.; Macdonald, G.J.; Sanderson, W.; Megens, A.; et al. Discovery of a Potent, Selective, and Orally Active Phosphodiesterase 10A Inhibitor for the Potential Treatment of Schizophrenia. *J. Med. Chem.* **2014**, *57*, 4196–4212. [CrossRef] [PubMed]

56. Reader, J.C.; Matthews, T.P.; Klair, S.; Cheung, K.-M.J.; Scanlon, J.; Proisy, N.; Addison, G.; Ellard, J.; Piton, N.; Taylor, S.; et al. Structure-Guided Evolution of Potent and Selective CHK1 Inhibitors through Scaffold Morphing. *J. Med. Chem.* **2011**, *54*, 8328–8342. [CrossRef] [PubMed]

57. Hutchings, M.G.; Meyrick, B.H.; Nelson, A.J. Colour and constitution of azo compounds derived from diaminoazines. *Tetrahedron* **1984**, *40*, 5081–5088. [CrossRef]

58. Miesel, J.L. Novel 1-(mono-o-substituted benzoyl)-3-(substituted pyrazinyl) ureas. US Patent No. 4,293,552, 31 July 1979.

59. Massaux, P.M.; Bernard, M.J.; Le Bihan, M.-T. Etude Structurale de CuBr.CH$_3$CN. *Acta Cryst.* **1972**, *B27*, 2419–2424. [CrossRef]

60. Rigaku Oxford Diffraction. *CrysAlisPro Version 38.46*; Rigaku: Tokyo, Japan, 2018.

61. Bruker AXS Inc. COLLECT. In *Bruker AXS BV, 1997-2004*; Bruker AXS Inc.: Madison, WI, USA.

62. Otwinowski, Z.; Minor, W. Processing of X-ray diffraction data collected in oscillation mode. *Methods Enzymol.* **1997**, *276*, 307–326. [CrossRef]

63. Sheldrick, G.M. *SADABS. Program for Empirical Absorption Correction*; University of Göttingen: Göttingen, Lower Saxony, Germany, 1996.

64. Sheldrick, G.M. Crystal structure refinement with SHELXL. *Acta Cryst.* **2015**, *C71*, 3–8. [CrossRef]

65. Dolomanov, O.V.; Bourhis, L.J.; Gildea, R.J.; Howard, J.A.K.; Puschmann, H.J.J. OLEX2: A complete structure solution, refinement and analysis program. *J. Appl. Cryst.* **2009**, *42*, 339–341. [CrossRef]

66. Gruene, T.; Hahn, H.W.; Luebben, A.V.; Meilleur, F.; Sheldrick, G.M.J. Refinement of macromolecular structures against neutron data with SHELXL2013. *Appl. Cryst.* **2014**, *47*, 462–466. [CrossRef]

67. Degen, T.; Sadki, M.; Bron, E.; König, U.; Nénert, G. The HighScore suite. Powder Diffraction (Supplement S2). 20 December 2014; 29, S13–S18.

68. Pawley, G.S. Unit-cell refinement from powder diffraction scans. *J. Appl. Cryst.* **1981**, *14*, 357–361. [CrossRef]

69. Mailman, A.; *unpublished results:* A preliminary structure of the CuBr$_2$ complex of 2-amino-5-chloropyrazine (**2**), *i.e.* **2**·CuBr$_2$ has been determined by single crystal X-ray diffraction methods: Triclinic, *P-1*, *a* = 6.002(5), 7.078 (5), 8.467 (7), α = 94.225 (6), β =97.238(7) γ = 102.105(7), V = 347.03 Å^3.

70. Kochi, J.K. The Reduction of Cupric Chloride by Carbonyl Compounds. *J. Am. Chem. Soc.* **1955**, *77*, 5274–5278. [CrossRef]

71. Yamada, Y.; Ohba, H.; Noboru, Y.; Daicho, S.; Nibu, Y. Solvation Effect on the NH Stretching Vibrations of Solvated Aminopyrazine, 2-Aminopyridine, and 3-Aminopyridine Clusters. *J. Phys. Chem. A* **2012**, *116*, 9271–9278. [CrossRef] [PubMed]

72. Peng, R.; Li, M.; Li, D. Copper(I) halides: A versatile family in coordination chemistry and crystal engineering. *Coord. Chem. Rev.* **2010**, *254*, 1–18. [CrossRef]

73. Knorr, M.; Bonnot, A.; Lapprand, A.; Khatyr, A.; Strohmann, C.; Kubicki, M.M.; Rousselin, Y.; Harvey, P.D. Reactivity of CuI and CuBr toward Dialkyl Sulfides RSR: From Discrete Molecular Cu4I4S4 and Cu8I8S6 Clusters to Luminescent Copper(I) Coordination Polymers. *Inorg. Chem.* **2015**, *54*, 4076–4093. [CrossRef] [PubMed]

74. Harisomayajula, N.V.S.; Makovetskyi, S.; Tsai, Y.-C. Cuprophilic Interactions in and between Molecular Entities. *Chem. A Eur. J.* **2019**, *25*, 8936–8954. [CrossRef] [PubMed]

75. Barclay, T.M.; Cordes, A.W.; Oakley, R.T.; Preuss, K.E.; Zhang, H. 2,5-Diamino-3,6-dichloropyrazine. *Acta Crystallogr. Sect. C* **1998**, *54*, 1018–1019. [CrossRef]

76. Chao, M.; Schempp, E.; Rosenstein, R.D. Aminopyrazine. *Acta Crystallogr. Sect. B* **1976**, *32*, 288–290. [CrossRef]
77. Näther, C.; Greve, J.; Jeß, I. Synthesis, crystal structures and thermal properties of new copper(I) halide coordination polymers. *Solid State Sci.* **2002**, *4*, 813–820. [CrossRef]
78. Conesa-Egea, J.; Gallardo-Martínez, J.; Delgado, S.; Martínez, J.I.; Gonzalez-Platas, J.; Fernández-Moreira, V.; Rodríguez-Mendoza, U.R.; Ocón, P.; Zamora, F.; Amo-Ochoa, P. Multistimuli Response Micro- and Nanolayers of a Coordination Polymer Based on Cu2I2 Chains Linked by 2-Aminopyrazine. *Small* **2017**, *13*, 1700965. [CrossRef]

Article

From Frustrated Packing to Tecton-Driven Porous Molecular Solids

Chamara A. Gunawardana [1], Abhijeet S. Sinha [1], Eric W. Reinheimer [2] and Christer B. Aakeröy [1,*]

[1] Department of Chemistry, Kansas State University, Manhattan, KS 66506, USA; chamara@ksu.edu (C.A.G.); sinha@ksu.edu (A.S.S.)
[2] Rigaku Americas Corporation, 9009 New Trails Drive, The Woodlands, TX 77381, USA; eric.reinheimer@rigaku.com
* Correspondence: aakeroy@ksu.edu; Tel.: +1-785-532-6096

Received: 24 February 2020; Accepted: 10 March 2020; Published: 13 March 2020

Abstract: Structurally divergent molecules containing bulky substituents tend to produce porous materials via frustrated packing. Two rigid tetrahedral cores, tetraphenylmethane and 1,3,5,7-tetraphenyladamantane, grafted peripherally with four (trimethylsilyl)ethynyl moieties, were found to have only isolated voids in their crystal structures. Hence, they were modified into tecton-like entities, tetrakis(4-(iodoethynyl)phenyl)methane [**I₄TEPM**] and 1,3,5,7-tetrakis(4-(iodoethynyl)phenyl)adamantane [**I₄TEPA**], in order to deliberately use the motif-forming characteristics of iodoethynyl units to enhance crystal porosity. **I₄TEPM** not only holds increased free volume compared to its precursor, but also forms one-dimensional channels. Furthermore, it readily co-crystallizes with Lewis basic solvents to afford two-component porous crystals.

Keywords: crystal engineering; porous material; molecular recognition; halogen bond; co-crystal; molecular tecton; binary solid; network structure; σ-hole; molecular electrostatic potential

---

1. Introduction

According to Kitaigorodskii's principle of close packing [1–5], molecules in crystals tend to dovetail and pack as efficiently as possible in order to maximize attractive dispersion forces and to minimize free energy. In other words, void space in crystals is always unfavorable. Thus, the construction of porous materials from discrete organic molecules (i.e., molecular porous materials (MPMs)) demands some special tactics [6–11]. For example, the packing of molecules specifically designed to bear sufficiently large and dimensionally fixed inner cavities or clefts (e.g., molecular cages and bowl-shaped compounds) can lead to porous structures [12–14].

Another viable synthetic strategy towards MPMs is to employ molecules with bulky, divergent and/or awkward shapes so that they no longer have the ability to pack tightly. Molecules such as 4-*p*-Hydroxyphenyl-2,2,4-trimethylchroman (Dianin's compound) [15,16], tris(*o*-phenylenedioxy) cyclotriphosphazene (TPP) [17–19] and 3,3′,4,4′-tetrakis(trimethylsilylethynyl)biphenyl (TTEB) [20] are well-known for producing MPMs merely as a consequence of frustrated packing, even though they do not have pre-fabricated molecular free volumes.

We have now expanded this idea to a family of tetrahedral molecules substituted at the four vertices with bulky groups. Here, we report the synthesis and structural investigation of tetrakis(4-((trimethylsilyl)ethynyl)phenyl)methane (**TMS₄TEPM**) and 1,3,5,7-tetrakis (4-((trimethylsilyl)ethynyl)phenyl)adamantane (**TMS₄TEPA**) (Scheme 1). By affixing large trimethylsilylethynyl (TMS-acetylenyl) moieties to the parent tetraphenylmethane (TPM) and 1,3,5,7-tetraphenyladamantane (TPA) core units, our aim was to disturb close-packing and to realize more open crystalline solids.

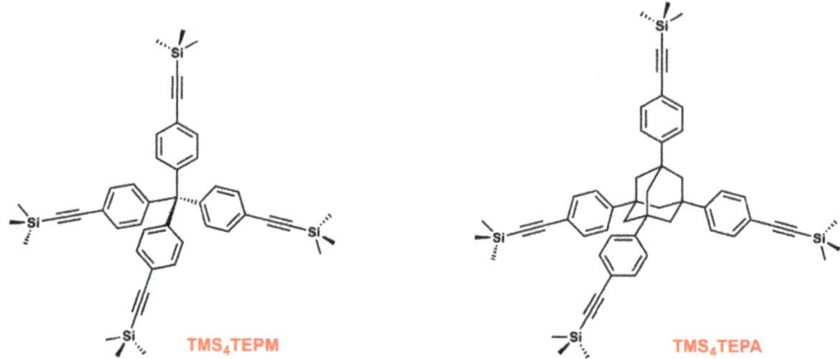

Scheme 1. Structural formulas of tetrakis(4-((trimethylsilyl)ethynyl)phenyl)methane (**TMS₄TEPM**) and 1,3,5,7-tetrakis(4-((trimethylsilyl)ethynyl)phenyl)adamantane (**TMS₄TEPA**).

Even though molecular shape is of primary importance in crystal packing, it is not the only structure-directing factor. The presence of functional units that can partake in directional and energetically significant non-covalent interactions has a major influence on molecular arrangement. With tectons (i.e., molecules featuring multiple peripheral binding sites) [21–24], the structure is built up so as to saturate the maximum amount of interactions, which is usually accompanied by compromises regarding dense-packing. Their association induces the assembly of networks where each molecule is positioned, through directional molecular recognition events, in a definite way with respect to its neighbors. Moreover, unlike van der Waals contacts, intermolecular point contacts consume only a limited amount of molecular surface, thereby leaving more usable surface. In this context, a great body of work has been done with hydrogen-bonding tectons to build so-called hydrogen-bonded organic frameworks (HOFs) [25–28]. Some notable examples include triptycenetrisbenzimidazolone (TTBI) [29], triaminotriazine-functionalized spirobifluorene [30,31] and polyfluorinated triphenylbenzene equipped with pyrazole [32].

Molecular tectonics based on halogen bonding (XB) is still in its infancy [33,34]. We therefore decided to modify the TPM and TPA scaffolds and transform them into new tecton-like entities, tetrakis(4-(iodoethynyl)phenyl)methane (**I₄TEPM**) and 1,3,5,7-tetrakis(4-(iodoethynyl)phenyl)adamantane (**I₄TEPM**) (Scheme 2). When iodine is directly bonded to an *sp*-hybridized carbon, it is strongly polarized, resulting in a more pronounced electron-deficient region (i.e., σ-hole) at the tip along the C–I bond axis [35–38]. The iodoethynyl functionality is, therefore, a perfect candidate for σ-hole/XB interactions. Although largely overlooked in molecular tectonics and crystal engineering, it can direct the assembly of network structures through C≡C–I···(C≡C) interactions (wherein the ethynyl π system acts as the XB acceptor) [39–41]. These T-shaped contacts frequently lead to *zigzag* chain motifs and are topologically parallel to those formed by C≡C–H···(C≡C) and C≡C–Br···(C≡C) contacts [37,42–52], but preferably serve as a stronger counterpart. Additional features that make the iodoethynyl unit well-suited for devising molecular building blocks include its structural rigidity, steric openness and core expanding ability.

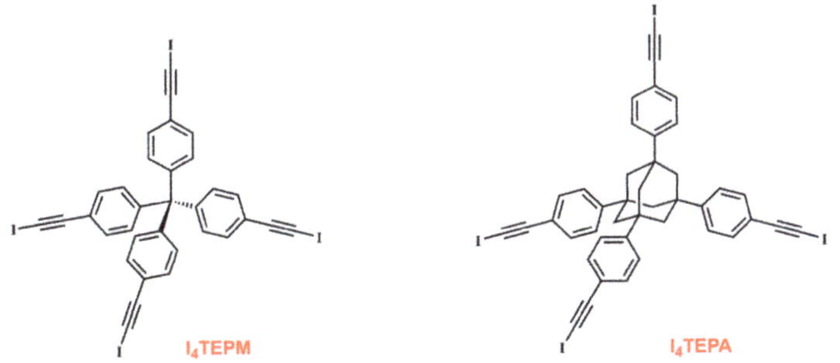

Scheme 2. Structural formulas of tetrakis(4-(iodoethynyl)phenyl)methane (**I₄TEPM**) and 1,3,5,7-tetrakis(4-(iodoethynyl)phenyl)adamantane (**I₄TEPM**).

2. Results and Discussion

The four molecules of interest were obtained according to the synthetic pathways shown in Schemes 3 and 4. Starting with commercially available tetraphenylmethane, **TMS₄TEPM** was prepared in two steps (tetra-*para*-bromination followed by coupling with trimethylsilylacetylene) with an overall yield of 78%. The synthesis of **TMS₄TEPA** required three steps (Friedel-Crafts reaction of 1-bromoadamantane and benzene, tetra-*para*-iodination followed by coupling with trimethylsilylacetylene), and the yield over these three steps was 50% (with respect to 1-bromoadamantane).

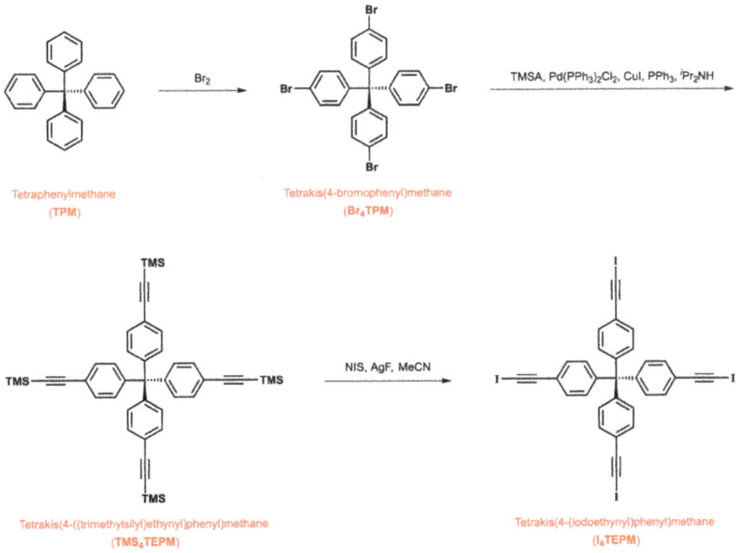

Scheme 3. Synthetic route to **TMS₄TEPM** and **I₄TEPM**.

Both **I₄TEPM** and **I₄TEPA** were accessible from the corresponding TMS derivatives, **TMS₄TEPM** and **TMS₄TEPA**, via one-pot/in situ desilylative iodination using silver(I) fluoride and *N*-iodosuccinimide. This direct trimethylsilyl-to-iodo transformation allowed us to avoid potentially unstable ethynyl intermediates and to achieve the target compounds in moderate yields (56% and 63%, respectively). Even though the ^{1}H and proton-decoupled ^{13}C-NMR spectra of these four-fold symmetric tetraiodoethynyl species are quite simple, the signals display considerable solvent dependency due to their XB-based complexation ability, with the alkynyl carbon bonded to iodine being most strongly affected (**I$_4$TEPM**: 7.0 ppm in CDCl$_3$ versus 18.4 ppm in DMSO-d_6, **I$_4$TEPA**: 6.2 ppm in CDCl$_3$ versus 17.0 ppm in DMSO-d_6). It is also worth mentioning that the ^{1}H-NMR spectrum of **I$_4$TEPA** exhibits conspicuous second order (leaning/roofing) effects.

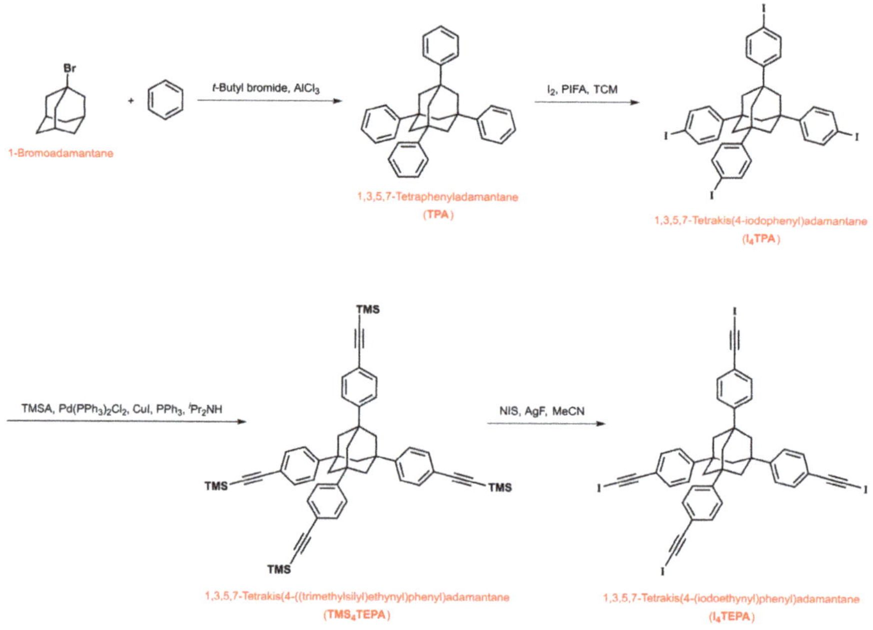

Scheme 4. Synthetic route to **TMS$_4$TEPA** and **I$_4$TEPA**.

Crystals of **TMS$_4$TEPM** suitable for single-crystal X-ray analysis were obtained by slow evaporation of either tetrahydrofuran/ethanol or chloroform/ethanol solution. For **TMS$_4$TEPA**, X-ray quality crystals could be harvested from hexane, heptane, heptane/dichloromethane or chloroform/ethanol. As anticipated, structural determination revealed that both are somewhat porous in nature (14.9% and 14.5% free volume, respectively). They, however, do not form empty-channel structures; instead, they have disconnected spatial voids or "porosity without pores", as described by Barbour (Figure 1) [53]. The overall packing is mainly mediated by extensive phenyl embraces.

In order to get some insight about the electron density/charge distribution over the free tetraiodoethynyl tectons and the degree of activation of XB donor sites (i.e., iodine atoms) delivered by *sp*-hybridized carbons [35–38], their molecular electrostatic potential (MEP) maps were generated (Figure 2). As expected, both **I$_4$TEPM** and **I$_4$TEPA** were found to have well-built σ-holes (+172.4 and +170.7 kJ/mol, respectively) on each iodine atom. Indeed, these σ-hole potential values are significantly higher than those of other closely-related tetra-halogenated molecules (see Supplementary Materials, Figure S33).

We then tried to grow crystals of **I$_4$TEPM** and **I$_4$TEPA** but were successful only with the former. The structural analysis of **I$_4$TEPM** crystals (harvested from hexanes) showed that the molecules are arranged in stacks which, in turn, are linked together by C≡C–I···(C≡C) halogen bonds, with near orthogonal approach of C–I donors towards C≡C triple bonds (detailed geometrical data are given in Table 1). In each **I$_4$TEPM** molecule, only two iodoethynyl arms participate in these T-shaped contacts, and the remaining two form weak C≡C–I···π(phenyl) interactions. The extended (and possibly cooperative) *zigzag* arrays of the C≡C–I···π(ethynyl) interactions ultimately make ladder-like motifs between individual molecular rows, leading to an infinite two-dimensional network (Figure 3 left).

I$_4$TEPM shares these packing features with its bromo analog, tetrakis(4-(bromoethynyl)phenyl)methane (**Br$_4$TEPM**) [42], but not with tetrakis(4-ethynylphenyl)methane (**TEPM**), which forms an interwoven diamondoid net [44].

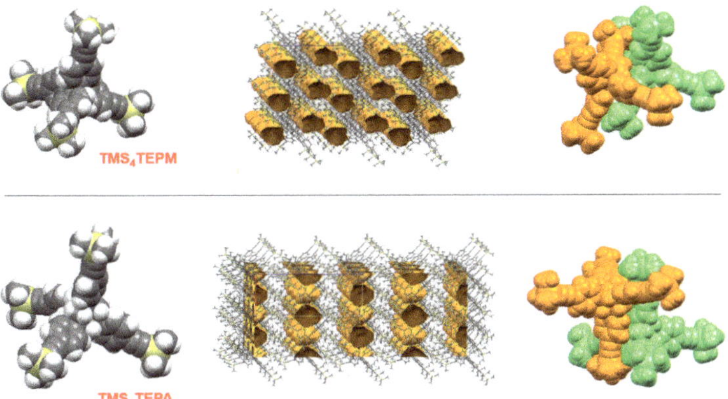

Figure 1. Crystal structures of **TMS$_4$TEPM** and **TMS$_4$TEPA**. (from left) Single molecules, overall packing and phenyl embraces (representative structures are shown from disordered structures).

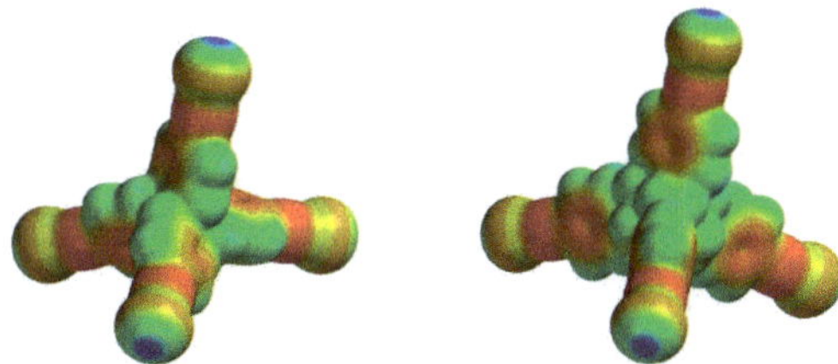

Figure 2. Molecular electrostatic potential (MEP) surfaces of the free tetraiodoethynyl tectons, **I$_4$TEPM** and **I$_4$TEPA**. Both plots have been set to the same color scale for visual comparison. Range: from −80 kJ/mol (red) to +175 kJ/mol (blue).

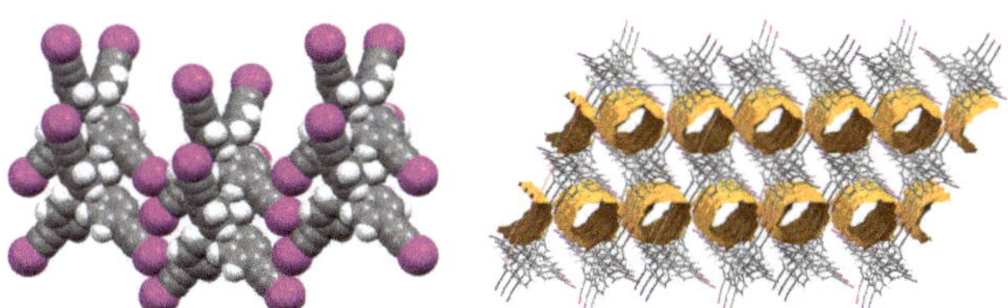

Figure 3. Crystal structure of **I$_4$TEPM**, showing halogen bonding (XB)-driven network formation (left) and void space in overall packing (right).

In contrast to the structure of **TMS$_4$TEPM** with isolated voids, **I$_4$TEPM** possesses one-dimensional channels along the crystallographic *b* axis (Figure 3 right). These channels account for 26.5% of the crystal volume, which is roughly twice as high as that of **TMS$_4$TEPM**. Another point worth emphasizing is that the precursor molecules, tetraphenylmethane (**TPM**), tetrakis(4-bromophenyl)methane (**Br$_4$TPM**) and tetrakis(4-iodophenyl)methane (**I$_4$TPM**), all form non-porous structures (see Supplementary Materials, Figure S34), highlighting the effectiveness of our strategy.

Since MPMs are usually held together by relatively weak interactions, they are not as rigid and robust as zeolites, metal-organic frameworks (MOFs) or covalent-organic frameworks (COFs). In most cases, attempts at activation (i.e., removal of entrapped guest molecules) cause structural disintegration. Hence, the real challenge lies in attaining permanently porous molecular materials that can behave analogously to framework-type solids. Most importantly, **I₄TEPM**, sustained primarily by the iodoethynyl catemer motif (i.e., the infinite C≡C–I···C≡C–I··· synthon), can maintain its structural integrity upon guest solvent loss, indicating its potential to exhibit permanent porosity.

In addition to tectonic construction, we also wanted to test the suitability of **I₄TEPM** in modular construction by co-crystallizing it with appropriate Lewis basic (i.e., XB-accepting) co-formers, in order to realize multicomponent architectures. With tetraphenylphosphonium halide salts ($Ph_4P^+X^-$; X^- = Cl^-, Br^-, I^-), it readily afforded diamondoid (**dia**) frameworks, but interpenetration and the inclusion of bulky Ph_4P^+ cations gave rise to highly compact arrangements within those solids [54]. As a charge-neutral co-crystallizing partner, our first choice was pyridine, one of the simplest XB acceptors, even though it cannot lead **I₄TEPM** to a polymeric assembly. We managed to get a binary crystalline material (confirmed by IR, NMR and TGA) but the structural characterization was not successful, as those crystals were quite fragile and rapidly deteriorated during data collection. This intrigued us to try out other Lewis basic/coordinating solvents with multiple bond forming ability. In three cases, with tetrahydrofuran (THF), dimethyl sulfoxide (DMSO) and 1,4-dioxane, **I₄TEPM** afforded crystalline binary solids.

Crystallization of **I₄TEPM** in THF/methanol afforded crystals of **I₄TEPM·2THF** where each THF molecule forms two halogen bonds in a bifurcated manner and connect adjacent **I₄TEPM** molecules together, thereby forming a one-dimensional twisted ribbon-like architecture (Figure 4a left). The resulting lattice comprises isolated voids that account for 14.4% of unit cell volume (Figure 4a right).

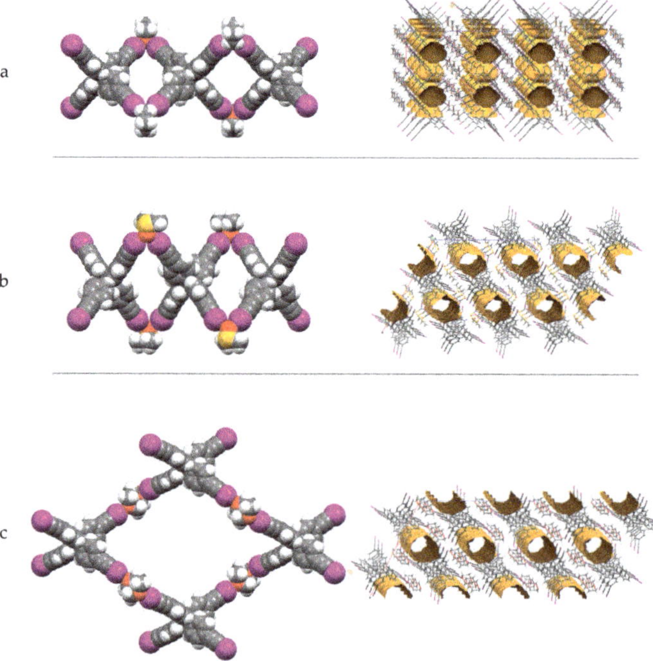

Figure 4. Crystal structures of (**a**) **I₄TEPM·2THF**, (**b**) **I₄TEPM·2DMSO** and (**c**) **I₄TEPM·2Dioxane**, showing XB-directed chain/net formation (left) and void space in overall packing (right).

Crystallization of I_4**TEPM** from neat DMSO or DMSO/methanol yielded crystals of I_4**TEPM**·2DMSO which has XB interactions analogous to those observed in I_4**TEPM**·2THF. Once again, the coordinating solvent acts as a bridging ligand and gives rise to a twisted-ribbon supramolecular chain (Figure 4b left), with one-dimensional channels of 21.0% free volume in the overall packing (Figure 4b right).

By using 1,4-dioxane/dichloromethane as the solvent system, crystals of I_4**TEPM**·2Dioxane could be obtained. As expected, dioxane serves as a linear ditopic ligand, so the structure propagates into two dimensions (Figure 4c left). As in I_4**TEPM**·2DMSO, the structure creates one-dimensional channels parallel to the crystallographic *c* axis, holding 21.0% free volume (Figure 4c right).

Unfortunately, as is the case with many other crystalline solvates, all these binary crystals are unstable at room temperature. Once removed from the mother liquor, they gradually become opaque because of the partial loss of halogen-bonded and freely-occupying solvent molecules. The DSC and TGA thermograms (Figure 5), however, show that the solvents are somewhat strongly attached to the crystal lattice. In particular, for I_4**TEPM**·2THF and I_4**TEPM**·2Dioxane, the removal temperatures are noticeably higher than their respective boiling points.

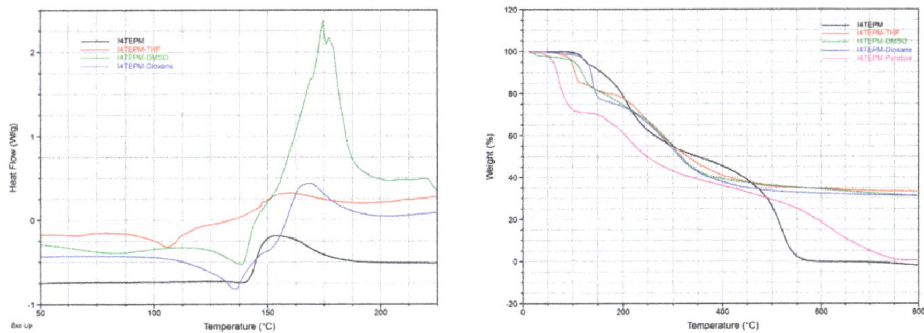

Figure 5. (**left**) DSC traces (Tzero aluminum pan, 1–2 mg sample size, 5 °C·min^{-1} heating rate, nitrogen atmosphere) and (**right**) TGA traces (platinum pan, 5–10 mg sample size, 10 °C·min^{-1} heating rate, nitrogen atmosphere).

Table 1 presents XB distances and angles of I_4**TEPM** and its binary crystals/solvates, along with the normalized distance (*ND*) and the percent radii reduction (*%RR*) values, which are two common indicators used as rough measures of the XB strength. In I_4**TEPM**, C≡C–I···(C≡C) interactions are not symmetric and the C–I donors reach more toward terminal acetylenic carbons. Consequently, one I···C separation is significantly longer (with a low *%RR* value) and deviates from linearity. The *%RR* values calculated for XBs observed in the three solvates are greater than 15% (except in one case), reflecting the moderate strength of those interactions. Moreover, all bonds have near-linear (> 170° angles, again one exception) arrangements, reflecting their high directionality.

Table 1. XB interaction parameters in the studied crystal structures.

Compound	C–I···O	d(I···O)/Å	ND [a]	$\%RR$ [b]	$\angle$(C–I···O)/°
I$_4$TEPM	C9–I10···C8(π) [c]	3.405(12)	0.925	7.47	165.5(4)
	C9–I10···C9(π) [c]	3.266(13)	0.888	11.2	173.6(5)
I$_4$TEPM·2THF	C9–I10···O11 [d]	2.965(5)	0.847	15.3	170.1(3)
I$_4$TEPM·2DMSO	C9–I10···O23 [e]	3.013(3)	0.861	13.9	162.0(14)
	C18–I19···O23 [f]	2.797(3)	0.799	20.1	170.0(14)
I$_4$TEPM·2Dioxane	C1–I1···O1	2.773(4)	0.792	20.8	174.3(11)
	C17–I2···O2 [g]	2.819(3)	0.805	19.5	174.4(9)

[a] Normalized distance, $ND = d_{xy}/(r_x + r_y)$, where d_{xy} is the crystallographically determined XB distance, and r_x and r_y are the van der Waals radii for the two involved atoms (I = 1.98 Å, C = 1.70 Å, O = 1.52 Å). [b] Percent radii reduction, $\%RR = (1 - ND) \times 100$. Symmetry transformations used to generate equivalent atoms: [c] $1-x$, $\frac{1}{2}+y$, $1.5-z$. [d] $\frac{1}{2}+x$, $1.5-y$, $1-z$. [e] $-\frac{1}{2}+x$, $-\frac{1}{2}-y$, $-\frac{1}{2}+z$. [f] x, $1+y$, z. [g] $-\frac{1}{2}+x$, $-2.5+y$, $-\frac{1}{2}+z$.

3. Conclusions

The solid-state packing behavior of tetrakis(4-((trimethylsilyl)ethynyl)phenyl)methane [**TMS$_4$TEPM**] and 1,3,5,7-tetrakis(4-((trimethylsilyl)ethynyl)phenyl)adamantane [**TMS$_4$TEPA**] showed some degree of extrinsic porosity. These two molecules were converted into tecton-like derivatives with XB capability, **I$_4$TEPM** and **I$_4$TEPA**, in order to investigate the power of iodoethynyl recognition sites in the context of solid-state packing and extrinsic porosity. Our results demonstrate that, even though **I$_4$TEPA** tends not to form crystalline unary or binary solids, **I$_4$TEPM** crystallizes into porous solids in its neat form as well as with suitable co-formers. The binary systems formed with coordinating solvents (i.e., **I$_4$TEPM**·4Pyridine, **I$_4$TEPM**·2THF, **I$_4$TEPM**·2DMSO and **I$_4$TEPM**·2Dioxane) are prone to collapse upon solvent removal. It is therefore rational to think that **I$_4$TEPM** would offer more stable crystals if the co-formers employed are solids at ambient conditions. Efforts to explore these new possibilities, especially utilizing molecules with tetrahedrally-disposed XB accepting sites (e.g., tetraazaadamantane, tetrakis(4-pyridyl)cyclobutane, tetrakis(4-pyridyloxymethyl)methane) are currently being undertaken in our lab.

4. Materials and Methods

Unless otherwise noted, all reagents, solvents and precursors (tetraphenylmethane and 1-bromoadamantane) were purchased from commercial sources and used as received, without further purification. Nuclear magnetic resonance (NMR) spectra were recorded at room temperature on a Varian Unity Plus (400 MHz) spectrometer (Agilent Technologies, Inc., Santa Clara, CA, USA). Chemical shifts for ^{1}H-NMR spectra were referenced to the residual protio impurity peaks in the deuterated solvents, while ^{13}C{^{1}H} NMR spectra were referenced against the solvent ^{13}C resonances. A Nicolet 380 FT-IR system (Thermo Fisher Scientific Inc., Waltham, MA, USA) was used for the infrared (IR) spectroscopic analysis. Differential scanning calorimetry (DSC) and thermogravimetric analysis (TGA) were performed on TA Q20 and TA Q50 (TA Instruments, New Castle, DE, USA), respectively. In order to calculate the molecular surface electrostatic potentials of tetra-halogenated TPM and TPA species, their geometries were optimized (using Spartan '14 software [55]) at hybrid functional B3LYP/6-311+G** and B3LYP/6-311++G** levels of theory, respectively, and potential values were subsequently mapped onto 0.002 au isosurface. Detailed crystallographic information about data collections, solutions, and refinements can be found in the Supplementary Materials. Structural visualizations and void mapping were done using Mercury software [56]. For free volume calculations, the voids function in Mercury (with contact surface, 1.2 Å probe radius and 0.2 Å approximate grid spacing) and/or the solvent-masking tool in Olex2 (with its default parameters) were employed [56,57].

4.1. Synthesis of Tetrakis(4-bromophenyl)methane (**Br₄TPM**)

The bromination of tetraphenylmethane was performed neat using an excess of molecular bromine. To a 100-mL round-bottom flask containing tetraphenylmethane (2.00 g, 6.24 mmol, 1 equiv.), bromine liquid (6.4 mL, 124.8 mmol, 20 equiv.) was added carefully at 0 °C. After attaching a water-cooled reflux condenser, the resultant dark reddish slurry was stirred vigorously at room temperature for one hour, and then cooled to −78 °C by using a dry ice/acetone bath. Ethanol (25 mL) was added slowly and the reaction mixture was allowed to warm to room temperature overnight. Then, to destroy excess/unreacted bromine, it was treated with 40% aqueous solution of sodium bisulfite (approximately 75 mL) and stirred for an additional 30 min until the orange color disappeared. The tan colored solid was collected by filtration, washed well with distilled water (100 mL) and oven-dried at 60 °C for five hours. This solid was further purified by re-crystallization from chloroform/ethanol (2:1), affording tetrakis(4-bromophenyl)methane, **Br₄TPM**, as an off-white crystalline material. Yield: 3.65 g (5.74 mmol, 92%). ^{1}H-NMR (400 MHz, CDCl$_3$) δ (ppm): 7.39 (d, 8H); 7.01 (d, 8H). ^{13}C-NMR (100 MHz, CDCl$_3$) δ (ppm): 144.64, 132.57, 131.30, 121.02, 63.84. ATR-FTIR (cm^{-1}): 3059, 1919, 1569, 1478, 1395, 1185, 1077, 1007, 948, 808, 753.

4.2. Synthesis of Tetrakis(4-((trimethylsilyl)ethynyl)phenyl)methane (**TMS₄TEPM**)

This step involved a Sonogashira cross-coupling reaction of tetrakis(4-bromophenyl)methane with trimethylsilylacetylene. Tetrakis(4-bromophenyl)methane (3.50 g, 5.50 mmol, 1 equiv.) and triphenylphosphine (462 mg, 1.76 mmol, 32 mol%) were placed in a 250-mL round-bottomed flask. Diisopropyl amine (100 mL) was added and the resulting solution was purged with dinitrogen gas for 30 min. Then, bis(triphenylphosphine)palladium(II) dichloride (618 mg, 0.88 mmol, 16 mol%), copper(I) iodide (168 mg, 0.88 mmol, 16 mol%) and trimethylsilylacetylene (6.2 mL, 44.0 mmol, 8 equiv.) were added. The reaction flask was fitted to a water-jacketed condenser, cooled to −78 °C, subjected to a brief vacuum/backfill cycle and refluxed for 24 h under nitrogen atmosphere. After removing volatile materials in vacuo, the residue was re-dissolved in chloroform (100 mL) and filtered through a pad of Celite, using an extra 50 mL portion of chloroform to wash the filter pad. The combined filtrate was then washed with distilled water (2 × 25 mL) and brine (25 mL), dried over anhydrous magnesium sulfate, and evaporated to dryness under vacuum. The crude product was flash-column-chromatographed on silica gel using pure hexanes followed by hexanes/ethyl acetate (4:1) as eluents to obtain the title compound, **TMS₄TEPM**, as a pale yellowish solid. Yield: 3.30 g (4.68 mmol, 85%). ^{1}H-NMR (400 MHz, CDCl$_3$) δ (ppm): 7.33 (d, 8H), 7.05 (d, 8H), 0.24 (s, 36H). ^{13}C-NMR (100 MHz, CDCl$_3$) δ (ppm): 146.21, 131.59, 130.95, 121.42, 104.82, 95.00, 64.98, 0.18. ATR-FTIR (cm^{-1}): 2957, 2157, 1496, 1405, 1247, 1187, 1019, 835, 758.

4.3. Synthesis of Tetrakis(4-(iodoethynyl)phenyl)methane (**I₄TEPM**)

The one-pot/in situ desilylative iodination (i.e., direct trimethylsilyl-to-iodo conversion) method was employed. Acetonitrile (150 mL) was transferred into a 250-mL round-bottom flask that contained tetrakis(4-((trimethylsilyl)ethynyl)phenyl)methane (2.50 g, 3.54 mmol, 1 equiv.). The flask was wrapped in aluminium foil, and then silver(I) fluoride (2.70 g, 21.3 mmol, 6 equiv.) and *N*-iodosuccinimide (4.78 g, 21.3 mmol, 6 equiv.) were added. It was then evacuated (while stirring), refilled with nitrogen and stirred at room temperature for 24 h. Distilled water (200 mL) was added and the resulting mixture was extracted with diethyl ether (4 × 50 mL). The combined organic layers were washed with saturated sodium bisulfite (40 mL), distilled water (40 mL) and brine (40 mL), and dried over anhydrous magnesium sulfate. The evaporation of the solvent under reduced pressure resulted in an orange colored residue. Additional cleanup by column chromatography (silica gel, hexanes/ethyl acetate = 9:1) gave the desired compound, **I₄TEPM**, as a yellow solid. Crystals suitable for single-crystal X-ray diffraction were grown from hexanes. Yield: 1.83 g (1.98 mmol, 56%). ^{1}H-NMR (400 MHz, CDCl$_3$) δ (ppm): 7.32 (d, 8H), 7.06 (d, 8H). ^{13}C-NMR (100 MHz, CDCl$_3$) δ (ppm): 146.34, 132.04, 130.87,

121.81, 93.87, 65.02, 7.03. ^{1}H-NMR (400 MHz, DMSO-d_6) δ (ppm): 7.37 (d, 8H), 7.04 (d, 8H). ^{13}C-NMR (100 MHz, DMSO-d_6) δ (ppm): 145.68, 131.66, 130.36, 121.08, 92.11, 64.26, 18.41. ATR-FTIR (cm^{-1}): 2944, 2167, 1490, 1400, 1186, 1112, 1016, 955, 898, 827, 722.

4.4. Synthesis of 1,3,5,7-Tetraphenyladamantane (TPA)

In a 250-mL round-bottom flask, *tert*-butyl bromide (3.9 mL, 34.9 mmol, 2.5 equiv.) was added to a solution of 1-bromoadamantane (3.00 g, 13.9 mmol, 1 equiv.) in anhydrous benzene (30 mL). The flask was placed in an ice bath and aluminium chloride (186 mg, 1.39 mmol, 10 mol%) was carefully charged to the chilled stirring solution. The mixture was then heated under reflux until the evolution of hydrogen bromide ceased (the top of the condenser was connected to a gas absorption trap containing 30% aqueous sodium hydroxide). The resultant heterogeneous mixture was allowed to cool to room temperature and filtered, and the residue was washed sequentially with chloroform (30 mL), water (50 mL) and chloroform (30 mL). The off-white solid was further purified by washing overnight with refluxing chloroform in a Soxhlet apparatus, which gave 1,3,5,7-tetraphenyladamantane, **TPA**, as a fine white powder. Yield: 5.04 g (11.4 mmol, 82%). Mp: > 300 °C. ATR-FTIR (cm^{-1}): 3055, 3020, 2918, 2849, 1597, 1493, 1442, 1355, 1263, 1078, 1030, 918, 889, 844, 788, 760, 746, 699.

4.5. Synthesis of 1,3,5,7-Tetrakis(4-iodophenyl)adamantane (I₄TPA)

To a 250-mL round-bottom flask containing a suspension of 1,3,5,7-tetraphenyladamantane (4.00 g, 9.08 mmol, 1 equiv.) in chloroform (100 mL) was added iodine (5.76 g, 22.7 mmol, 2.5 equiv.). This mixture was stirred vigorously at room temperature until the iodine fully dissolved. The flask was flushed with nitrogen gas and bis(trifluoroacetoxy)iodo)benzene (9.76 g, 22.7 mmol, 2.5 equiv.) was added. The resulting mixture was stirred at room temperature for 12 h. It was then filtered off, and the collected solid was washed with an excess amount of chloroform (200 mL). The combined dark purple filtrate was washed with 5% sodium bisulfite solution twice (2 × 50 mL), followed by distilled water (100 mL) and saturated sodium chloride solution (100 mL). It was dried with anhydrous magnesium sulfate and the solvent was removed under reduced pressure, which resulted in a pale-yellow solid. After refluxing in methanol (200 mL) for 12 h, the pure compound, **I₄TPA**, was isolated as a white solid by filtration and air-drying. Yield: 5.91 g (6.26 mmol, 69%). ^{1}H-NMR (400 MHz, CDCl$_3$) δ (ppm): ^{1}H-NMR (400 MHz, CDCl$_3$) δ (ppm): 7.67 (d, 8H), 7.18 (d, 8H), 2.06 (s, 12H). ^{13}C-NMR (100 MHz, CDCl$_3$) δ (ppm): 148.63, 137.75, 127.34, 91.96, 46.92, 39.29. ATR-FTIR (cm^{-1}): 3046, 2928, 2898, 2851, 1900, 1782, 1647, 1579, 1483, 1441, 1390, 1355, 1180, 1120, 1064, 1001, 888, 819, 775, 701, 659.

4.6. Synthesis of 1,3,5,7-Tetrakis(4-((trimethylsilyl)ethynyl)phenyl)adamantane (TMS₄TEPA)

As in the synthesis of **TMS₄TEPM**, this step involved a four-fold Sonogashira cross-coupling reaction of 1,3,5,7-tetrakis(4-iodophenyl)adamantane (**I₄TPA**) with trimethylsilylacetylene. Yield: 88%. ^{1}H-NMR (400 MHz, CDCl$_3$) δ (ppm): ^{1}H-NMR (400 MHz, CDCl$_3$) δ (ppm): 7.45 (d, 8H), 7.38 (d, 8H), 2.09 (s, 12H), 0.24 (s, 36H). ^{13}C-NMR (100 MHz, CDCl$_3$) δ (ppm): 149.63, 132.29, 125.13, 121.32, 105.19, 94.20, 46.97, 39.53, 0.25. ATR-FTIR (cm^{-1}): 3033, 2958, 2897, 2852, 2155, 1604, 1502, 1445, 1398, 1355, 1248, 1115, 1016, 859, 835, 758.

4.7. Synthesis of 1,3,5,7-Tetrakis(4-(iodoethynyl)phenyl)adamantane (I₄TEPA)

The same one-pot desilylative iodination method described above for the synthesis of **I₄TEPM** (i.e., the direct trimethylsilyl-to-iodo transformation using silver(I) fluoride and *N*-iodosuccinimide) was employed. Yield: 63%. ^{1}H-NMR (400 MHz, CDCl$_3$) δ (ppm): 7.42 (d, 8H), 7.39 (d, 8H), 2.09 (s, 12H). ^{13}C-NMR (100 MHz, CDCl$_3$) δ (ppm): 149.82, 132.64, 125.16, 121.57, 94.16, 46.88, 39.50, 6.18. ^{1}H-NMR (400 MHz, DMSO-d_6) δ (ppm): 7.51 (d, 8H), 7.37 (d, 8H), 2.00 (s, 12H). ^{13}C-NMR (100 MHz, DMSO-d_6) δ (ppm): 150.14, 131.74, 125.48, 120.50, 92.59, 45.48, 38.95, 17.02. ATR-FTIR (cm^{-1}): 3033, 2919, 2896, 2849, 2165, 1908, 1701, 1603, 1501, 1439, 1355, 1241, 1176, 1115, 1016, 837, 822, 769, 693.

4.8. Synthesis of **I₄TEPM**·*4pyridine*

In a 2-dram glass vial, **I₄TEPM** (10 mg, 0.011 mmol) was dissolved in 0.5 mL of pyridine. This open vial was placed in a second larger container (50-mL glass jar) containing 10 mL of pyridine/methanol (1:4) mixture. The outer container was then closed/sealed, and the apparatus was kept at ambient conditions to allow the vapor from methanol (anti-solvent) to diffuse into the sample solution. When the total volume of the inner vial became ~3 mL, it was taken out and, after partially tightening the lid, left undisturbed at ambient conditions to allow the solvents to evaporate slowly. Colorless/pale-yellow crystals were observed after few days. ATR-FTIR (cm^{-1}): 3032, 2923, 2851, 2158, 1909, 1587, 1493, 1438, 1405, 1210, 1185, 1147, 1066, 1017, 997, 955, 827, 745, 699.

4.9. Synthesis of **I₄TEPM**·*2THF*

In a 2-dram glass vial, **I₄TEPM** (10 mg, 0.011 mmol) was dissolved in 1 mL of tetrahydrofuran. After adding 1 mL of methanol, the vial (with a partially-tightened screw cap) was left undisturbed at ambient conditions to allow the solvents to evaporate slowly. Colorless/pale-yellow crystals suitable for single-crystal X-ray diffraction were observed after few days. ATR-FTIR (cm^{-1}): 2974, 2865, 2165, 1684, 1588, 1494, 1423, 1404, 1365, 1190, 1115, 1044, 1018, 884, 830, 809.

4.10. Synthesis of **I₄TEPM**·*2DMSO*

In a 2-dram glass vial, **I₄TEPM** (10 mg, 0.011 mmol) was dissolved in 0.5 mL of dimethyl sulfoxide. The vial (with a partially-tightened screw cap) was then allowed to stand at room temperature for one week, during which time colorless/pale-yellow crystals suitable for single-crystal X-ray diffraction were appeared. ATR-FTIR (cm^{-1}): 3032, 2986, 2908, 2160, 1494, 1429, 1398, 1308, 1186, 1113, 1039, 1014, 945, 826, 697.

4.11. Synthesis of **I₄TEPM**·*2dioxane*

In a 2-dram glass vial, **I₄TEPM** (10 mg, 0.011 mmol) was suspended in 0.5 mL 1,4-dioxane. After adding a few drops of methylene chloride, the vial was sealed and heated to obtain a clear solution. Colorless/pale-yellow crystals suitable for single-crystal X-ray diffraction were harvested by slow evaporation. ATR-FTIR (cm^{-1}): 2958, 2906, 2851, 2171, 1490, 1448, 1401, 1369, 1288, 1252, 1186, 1113, 1077, 1016, 976, 866, 829, 735.

Supplementary Materials: NMR and IR spectra, and crystallographic data are available online at http://www.mdpi.com/2624-8549/2/1/11/s1. The crystallographic data for this paper (CCDC 1971906–1971911) can also be obtained free of charge via www.ccdc.cam.ac.uk/data_request/cif, or by emailing data_request@ccdc.cam.ac.uk, or by contacting The Cambridge Crystallographic Data Centre, 12 Union Road, Cambridge CB2 1EZ, UK; fax: +44 1223 336033.

Author Contributions: C.A.G. and C.B.A. conceived and designed the experiments; C.A.G. performed the experiments; A.S.S. and E.W.R. performed the single-crystal X-ray crystallography; C.A.G. and C.B.A. analyzed the data and wrote the paper. All authors have read and agreed to the published version of the manuscript.

Funding: This research was funded by the U. S. Army Research Laboratory and the U. S. Army Research Office, grant number W911NF-13-1-0387.

Acknowledgments: We are grateful to Victor W. Day at the University of Kansas for collecting some single-crystal X-ray data. He, in turn, acknowledges the NSF-MRI grant CHE-0923449 which was used to purchase an X-ray diffractometer and software used in this study.

Conflicts of Interest: The authors declare no conflict of interest. The funders had no role in the design of the study; in the collection, analyses, or interpretation of data; in the writing of the manuscript, or in the decision to publish the results.

References

1. Kitaigorodskii, A.I. *Molecular Crystals and Molecules*; Academic Press: New York, NY, USA, 1973.
2. Kitaigorodskii, A.I. *Organic Chemical Crystallography*; Consultants Bureau: New York, NY, USA, 1961.

3.   Kitaigorodskii, A.I. Non-bonded interactions of atoms in organic crystals and molecules. *Chem. Soc. Rev.* **1978**, *7*, 133–163. [CrossRef]

4.   Kitaigorodskii, A.I. The principle of close packing and the condition of thermodynamic stability of organic crystals. *Acta Crystallogr.* **1965**, *18*, 585–590. [CrossRef]

5.   Kitaigorodskii, A.I. The close-packing of molecules in crystals of organic compounds. *J. Phys. (USSR)* **1945**, *9*, 351–352.

6.   Lü, J.; Cao, R. Porous organic molecular frameworks with extrinsic porosity: A platform for carbon storage and separation. *Angew. Chem. Int. Ed.* **2016**, *55*, 9474–9480. [CrossRef] [PubMed]

7.   Hashim, M.I.; Hsu, C.-W.; Le, H.T.M.; Miljanić, O.Š. Organic molecules with porous crystal structures. *Synlett* **2016**, *27*, 1907–1918.

8.   Tian, J.; Thallapally, P.K.; McGrail, B.P. Porous organic molecular materials. *CrystEngComm* **2012**, *14*, 1909–1919. [CrossRef]

9.   Mastalerz, M. Permanent porous materials from discrete organic molecules—towards ultra-high surface areas. *Chem. Eur. J.* **2012**, *18*, 10082–10091. [CrossRef]

10.   McKeown, N.B. Nanoporous molecular crystals. *J. Mater. Chem.* **2010**, *20*, 10588–10597. [CrossRef]

11.   Holst, J.R.; Trewin, A.; Cooper, A.I. Porous organic molecules. *Nat. Chem.* **2010**, *2*, 915–920. [CrossRef]

12.   Hasell, T.; Cooper, A.I. Porous organic cages: Soluble, modular and molecular pores. *Nat. Rev. Mater.* **2016**, *1*, 16053. [CrossRef]

13.   Evans, J.D.; Sumby, C.J.; Doonan, C.J. Synthesis and applications of porous organic cages. *Chem. Lett.* **2015**, *44*, 582–588. [CrossRef]

14.   Zhang, G.; Mastalerz, M. Organic cage compounds—from shape-persistency to function. *Chem. Soc. Rev.* **2014**, *43*, 1934–1947. [CrossRef] [PubMed]

15.   Imashiro, F.; Yoshimura, M.; Fujiwara, T. 'Guest-free' Dianin's compound. *Acta Crystallogr.* C **1998**, *54*, 1357–1360. [CrossRef]

16.   Barrer, R.M.; Shanson, V.H. Dianin's compound as a zeolitic sorbent. *J. Chem. Soc. Chem. Commun.* **1976**, 333–334. [CrossRef]

17.   Kaleta, J.; Bastien, G.; Wen, J.; Dračínský, M.; Tortorici, E.; Císařová, I.; Beale, P.D.; Rogers, C.T.; Michl, J. Bulk inclusions of double pyridazine molecular rotors in hexagonal tris(*o*-phenylene)cyclotriphosphazene. *J. Org. Chem.* **2019**, *84*, 8449–8467. [CrossRef]

18.   Sozzani, P.; Bracco, S.; Comotti, A.; Ferretti, L.; Simonutti, R. Methane and carbon dioxide storage in a porous van der Waals crystal. *Angew. Chem. Int. Ed.* **2005**, *44*, 1816–1820. [CrossRef]

19.   Allcock, H.R.; Siegel, L.A. Phosphonitrilic compounds. III. Molecular inclusion compounds of tris(*o*-phenylenedioxy)phosphonitrile trimer. *J. Am. Chem. Soc.* **1964**, *86*, 5140–5144. [CrossRef]

20.   Msayib, K.J.; Book, D.; Budd, P.M.; Chaukura, N.; Harris, K.D.M.; Helliwell, M.; Tedds, S.; Walton, A.; Warren, J.E.; Xu, M.C.; et al. Nitrogen and hydrogen adsorption by an organic microporous crystal. *Angew. Chem. Int. Ed.* **2009**, *48*, 3273–3277. [CrossRef]

21.   Wuest, J.D. Engineering crystals by the strategy of molecular tectonics. *Chem. Commun.* **2005**, 5830–5837. [CrossRef]

22.   Hosseini, M.W. Molecular tectonics: From simple tectons to complex molecular networks. *Acc. Chem. Res.* **2005**, *38*, 313–323. [CrossRef]

23.   Hosseini, M.W. Reflexion on molecular tectonics. *CrystEngComm* **2004**, *6*, 318–322. [CrossRef]

24.   Su, D.; Wang, X.; Simard, M.; Wuest, J.D. Molecular tectonics. *Supramol. Chem.* **1995**, *6*, 171–178. [CrossRef]

25.   Lin, R.-B.; He, Y.; Li, P.; Wang, H.; Zhou, W.; Chen, B. Multifunctional porous hydrogen-bonded organic framework materials. *Chem. Soc. Rev.* **2019**, *48*, 1362–1389. [CrossRef] [PubMed]

26.   Hisaki, I.; Xin, C.; Takahashi, K.; Nakamura, T. Designing hydrogen-bonded organic frameworks (HOFs) with permanent porosity. *Angew. Chem. Int. Ed.* **2019**, *58*, 11160–11170. [CrossRef] [PubMed]

27.   Luo, J.; Wang, J.-W.; Zhang, J.-H.; Lai, S.; Zhong, D.-C. Hydrogen-bonded organic frameworks: Design, structures and potential applications. *CrystEngComm* **2018**, *20*, 5884–5898. [CrossRef]

28.   Han, Y.-F.; Yuan, Y.-X.; Wang, H.-B. Porous hydrogen-bonded organic frameworks. *Molecules* **2017**, *22*, 266. [CrossRef]

29.   Mastalerz, M.; Oppel, I.M. Rational construction of an extrinsic porous molecular crystal with an extraordinary high specific surface area. *Angew. Chem. Int. Ed.* **2012**, *51*, 5252–5255. [CrossRef]

30. Fournier, J.H.; Maris, T.; Wuest, J.D. Molecular tectonics. Porous hydrogen-bonded networks built from derivatives of 9,9′-spirobifluorene. *J. Org. Chem.* **2004**, *69*, 1762–1775. [CrossRef]

31. Demers, E.; Maris, T.; Wuest, J.D. Molecular tectonics. Porous hydrogen-bonded networks built from derivatives of 2,2′,7,7′-tetraphenyl-9,9′-spirobi[9 *H*-fluorene]. *Cryst. Growth Des.* **2005**, *5*, 1227–1235. [CrossRef]

32. Chen, T.H.; Popov, I.; Kaveevivitchai, W.; Chuang, Y.C.; Chen, Y.S.; Daugulis, O.; Jacobson, A.J.; Miljanic, O.S. Thermally robust and porous noncovalent organic framework with high affinity for fluorocarbons and CFCs. *Nat. Commun.* **2014**, *5*. [CrossRef]

33. Shankar, S.; Chovnik, O.; Shimon, L.J.W.; Lahav, M.; van der Boom, M.E. Directed molecular structure variations of three-dimensional halogen-bonded organic frameworks (XBOFs). *Cryst. Growth Des.* **2018**, *18*, 1967–1977. [CrossRef]

34. Nikolayenko, V.I.; Castell, D.C.; van Heerden, D.P.; Barbour, L.J. Guest-induced structural transformations in a porous halogen-bonded framework. *Angew. Chem. Int. Ed.* **2018**, *57*, 12086–12091. [CrossRef] [PubMed]

35. Nguyen, S.T.; Ellington, T.L.; Allen, K.E.; Gorden, J.D.; Rheingold, A.L.; Tschumper, G.S.; Hammer, N.I.; Watkins, D.L. Systematic experimental and computational studies of substitution and hybridization effects in solid-state halogen bonded assemblies. *Cryst. Growth Des.* **2018**, *18*, 3244–3254. [CrossRef]

36. González, L.; Gimeno, N.; Tejedor, R.M.; Polo, V.; Ros, M.B.; Uriel, S.; Serrano, J.L. Halogen-bonding complexes based on bis(iodoethynyl)benzene units: A new versatile route to supramolecular materials. *Chem. Mater.* **2013**, *25*, 4503–4510. [CrossRef]

37. Aakeröy, C.B.; Baldrighi, M.; Desper, J.; Metrangolo, P.; Resnati, G. Supramolecular hierarchy among halogen-bond donors. *Chem. Eur. J.* **2013**, *19*, 16240–16247. [CrossRef]

38. Sarwar, M.G.; Dragisic, B.; Salsberg, L.J.; Gouliaras, C.; Taylor, M.S. Thermodynamics of halogen bonding in solution: Substituent, structural, and solvent effects. *J. Am. Chem. Soc.* **2010**, *132*, 1646–1653. [CrossRef]

39. Baldrighi, M.; Bartesaghi, D.; Cavallo, G.; Chierotti, M.R.; Gobetto, R.; Metrangolo, P.; Pilati, T.; Resnati, G.; Terraneo, G. Polymorphs and co-crystals of haloprogin: An antifungal agent. *CrystEngComm.* **2014**, *16*, 5897–5904. [CrossRef]

40. Lemouchi, C.; Vogelsberg, C.S.; Zorina, L.; Simonov, S.; Batail, P.; Brown, S.; Garcia-Garibay, M.A. Ultra-fast rotors for molecular machines and functional materials via halogen bonding: Crystals of 1,4-bis(iodoethynyl)bicyclo[2.2.2]octane with distinct gigahertz rotation at two sites. *J. Am. Chem. Soc.* **2011**, *133*, 6371–6379. [CrossRef]

41. Dunitz, J.D.; Gehrer, H.; Britton, D. The crystal structure of diiodacetylene: An example of pseudosymmetry. *Acta Crystallogr. B* **1972**, *28*, 1989–1994. [CrossRef]

42. Guo, W.Z.; Galoppini, E.; Gilardi, R.; Rydja, G.I.; Chen, Y.H. Weak intermolecular interactions in the crystal structures of molecules with tetrahedral symmetry: Diamondoid nets and other motifs. *Cryst. Growth Des.* **2001**, *1*, 231–237. [CrossRef]

43. Kaleta, J.; Bastien, G.; Císařová, I.; Batail, P.; Michl, J. Molecular rods: Facile desymmetrization of 1,4-diethynylbicyclo[2.2.2]octane. *Eur. J. Org. Chem.* **2018**, *2018*, 5137–5142. [CrossRef]

44. Galoppini, E.; Gilardi, R. Weak hydrogen bonding between acetylenic groups: The formation of diamondoid nets in the crystal structure of tetrakis(4-ethynylphenyl)methane. *Chem. Commun.* **1999**, 173–174. [CrossRef]

45. Dikundwar, A.G.; Sathishkumar, R.; Guru Row, T.N.; Desiraju, G.R. Structural variability in the monofluoroethynylbenzenes mediated through interactions involving "organic" fluorine. *Cryst. Growth Des.* **2011**, *11*, 3954–3963. [CrossRef]

46. Thakur, T.S.; Sathishkumar, R.; Dikundwar, A.G.; Guru Row, T.N.; Desiraju, G.R. Third polymorph of phenylacetylene. *Cryst. Growth Des.* **2010**, *10*, 4246–4249. [CrossRef]

47. Dziubek, K.; Podsiadło, M.; Katrusiak, A. Nearly isostructural polymorphs of ethynylbenzene: Resolution of ≡CH···π(arene) and cooperative ≡CH···π(C≡C) interactions by pressure freezing. *J. Am. Chem. Soc.* **2007**, *129*, 12620–12621. [CrossRef] [PubMed]

48. Saha, B.K.; Nangia, A. Ethynyl group as a supramolecular halogen and C≡C–H···C≡C trimer synthon in 2,4,6-tris(4-ethynylphenoxy)-1,3,5-triazine. *Cryst. Growth Des.* **2007**, *7*, 393–401. [CrossRef]

49. Ohkita, M.; Suzuki, T.; Nakatani, K.; Tsuji, T. Polar assembly of 2,6-diethynylpyridine through C(sp^2)–H···N, C(sp)–H···π and π–π stacking interactions: Crystal structure and nonlinear optical properties. *Chem. Lett.* **2001**, *30*, 988–989. [CrossRef]

50. Robinson, J.M.; Kariuki, B.M.; Harris, K.D.; Philp, D. Interchangeability of halogen and ethynyl substituents in the solid state structures of di- and tri-substituted benzenes. *J. Chem. Soc. Perkin Trans.* **1998**, *2*, 2459–2470. [CrossRef]

51. Weiss, H.-C.; Bläser, D.; Boese, R.; Doughan, B.; Haley, M. C–H⋯π interactions in ethynylbenzenes: The crystal structures of ethynylbenzene and 1,3,5-triethynylbenzene, and a redetermination of the structure of 1,4-diethynylbenzene. *Chem. Commun.* **1997**, 1703–1704. [CrossRef]

52. Steiner, T.; Starikov, E.B.; Amado, A.M.; Teixeira-Dias, J.J.C. Weak hydrogen bonding. Part 2. The hydrogen bonding nature of short C–H⋯π contacts: Crystallographic, spectroscopic and quantum mechanical studies of some terminal alkynes. *J. Chem. Soc. Perkin Trans.* **1995**, *2*, 1321–1326. [CrossRef]

53. Barbour, L.J. Crystal porosity and the burden of proof. *Chem. Commun.* **2006**, *11*, 1163–1168. [CrossRef] [PubMed]

54. Gunawardana, C.A.; Đaković, M.; Aakeröy, C.B. Diamondoid architectures from halogen-bonded halides. *Chem. Commun.* **2018**, *54*, 607–610. [CrossRef] [PubMed]

55. Wavefunction, Inc. Irvine, CA 92612, USA. Available online: https://www.wavefun.com/spartan (accessed on 8 March 2020).

56. Macrae, C.F.; Sovago, I.; Cottrell, S.J.; Galek, P.T.A.; McCabe, P.; Pidcock, E.; Platings, M.; Shields, G.P.; Stevens, J.S.; Towler, M.; et al. Mercury 4.0: From visualization to analysis, design and prediction. *J. Appl. Crystallogr.* **2020**, *53*, 226–235. [CrossRef] [PubMed]

57. Dolomanov, O.V.; Bourhis, L.J.; Gildea, R.J.; Howard, J.A.K.; Puschmann, H. Olex2: A complete structure solution, refinement and analysis program. *J. Appl. Crystallogr.* **2009**, *42*, 339–341. [CrossRef]

Article

Solvatochromism and Selective Sorption of Volatile Organic Solvents in Pyridylbenzoate Metal-Organic Frameworks

Christophe A. Ndamyabera, Savannah C. Zacharias, Clive L. Oliver and Susan A. Bourne *

Centre for Supramolecular Chemistry Research, Department of Chemistry, University of Cape Town, Rondebosch 7701, South Africa
* Correspondence: susan.bourne@uct.ac.za

Received: 20 July 2019; Accepted: 13 August 2019; Published: 15 August 2019

Abstract: Using cobalt(II) as a metal centre with different solvent systems afforded the crystallization of isomorphous metal-organic frameworks {[Co(34pba)(44pba)]·DMF}$_n$ (**1**) and {[Co(34pba)(44pba)]·(C$_3$H$_6$O)}$_n$ (**2**) from mixed 4-(4-pyridyl)benzoate (44pba) and 3-(4-pyridyl)benzoate (34pba) ligands. Zinc(II) under the same reaction conditions that led to the formation of **1** formed an isostructural {[Zn(34pba)(44pba)]·DMF}$_n$ framework (**3**). Crystal structures of all three MOFs were elucidated and their thermal stabilities were determined. The frameworks of **1**, **2**, and **3** were activated under vacuum to form the desolvated forms **1d**, **2d**, and **3d**, respectively. PXRD results showed that **1d** and **2d** were identical, consequently, **1d** and **3d** were then investigated for sorption of volatile organic compounds (VOCs) containing either chloro or amine moieties. Thermogravimetric analysis (TGA) and nuclear magnetic resonance (NMR) were used to determine the sorption capacity and selectivity for the VOCs. Some sorption products of **1d** with amines became amorphous, but the crystalline framework could be recovered on desorption of the amines. Investigation of the sorption of water (H$_2$O) and ammonia (NH$_3$) in **1d** gave rise to new phases identifiable by means of a colour change (solvatochromism). The kinetics of desorption of DMF, water and ammonia from frameworks **1d** and **3d** were studied using non-isothermal TGA. Activation energies for both cobalt(II) and zinc(II) frameworks are in the order NH$_3$ < H$_2$O < DMF, with values for the **1d** analogue always higher than those for **3d**.

Keywords: metal-organic frameworks; vapour sorption; solvatochromism; desorption kinetics

---

1. Introduction

Volatile organic compounds (VOCs) are organic compounds with an appreciable vapour pressure at ambient temperature. They include naturally occurring and synthetic compounds and range in effect from harmless to toxic. Some VOCs have been shown to have malodorous, mutagenic or carcinogenic properties [1–3] and some have been implicated in causing air pollution, particularly in developing countries [4], and are partly responsible for the generation of photochemical ozone and smog precursors. They are thus considered as harmful pollutants [2,3]. Some industrial manufacturing processes, as well as the use of manufactured materials, can increase the emission of VOCs into the local environment [5,6]. As a consequence, the development of effective technologies to mitigate the emission of VOCs has received increasing attention [1]. Some reports have shown promising removal and recovery methods of VOCs from air and water through adsorption processes [7–9]. Solid adsorbents have been shown to be superior compared to other techniques of decontamination of air or water, owing to their relative low cost, wide range of applications, simplicity of design, easy operation, low harmful secondary products and the feasible regeneration of these solid adsorbents [10]. Traditional solids adsorbents such as zeolites and activated carbon (AC) can be used for sorption purposes but they have shortcomings such as low surface area and the requirement of high temperature for their synthesis and regeneration [11,12]. Recent reports have shown that metal-organic frameworks (MOFs) have higher adsorption capacities and lower energy costs for regeneration [4,10,12].

Porous MOFs are crystalline frameworks with a wide range of possible configurations arising from the coordination of metal centres or clusters and organic linkers. MOFs can be designed to have high surface areas [10,13–15], easy functionalization, and tunable porosities, making them preferable to zeolites and activated carbon for many applications. Additionally, the coexistence of inorganic (hydrophilic) and organic (hydrophobic) moieties in MOFs structure may offer control of their interaction with guest molecules [4]. Thus, MOFs are of interest for a wide range of applications such as gas sorption [4,7,16], storage [17], separation [18–21], and sensing [22–25]. The choice of organic linker is key to MOF properties. The most commonly used linkers are those that can coordinate to metal ions via oxygen or nitrogen donors. Prior studies in our laboratories [19] and elsewhere [26] have shown that combining carboxylate and pyridyl or triazole aromatic rings allows dynamic rotation between the aromatic rings which in turn generates flexible MOFs. This is a key feature for their selective sorption capacity [19,26]. The recognition of chemical information by an adsorbent MOF may be characterised by colour change known as chromism [27,28], or reversible change in structure size known as a breathing phenomenon [19,23]. The latter has been observed in both single ligand MOFs such as $[Zn(34pba)_2]_n$ as well as in a mixed ligand MOF $[Cd(34pba)(44pba)]_n$; where the channels react to stimuli caused by the temperature and size of the entering molecules such as alkyl alcohols, N,N-dimethylformamide (DMF) and N,N-dimethylacetamide (DMA) [19,26]. However, it can be difficult to characterise the sorbed product due to a loss of crystallinity after removal or inclusion of guests [27,29–31]. Furthermore, the selective sorption capacity for VOCs such as chlorinated solvents and amines are rarely investigated. In this paper we report the synthesis of three-dimensional isomorphous and isostructural MOFs from cobalt(II) and zinc(II) with two related ligands, 3-(4-pyridylbenzoate) (34pba) and 4-(4-pyridylbenzoate) (44pba). These non-interpenetrated frameworks retain the framework structure and crystallinity on activation under vacuum. Their sorption capacity for amines and chlorinated solvents was investigated, as was their relative selectivity for sorption of chlorinated VOCs.

2. Materials and Methods

All chemicals were obtained from commercial sources and were used without further purification. $\{[Co(34pba)(44pba)]\cdot DMF\}_n$ (**1**), $\{[Co(34pba)(44pba)]\cdot(C_3H_6O)\}_n$ (**2**), and $\{[Zn(34pba)(44pba)]\cdot DMF\}_n$ (**3**) (44pba = 4-(4-pyridyl)benzoate and 34pba = 3-(4-pyridyl)benzoate) were solvothermally synthesized as detailed in Table 1. Compounds **1**, **2**, and **3** were activated at 210 °C under vacuum for 6 h which resulted in **1d**, **2d**, and **3d**, respectively. The activated samples were placed in narrow vials which were placed into larger vials containing VOCs and sealed to allow vapour sorption at room temperature (r.t., ca. 25 °C) for between one day and two weeks. The VOCs selected for study were dichloromethane (DCM), chloroform (CHCl$_3$), chlorobenzene (ClBenz), water, ammonia, methylamine (MeNH$_2$), 1-propylamine (PropNH$_2$), 1-butylamine (ButNH$_2$), benzylamine (BzNH$_2$), and 1-phenylethylamine(PhEtNH$_2$). The regeneration of the activated sorbents was carried out using the same conditions as for activation.

Table 1. Experimental conditions for the synthesis of **1**, **2**, and **3**.

	Metal Salt	Ligands	Solvent System	Conditions
1	CoCl$_2$·6H$_2$O (6 mg, 0.03 mmol)	34pba/44pba (10 mg, 0.050 mmol each)	DMF(6 mL)/Ethanol (2 mL)	120 °C for 3 days
2	CoCl$_2$·6H$_2$O (6 mg, 0.03 mmol)	34pba/44pba (10 mg, 0.050 mmol each)	Acetonitrile(6 mL)/water (2 mL)	120 °C for 3 days
3	Zn(NO$_3$)·6H$_2$O (30 mg, 0.13 mmol)	34pba/44pba (40 mg, 0.20 mmol each)	DMF(6 mL)/Ethanol (2 mL)	120 °C for 3 days

Competitive sorption for chlorinated solvents was performed by placing equivalent volumes of two different solvents into a large vial and the relevant activated MOF into a small vial. The latter was then placed into the large vial and sealed for two days for the sorption of the vapours.

2.1. Thermogravimetric Analysis (TGA) and Differential Scanning Calorimetry (DSC)

Thermogravimetric analysis (TGA) was performed using a TA Instrument TA-Q500 on 1–2 mg samples in open platinum pans under nitrogen gas flow (50 mL min^{-1}) at a heating rate of 10 °C min^{-1} within the temperature range 25–500 °C. The onset temperature for guest loss was determined using Differential scanning calorimetry (DSC). Samples of mass 1–2 mg were placed in aluminium pans with ventilated lids and heated at 10 °C min^{-1} using a TA Instrument DSC-Q200 under nitrogen gas flow (50 mL min^{-1}).

2.2. Infrared Spectroscopy

IR spectra were measured on a PerkinElmer Spectrum Two FTIR spectrometer equipped with an ATR Diamond accessory for powder samples. Samples were scanned over a range of 400–4000 cm^{-1}.

2.3. Nuclear Magnetic Resonance (NMR)

Solids containing the guest species were soaked into DMSO-d_6 and heated in order to release the guests into the solution for the NMR analysis. ^{1}H NMR spectra were recorded in DMSO-d_6 solution using a BRUKER 300 MHz spectrometer at 303 K. Appropriate signals were integrated to determine the ratio of the respective guests in the MOFs.

2.4. Powder X-ray Diffraction (PXRD)

Powder X-ray diffraction (PXRD) patterns were measured on a Bruker D8 Advance X-ray diffractometer operating in a DaVinci geometry equipped with a Lynxeye detector using CuKα-radiation (λ = 1.5406 Å). X-rays were generated at 30 kV and 40 mA. Samples were placed on a zero-background sample holder and scanned over a range of 4–40° in 2θ.

2.5. Crystal Structure Determination

Single crystals of good quality were selected using optical microscopy under plane-polarized light. Intensity data were recorded on a Bruker KAPPA APEX II DUO diffractometer using graphite monochromated Mo-Kα radiation (λ = 0.71073 Å) at 100 or 173 K. Data were corrected for Lorentz-polarization effects and for absorption (SADABS) [32]. The structures were solved by direct methods in SHELXS and refined by full-matrix least-squares on F^2 using SHELXL [33] within the XSEED [34] interface. The non-hydrogen atoms were located in difference electron density maps and were refined anisotropically while hydrogen atoms were placed in calculated positions and refined with isotropic temperature factors. Details of crystal structure refinements are given in Table 2 and Table S2.

Table 2. Crystallographic information for compounds **1**, **2**, and **3**.

Compound	1	2	3
Formula	$C_{27}H_{23}CoN_3O_5$	$C_{27}H_{22}CoN_2O_5$	$C_{27}H_{23}N_3O_5Zn$
Mass (g·mol^{-1})	528.41	513.39	534.85
Crystal size (mm^3)	$0.080 \times 0.10 \times 0.11$	$0.030 \times 0.060 \times 0.090$	$0.030 \times 0.030 \times 0.090$
Crystal system	Monoclinic	Monoclinic	Monoclinic
Space group	$P2_1/c$	$P2_1/c$	$P2_1/c$
a (Å)	9.203(2)	10.068(4)	9.339(1)
b (Å)	17.823(4)	15.632(5)	17.678(3)
c (Å)	14.718(3)	15.399(5)	14.735(2)
β (°)	92.75(3)	98.588(7)	93.189(5)
V (Å^3)	2411.3(8)	2396.4(1)	2428.84(7)
T (K)	100(2)	100(2)	173(2)
Z	4	4	4
D_c (g·cm^{-3})	1.456	1.423	1.463
μ(Mo–Kα) (mm^{-1})	0.756	0.757	1.055
F(000)	1092	1060	1104
Range scanned, θ (°)	1.80–28.34	1.87–25.09	1.80–27.58
No. reflections collected	22,928	18,219	22,013
No. unique reflection	5981	4250	5584
No. reflections with I ≥ 2σ(I)	4089	2860	3713
Parameters/restraints	327/0	318/0	327/0
Goodness of fit, S	1.034	1.024	1.006
Final R indices (I ≥ 2σ(I))	0.0859	0.0899	0.0867
Final wR2 (all data)	0.1198	0.1248	0.1107
Min, max e$^-$ density (e Å^{-3})	0.414, −0.417	0.653, −0.455	0.421, −0.443

3. Results and Discussion

The frameworks in **1**, **2**, and **3** are identical in terms of connectivity and geometry, with the asymmetric unit consisting of a metal ion (Co^{2+} in **1** and **2**, Zn^{2+} in **3**) bound to one 34pba and one 44pba linker. A centre of inversion generates a dinuclear secondary building unit (SBU) in which the two metal ions are connected by two bridging 34pba linkers through carboxylate groups while each metal ion is also coordinated to one 34pba and one 44pba through the pyridyl-N and to a 44pba through a bidentate carboxylate. The extension of this SBU through space gives rise to a double-walled network of *bcu* topology where each side of the square channels consists of a 34pba and a 44pba linker (Figure 1 and Table 2) [26]. Hour-glass shaped channels running parallel to [100] contain DMF (**1** and **3**) or acetone (**2**) guest molecules. The presence of acetone in **2** was unexpected as a mixture of acetonitrile and water had been used to prepare this compound. Conversion of acetonitrile to acetone is likely to proceed via hydrolysis to acetic acid [35] followed by ketonization to form acetone [36,37]. There are weak hydrogen bonds between the guest oxygens and the aromatic walls of the MOF. While **1** and **3** are isostructural, the structure of **2** is subtly different. Torsion angles indicate that the rings of both linkers are twisted slightly more away from coplanar in **2** than in **1** or **3**, while the orientation of the carboxylate groups is closer to coplanar with the aromatic ring in **2** than in the other compounds (see Figure S1 and Table S1 in ESI). The effect of these small changes is a lengthening of unit cell axes *a* and *c* while axis *b* shortens, but without changing the symmetry or space group. It is likely that the guest influences this change through the flexibility of the bent 34pba and linear 44pba linkers which allow a hinge-like expansion or contraction of the guest-accessible void [26].

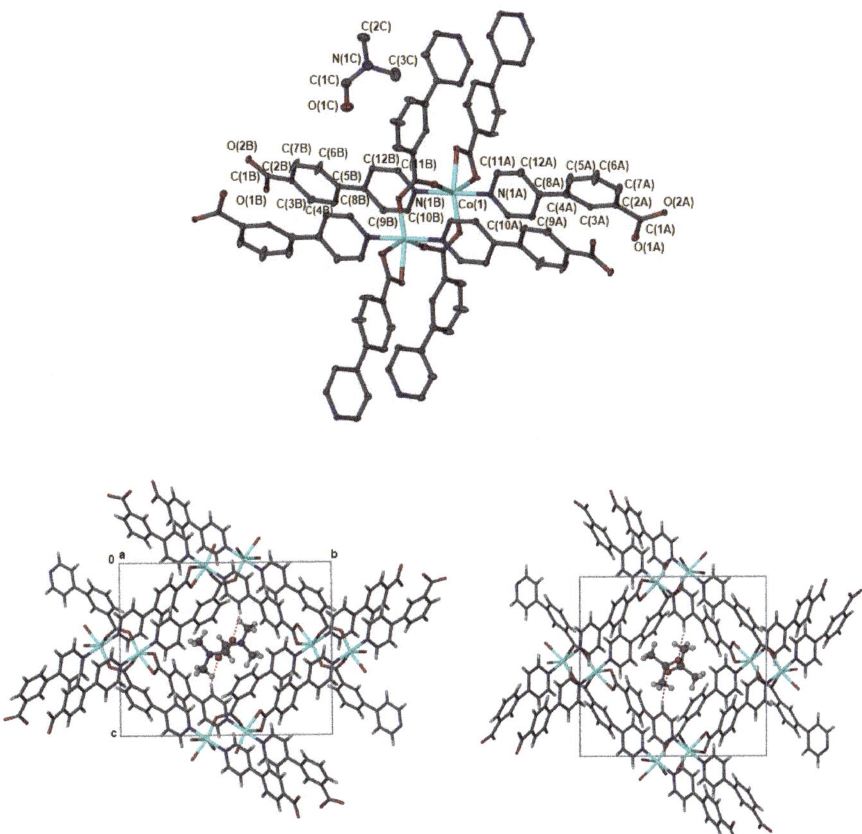

Figure 1. (**Top**) Coordination geometry and SBU in **1**. (**Bottom**) Packing diagrams of **1** (**left**) and **2** (**right**) showing the interactions between guest molecules and walls of the metal-organic frameworks (MOF).

The measured PXRD patterns in Figure 2 show the similarity of **1**, **2**, and **3** frameworks which matched well to the patterns calculated from single crystal structures. However, compound **2** had a small peak at 8.9° instead of 9.4° as for **1**. There are subtle differences in the pattern for **2** compared to those for **1** and **3**, for example, the shift in peaks at positions 12° and 21°. This dissimilarity could reflect the difference in the crystallographic data explained above. However, the activated forms of both **1d** and **2d** were the same after the removal of guest solvents. All activated forms **1d**, **2d**, and **3d** (**d**: Activated) retained their crystallinities with a slight shift of peaks (except **3d**) to higher 2θ values which corresponds to a small decrease in interplanar spacing in the frameworks after guest removal. Hence, these compounds were stable after removal of guest molecules which is not observed in all MOFs [27,29,30].

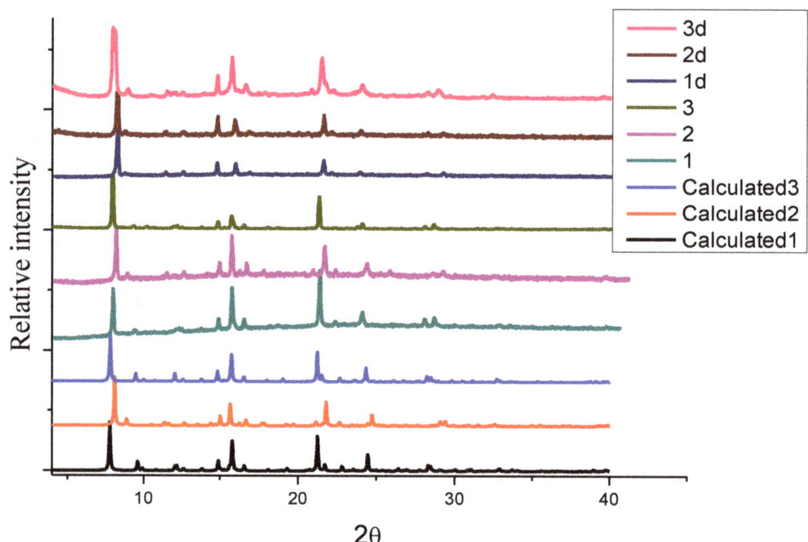

Figure 2. PXRD patterns for **1**, **2**, **3**, **1d**, **2d** and **3d** and their corresponding dry forms compared to their calculated patterns.

Carbonyl stretches in the FTIR spectra (Figure 3) confirm the presence of DMF (in **1** and **3**) at 1678 cm^{-1} and acetone (in **2**) at 1713 cm^{-1}. The removal of these guest solvents was confirmed by the absence of these bands in the spectra of **1d** and **3d**. The spectra of the activated forms were similar to one another as expected from the PXRD analysis.

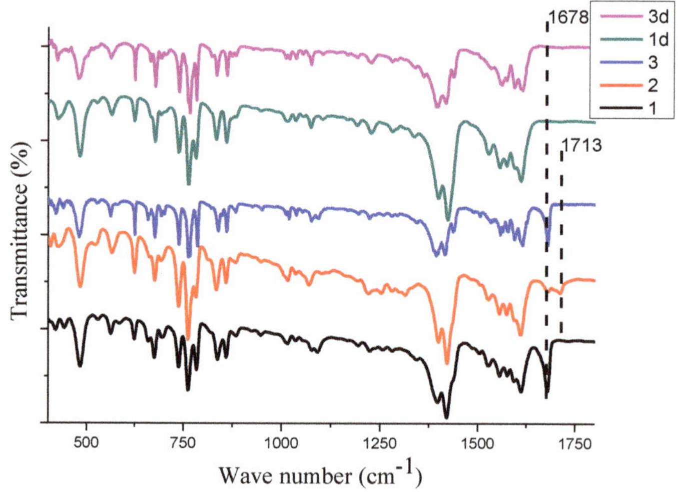

Figure 3. Infrared spectra of **1**, **2**, **3**, **1d**, and **3d** showing functional groups of guest molecules and coordination modes.

Thermogravimetric analysis (TGA) and DSC are shown in Figure 4. The weight loss of 14.1% between 120 and 216 °C in **1** was assigned to the removal of one DMF molecule (calculated 13.8%). This was characterised by a broad endothermic peak from 115–280 °C in the DSC. MOF **2** shows a total complex weight loss of 24.5%. The corresponding DSC trace shows an endothermic peak between 110 and 150 °C, followed by a small exotherm and a broad endothermic peak between 160 and 250 °C. It is possible that the removal of the acetone guest overlaps with the decomposition of the framework. This is contradictory to the PXRD evidence that the framework is robust. It is more likely therefore that the bulk sample selected for thermal analysis may contain a mixture of crystalline forms, only some of which correspond to the MOF characterised by crystal structure elucidation. An observed weight loss of 12.7% for **3** in the range of 120 and 216 °C was attributed to the removal of one DMF molecule (calculated 13.7%). The corresponding DSC curve shows a broad endothermic process between 130 and 260 °C. The TGA traces for **1d**, **2d**, and **3d** show no mass loss before 300 °C, indicating the solvent has been removed from the framework.

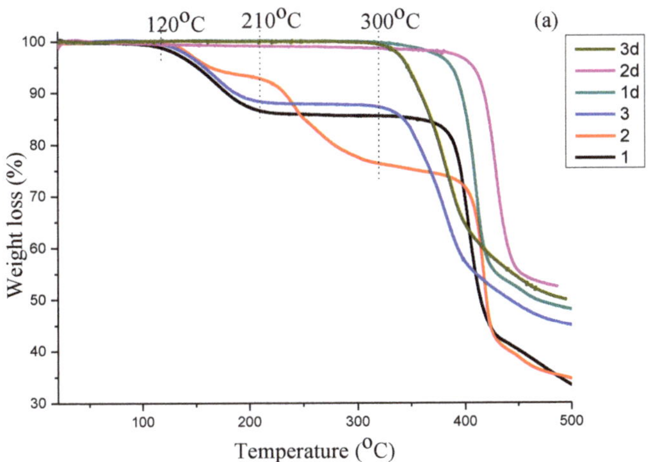

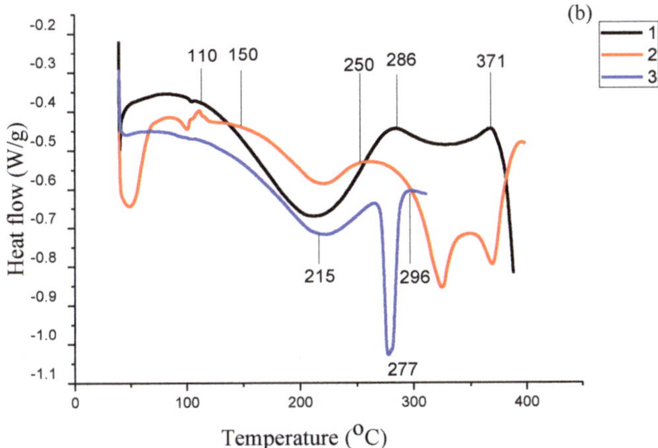

Figure 4. (a) TGA curves for **1, 2, 3, 1d, 2d,** and **3d** (b) DSC curves showing the process of the removal of guest molecules and decomposition of the framework.

3.1. Sorption of VOCs by Activated MOFs

To test the potential of these MOFs to serve as sorbents for pollutants, we carried out vapour sorption experiments using a series of chlorinated volatile organic compounds (VOCs) and another series of volatile amines. Sorption of water and of ammonia were also studied. Sorption experiments were carried out using activated samples of the Co-MOF (**1d**) and Zn-MOF (**3d**).

Sorption of chlorinated VOCs dichloromethane (DCM), chloroform ($CHCl_3$) and chlorobenzene (ClBenz) were achieved in a single crystal to single crystal manner, which allowed the elucidation of these crystal structures (Table S2 and Figure 5). The guests are stabilized in place by a number of weak interactions, including Cl···π, and C−H···π interactions and, in the case of chlorobenzene, through π···π interactions with the walls of the MOF. Comparable interactions have been observed in similar systems [38,39]. PXRD patterns (Figure S2a,b) of the phases obtained by vapour sorption of all tested chlorinated VOCs into **1d** or **3d** are unchanged from the starting activated phases, thus confirming the robustness of the retained framework structure [40].

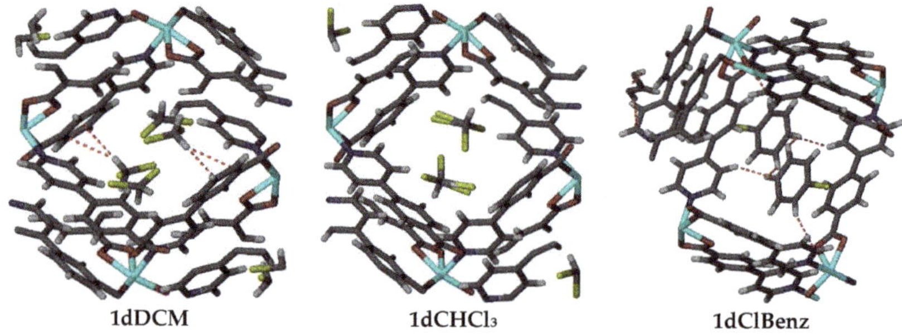

Figure 5. Inclusion of dichloromethane, chloroform and chlorobenzene into MOF **1d**.

The extent of selectivity in **1d** and **3d** was investigated from binary mixtures of the same chlorinated VOCs. Table 3 presents the solvent ratios obtained from the integration of relevant NMR peaks (Figure S5) from the competition studies. For **1d**, a mixture of DCM and chloroform were taken up without selectivity, while **3d** exposed to the same mixture selectivity absorbed DCM. Both MOFs selected DCM and chloroform over chlorobenzene from these respective binary mixtures. On the other hand, DCM was selectively sorbed 8.3 times over chlorobenzene. It should be noted that no attempt was made to compensate for differences in vapour pressure, and that the more volatile solvent was absorbed in each case, in contrast to a previous study carried out in our laboratory [19].

Table 3. Selectivity of **1d** and **3d** for chlorinated volatile organic compounds (VOCs).

1d	**Mole Ratio of VOCs in 1d** [a]	**Selectivity (Major Component)**
DCM/Chloroform	1:1	none
DCM/Chlorobenzene	8.3:1	DCM
Chloroform/Chlorobenzene	10:1	Chloroform
3d	**Mole Ratio of VOCs in 3d**	**Selectivity (Major Component)**
DCM/Chloroform	1.3:1	DCM
DCM/Chlorobenzene	1:0	DCM
Chloroform/Chlorobenzene	3:1	Chloroform

[a] Determined by NMR (Figure S5).

1d and **3d** show similar sorption trends for chlorinated VOCs as well as a series of volatile amines (Figure S3). Table 4 lists the VOC sorption results for **1d** and **3d**. PXRD traces for sorbed complexes are shown in Figure S2. The loading values were calculated from TGA analysis (Figure S4) and compared to theoretical maximum loading capacities. The loading capacity (L_c) is calculated from the crystallographically derived void volume and the liquid density of the respective solvents. The maximum loading capacity (ML_c) for the empty networks was estimated from

$$ML_c = (\text{solvent accessible void volume})/(Z \times \text{molecular volume}). \qquad (1)$$

Table 4. Uptake of selected solvents by the activated phases **1d** and **3d**.

VOC	Experimental Mass Loss, TGA (%)	Temperature Range of Mass Loss (°C)	Loading Capacity, L_c (x in Proposed Formula: {[M(34pba)(44pba)]·x Solvent}$_n$)	ML_c	% Loading Capacity
1d					
DCM	14.0	60–154	0.9	1.3	69
CHCl$_3$	17.1	118–285	0.8	1.0	80
ClBenz	13.0	87–264	0.6	0.8	75
H$_2$O	15.4	60–134	4.6	4.6	100
NH$_3$	12.9	60–150	4.0	3.5	114
MeNH$_2$	26.1	30–220	5.2	1.9	273
PropNH$_2$	33.4	30–220	3.9	1.0	390
ButNH$_2$	31.0	30–220	2.8	0.8	350
BzNH$_2$	52.0	65–260	4.6	0.8	575
PhEtNH$_2$	9.7	170–310	0.4	0.7	57
3d					
DCM	11.0	88–220	0.7	1.4	50
CHCl$_3$	13.3	110–232	0.6	1.1	55
ClBenz	11.0	61–252	0.5	0.8	63
H$_2$O	12.9	73–155	3.8	4.8	79
NH$_3$	12.5	59–127	3.9	3.6	108
MeNH$_2$	18.2	30–280	3.3	1.9	174
PropNH$_2$	18.4	30–263	1.8	1.0	180
ButNH$_2$	29.2	50–290	2.6	0.9	289
BzNH$_2$	36.0	88–290	2.4	0.8	300
PhEtNH$_2$	8.4	77–290	0.3	0.7	43

The solvent-accessible void volume of **1d** and **3d** were estimated using Mercury with a probe radius of 1.2 Å and a grid step of 0.2 Å and were found to be 549.0 and 571.4 Å^3 per unit cell respectively [41].

For the chlorinated solvents, the loading capacity (L_c) in the proposed formula {[M(34pba)(44pba)]·x solvent}$_n$ for both systems is lower than the maximum loading capacity. For each individual solvent, the sorption is higher for **1d** than for **3d**.

Water is taken up to near full capacity by both **1d** and **3d**, with little disruption of the framework.

To test the potential of these compounds as sorbents for amines, the activated MOFs **1d** and **3d** were exposed to the vapours of a series of amines, *viz.* ammonia (NH$_3$), methylamine (MeNH$_2$), propylamine (PropNH$_2$), 1-butylamine (ButNH$_2$), benzylamine (BzNH$_2$) and phenylethylamine (PhEtNH$_2$), Table 4, Figures S2 and S3. The crystal quality of the resultant compounds was too poor to allow full structural characterisation.

For all amines except phenylamine, the loading capacity of **1d** exceeds the maximum calculated from simple molecular volumes. Complexes also become amorphous. To further understand this, we exposed **1d** to benzylamine (BzNH$_2$) and found that the material remained crystalline until a mass loss of 40% was recorded. Subsequent desorption of the BzNH$_2$ from amorphous **1dBzNH$_2$** under vacuum recovered crystalline **1d** (Figure S9). In **3d** on the other hand, while the loading values obtained were again higher than the calculated maximum, the compounds retained their crystallinity but show some differences in phase in their PXRD traces. As with the chlorinated solvents, the amount sorbed by **1d** is greater than that for **3d**.

Amines are capable of hydrogen bonding, hence stronger intermolecular interactions, than chlorinated VOCs, which may allow them to pack more compactly into the channels, and to interact strongly with the internal surfaces of the MOFs, leading to higher loading values [39,42] and phase changes [43–45]. For benzylamine in particular, the MOFs took up a large amount, which could be attributed to aromatic stacking between $BzNH_2$ and the aromatic rings in the MOF walls [46]. The lower sorption capacity for $PhEtNH_2$ is the result of steric effects and lower polarity. No tests for selectivity among amine VOCs were performed.

3.2. Solvatochromism

Their PXRD patterns (Figure 6) show that the sorption of H_2O and NH_3 by **1d** formed new phases (**1dw** and **1dNH₃**, respectively) with noticeable colour changes from red to khaki (Figure 7). Upon desorption, both **1dw** and **1dNH₃** resulted in purple powder phases, which are amorphous (**1dwTG** and **1dNH₃TG**). However, the crystallinity, as well as their khaki colours, were restored after reabsorption (**1dwTGw** and **1dNH3TGNH₃**). Solvatochromism in MOFs has been reported to be the result of the supramolecular interactions such as hydrogen bonding and/or the coordination of the solvent molecules to the metal centres in the frameworks [27,44,47]. These interactions affect the energy associated with d-d transitions resulting in visible colour changes [27,39].

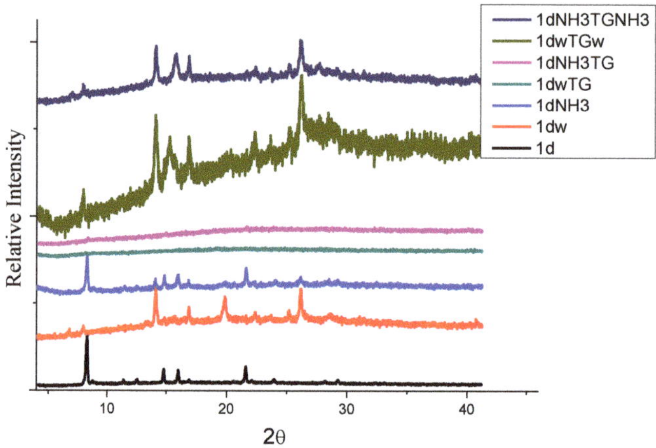

Figure 6. PXRD patterns for reversible sorption for ammonia, and H_2O by **1d**.

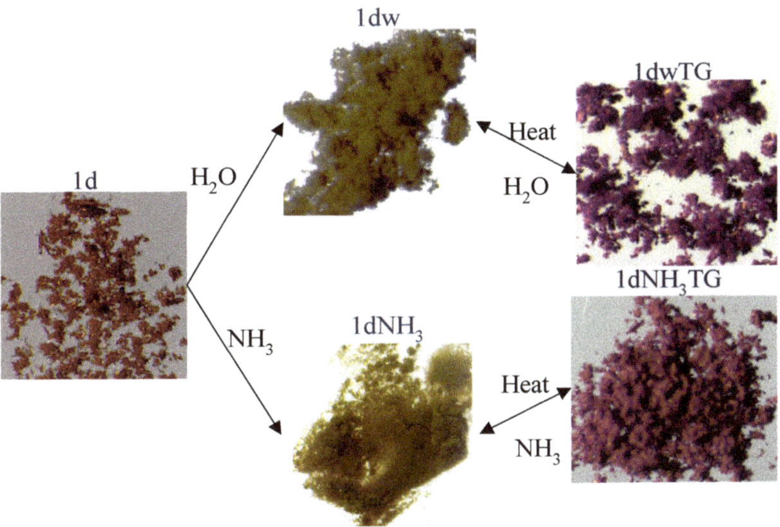

Figure 7. Reversible sorption of H_2O and NH_3 in **1d** and associated colour changes.

3.3. Kinetics of Desorption from 1 and 3

TGA may be used to determine the activation energy (Ea) of the guest desorption process. We used the Ozawa model-free method [48] to study the removal of guests DMF, NH_3, and H_2O for both systems reported here. Samples of mass 1–2 mg were heated at different heating rates (5, 10, 20, and 30 °C min^{-1}) in order to determine the activation energy associated with the removal of guest molecules from **1**, **3**, **1dw**, **3dw**, **1dNH$_3$**, and **3dNH$_3$** (Figure S7). Percentage mass losses along with the corresponding temperature at each heating rate were used to determine the activation energy (E_a) according to the equation:

$$\log\beta_\alpha = \log(A_\alpha\, E_{a\alpha}/g(\alpha)R) - 2.315 - 0.457(E_{a\alpha}/RT_\alpha) \tag{2}$$

where β_α is the heating rate, A_α is the frequency factor, $E_{a\alpha}$ is the activation energy, T_α is the temperature at each conversion level, and $g(\alpha)$ refers to the kinetic model. Figure S8 presents the plots of $\log\beta_\alpha$ versus reciprocal absolute temperature (in the form of 1000/T K^{-1}). Equating the slope to $-0.457(E_a/RT)$ allows one to calculate the activation energies, which are given in Table 5.

Table 5. Activation energy associated with removal of guest molecules.

Mass Loss (%)	E_a (kJ mol^{-1})					
	DMF from 1d	DMF from 3d	H_2O from 1dW	H_2O from 3dW	NH_3 from 1dNH$_3$	NH_3 from 3dNH$_3$
20	74.77	68.77	77.3	64.78	65. 10	58.46
40	75.31	66.50	72.59	57.35	67.8	59.39
60	72.77	70.57	75.24	65.23	68.61	62.01
80	77.30	64.08	74.75	68.38	68.77	62.01
Mean	**75.04 ± 1.68**	**67.48 ± 2.81**	**74.97 ± 1.93**	**63.94 ± 4.67**	**67.57 ± 1.70**	**60.47 ± 1.82**

The activation energies determined for desorption from **1d** are higher than the corresponding desorption from **3d**. This may be attributed to the difference in the metal centre as well as the solvent-accessible volume of the channels, *viz.* 549.0 Å^3 in **1d** and 571.4 Å^3 in **3d**, as the size of the cavities influences the supramolecular interactions possible between host and guest [47,49]. Activation energies associated with the desorption of DMF and H_2O are similar to one another but are higher than that of NH_3. Higher activation energies are generally associated with stronger host-guest interactions. The activation energies for desorption of DMF from **1d** and **3d** are comparable to those reported for the related MOF {[Co(34pba)$_2$]·DMF}$_n$ [47], while the average activation energies for the desorption of H_2O for **1d** and **3d** are also comparable to those reported for [Co(34pba)$_2$] isomers and chromium(III) terephthalate (MIL-101) [27,50]. There are no previous reports of desorption of ammonia from MOFs, so we compared our values to those reported for the desorption of NH_3 from Brønsted acid sites in zeolite ZSM-5 derivatives [51], which were found to have activation energies between 50 and 60 kJ mol^{-1}. Activation energies determined in this study are of the same order of magnitude, suggesting that intermolecular interactions such as hydrogen bonding with the channel walls are of approximately the same strength as those in the zeolite.

4. Conclusions

The coordination of two pyridylbenzoate ligands to cobalt(II) and zinc(II) metal centres formed isostructural {[Co(34pba)(44pba)]·DMF}$_n$ (**1**) and {[Zn(34pba)(44pba)]·DMF}$_n$ (**3**) compounds. Using an acetonitrile/water mixture instead of a DMF/ethanol solvent system led to a framework isomorphous to **1**, {[Co(34pba)(44pba)]·(C$_3$H$_6$O)}$_n$ (**2**) where acetonitrile had undergone hydrolysis and ketonization to produce the guest acetone (C$_3$H$_6$O). These MOFs retain their phase and crystallinity (**1d** and **3d**) after the removal of guest molecules under vacuum. Both **1d** and **3d** took up chlorinated and amine VOCs and showed potential selectivity in the sorption of binary chlorinated solvent mixtures, with preference for dichloromethane and chloroform. The activated MOFs had a higher sorption capacity for amine VOCs, which was attributed to their stronger intermolecular interactions with the framework. The sorption of chlorinated VOCs did not affect the crystallinity of the frameworks while some amine VOCs led to new phases in **3d** and amorphous phases in **1d**. The crystalline phase **1d** could be recovered from these amorphous phases on desolvation under vacuum. Characteristic solvatochromism was observed in **1d** on sorption of water or ammonia. The desorption of these two guests led to a new phase, which was reversible for both colour and crystallinity. The activation energy associated with the removal of DMF, H_2O, and NH_3 from MOFs **1d** and **3d** was determined and found to be comparable with previous systems studied. This study shows the potential of the synthesised MOFs having selectivity to take up guest molecules and undergo colour changes depending on the chemical and physical properties of the guest molecules. Therefore, studies on these MOFs for sensing and separation applications are ongoing.

Supplementary Materials: Supplementary data (crystal structure data, PXRD, thermal analysis, NMR) are available online at http://www.mdpi.com/2624-8549/1/1/9/s1. Crystallographic data for this paper have been deposited with the CCDC, accession numbers 1935229-1935234. These data can be obtained free of charge via www.ccdc.cam.ac.uk/data_request/cif or by emailing data_request@ccdc.cam.ac.uk.

Author Contributions: Conceptualization S.A.B.; synthesis, crystallography, thermal analysis C.A.N.; assistance with crystallography and analysis S.C.Z. and C.L.O.; writing—original draft preparation C.A.N.; writing—review and editing S.A.B., S.C.Z. and C.L.O.; supervision S.A.B. and C.L.O.

Funding: This research was funded by the National Research Foundation of South Africa, grant number 111699.

Conflicts of Interest: The authors declare no conflict of interest. The funders had no role in the design of the study; in the collection, analyses, or interpretation of data; in the writing of the manuscript, or in the decision to publish the results.

References

1. Lewandowski, D.A. *Design of Thermal Oxidation Systems for Volatile Organic Compounds*; CRC Press LLC: Boca Raton, FL, USA, 2000.

2. Martin, L.; Ognier, S.; Gasthauer, E.; Cavadias, S.; Dresvin, S.; Amouroux, J. Destruction of Highly Diluted Volatile Organic Components (VOCs) in Air by Dielectric Barrier Discharge and Mineral Bed Adsorption. *Energy Fuels* **2008**, *22*, 576–582. [CrossRef]

3. Hinojosa-Reyes, M.; Arriaga, S.; Diaz-Torres, L.A.; Rodríguez-González, V. Gas-Phase Photocatalytic Decomposition of Ethylbenzene over Perlite Granules Coated with Indium Doped TiO_2. *Chem. Eng. J.* **2013**, *224*, 106–113. [CrossRef]

4. Zhao, Z.; Wang, S.; Yang, Y.; Li, X.; Li, J.; Li, Z. Competitive Adsorption and Selectivity of Benzene and Water Vapor on the Microporous Metal Organic Frameworks (HKUST-1). *Chem. Eng. J.* **2015**, *259*, 79–89. [CrossRef]

5. Braek, A.M.; Almehaideb, R.A.; Darwish, N.; Hughes, R. Optimization of Process Parameters for Glycol Unit to Mitigate the Emission of BTEX/VOCs. *Process Saf. Environ. Prot.* **2001**, *79*, 218–232. [CrossRef]

6. Lee, Y.K.; Kim, H.J. The Effect of Temperature on VOCs and Carbonyl Compounds Emission from Wooden Flooring by Thermal Extractor Test Method. *Build. Environ.* **2012**, *53*, 95–99. [CrossRef]

7. Yang, K.; Sun, Q.; Xue, F.; Lin, D. Adsorption of Volatile Organic Compounds by Metal–Organic Frameworks MIL-101: Influence of Molecular Size and Shape. *J. Hazard. Mater.* **2011**, *195*, 124–131. [CrossRef] [PubMed]

8. Long, C.; Li, Q.; Li, Y.; Liu, Y.; Li, A.; Zhang, Q. Adsorption Characteristics of Benzene-Chlorobenzene Vapor on Hypercrosslinked Polystyrene Adsorbent and a Pilot-Scale Application Study. *Chem. Eng. J.* **2010**, *160*, 723–728. [CrossRef]

9. Bacchi, A.; Bourne, S.; Cantoni, G.; Cavallone, S.A.M.; Mazza, S.; Mehlana, G.; Pelagatti, P.; Righi, L. Reversible Guest Removal and Selective Guest Exchange with a Covalent Dinuclear Wheel-and-Axle Metallorganic Host Constituted by Half-Sandwich Ru(II) Wheels Connected by a Linear Diphosphine Axle. *Cryst. Growth Des.* **2015**, *15*, 1876–1888. [CrossRef]

10. Khan, N.A.; Hasan, Z.; Jhung, S.H. Adsorptive Removal of Hazardous Materials Using Metal-Organic Frameworks (MOFs): A Review. *J. Hazard. Mater.* **2013**, *244–245*, 444–456. [CrossRef]

11. Zhou, L.; Chen, Y.; Zhang, X.; Tian, F.; Zu, Z. Zeolites Developed from Mixed Alkali Modified Coal Fly Ash for Adsorption of Volatile Organic Compounds. *Mater. Lett.* **2014**, *119*, 140–142. [CrossRef]

12. Jhung, S.H.; Lee, J.H.; Yoon, J.W.; Serre, C.; Férey, G.; Chang, J.S. Microwave Synthesis of Chromium Terephthalate MIL-101 and Its Benzene Sorption Ability. *Adv. Mater.* **2007**, *19*, 121–124. [CrossRef]

13. Kobalz, M.; Lincke, J.; Kobalz, K.; Erhart, O.; Bergmann, J.; Lässig, D.; Lange, M.; Möllmer, J.; Gläser, R.; Staudt, R.; et al. Paddle Wheel Based Triazolyl Isophthalate MOFs: Impact of Linker Modification on Crystal Structure and Gas Sorption Properties. *Inorg. Chem.* **2016**, *55*, 3030–3039. [CrossRef] [PubMed]

14. Lincke, J.; Lässig, D.; Kobalz, M.; Bergmann, J.; Handke, M.; Möllmer, J.; Lange, M.; Roth, C.; Möller, A.; Staudt, R.; et al. An Isomorphous Series of Cubic, Copper-Based Triazolyl Isophthalate MOFs: Linker Substitution and Adsorption Properties. *Inorg. Chem.* **2012**, *51*, 7579–7586. [CrossRef] [PubMed]

15. Llewellyn, P.L.; Bourrelly, S.; Serre, C.; Vimont, A.; Daturi, M.; Hamon, L.; De Weireld, G.; Chang, J.-S.; Hong, D.-Y.; Hwang, Y.K.; et al. High Uptakes of CO2 and CH4 in Mesoporous Metal–Organic Frameworks MIL-100 and MIL-101. *Langmuir* **2008**, *24*, 7245–7250. [CrossRef] [PubMed]

16. Junghans, U.; Kobalz, M.; Erhart, O.; Preißler, H.; Lincke, J.; Möllmer, J.; Krautscheid, H.; Gläser, R. A Series of Robust Copper-Based Triazolyl Isophthalate Mofs: Impact of Linker Functionalization on Gas Sorption and Catalytic Activity. *Materials (Basel)* **2017**, *10*, 338. [CrossRef] [PubMed]

17. Xiang, Z.; Hu, Z.; Yang, W.; Cao, D. Lithium Doping on Metal-Organic Frameworks for Enhancing H2 Storage. *Int. J. Hydrogen Energy* **2012**, *37*, 946–950. [CrossRef]

18. Lv, X.; Shi, L.; Li, K.; Li, B.; Li, H. An Unusual Porous Cationic Metal – Organic Framework Fast and Highly Efficient Dichromate Trapping through a Single-Crystal to Single-Crystal Process. *Chem. Commun.* **2017**, *53*, 1860–1863. [CrossRef] [PubMed]

19. Mehlana, G.; Bourne, S.; Ramon, G. The Role of C–H$\cdots\pi$ Interactions in Modulating the Breathing Amplitude of a 2D Square Lattice Net: Alcohol Sorption Studies. *CrystEngComm* **2014**, *16*, 8160. [CrossRef]

20. Bhadra, B.N.; Ahmed, I.; Jhung, S.H. Remarkable Adsorbent for Phenol Removal from Fuel: Functionalized Metal–Organic Framework. *Fuel* **2016**, *174*, 43–48. [CrossRef]

21. Zhang, X.; Zhang, Y.Z.; Zhang, D.S.; Zhu, B.; Li, J.R. A Hydrothermally Stable Zn(II)-Based Metal-Organic Framework: Structural Modulation and Gas Adsorption. *Dalton Trans.* **2015**, *44*, 15697–15702. [CrossRef] [PubMed]

22. Tang, Y.-Y.; Wang, C.-J.; Chen, S.; Dai, H.-Y. A Terbium(III) Organic Framework as a Fluorescent Probe for Selectively Sensing of Organic Small Molecules and Metal Ions Especially Nitrobenzene and Fe^{3+}. *J. Coord. Chem.* **2017**, *70*, 3996–4007. [CrossRef]

23. Mahata, P.; Mondal, S.K.; Singha, D.K.; Majee, P. Luminescent Rare-Earth-Based MOFs as Optical Sensors. *Dalton Trans.* **2017**, *46*, 301–328. [CrossRef] [PubMed]

24. Wu, Y.; Yang, G.-P.; Zhao, Y.; Wu, W.-P.; Liu, B.; Wang, Y.-Y. Three New Solvent-Directed Cd(II)-Based MOFs with Unique Luminescent Properties and Highly Selective Sensors for Cu(2+) Cations and Nitrobenzene. *Dalton Trans.* **2015**, *44*, 3271–3277. [CrossRef] [PubMed]

25. Hulanicki, A.; Glab, S.; Ingman, F. Chemical Sensors Definitions and Classification. *Int. Union Pure Appl. Chem.* **1991**, *63*, 1247–1250. [CrossRef]

26. Zhou, H.L.; Zhang, Y.B.; Zhang, J.P.; Chen, X.M. Supramolecular-Jack-like Guest in Ultramicroporous Crystal for Exceptional Thermal Expansion Behaviour. *Nat. Commun.* **2015**, *6*, 1–7. [CrossRef] [PubMed]

27. Dzesse T, C.N.; Nfor, E.N.; Bourne, S.A. Vapor Sorption and Solvatochromism in a Metal-Organic Framework of an Asymmetric Pyridylcarboxylate. *Cryst. Growth Des.* **2018**, *18*, 416–423. [CrossRef]

28. Mehlana, G.; Ramon, G.; Bourne, S.A. Methanol Mediated Crystal Transformations in a Solvatochromic Metal Organic Framework Constructed from Co(II) and 4-(4-Pyridyl) Benzoate. *CrystEngComm* **2013**, *15*, 9521–9529. [CrossRef]

29. Davies, K.; Bourne, S.A.; Öhrström, L.; Oliver, C.L. Anionic Zinc-Trimesic Acid MOFs with Unusual Topologies: Reversible Hydration Studies. *Dalton Trans.* **2010**, *39*, 2869–2874. [CrossRef]

30. Lässig, D.; Lincke, J.; Moellmer, J.; Reichenbach, C.; Moeller, A.; Gläser, R.; Kalies, G.; Cychosz, K.A.; Thommes, M.; Staudt, R.; et al. A Microporous Copper Metal-Organic Framework with High H^2 and CO_2 Adsorption Capacity at Ambient Pressure. *Angew. Chem. Int. Ed.* **2011**, *50*, 10344–10348. [CrossRef]

31. Kubo, M.; Ushiyama, H.; Shimojima, A.; Okubo, T. Investigation on Specific Adsorption of Hydrogen on Lithium-Doped Mesoporous Silica. *Adsorption* **2011**, *17*, 211–218. [CrossRef]

32. Sheldrick, G.M. *SADABS, Version 2.05*; University of Göttingen: Göttingen, Germany, 2007.

33. Sheldrick, G.M. A short history of SHELX. *Acta Crystallogr. A* **2008**, *64*, 112–122. [CrossRef] [PubMed]

34. Barbour, L.J. X-Seed—A software tool for supramolecular crystallography. *J. Supramol. Chem.* **2001**, *1*, 189–191. [CrossRef]

35. Wang, Z.; Richter, S.M.; Rozema, M.J.; Schellinger, A.; Smith, K.; Napolitano, J.G. Potential Safety Hazards Associated with Using Acetonitrile and a Strong Aqueous Base. *Org. Process Res. Dev.* **2017**, *21*, 1501–1508. [CrossRef]

36. Bennett, J.A.; Parlett, C.M.A.; Isaacs, M.A.; Durndell, L.J.; Olivi, L.; Lee, A.F.; Wilson, K. Acetic Acid Ketonization over Fe_3O_4/SiO_2 for Pyrolysis Bio-Oil Upgrading. *ChemCatChem* **2017**, *9*, 1648–1654. [CrossRef] [PubMed]

37. Lukevics, E.; Stonkus, V.; Liepina, I.; Edolfa, K.; Jansone, D.; Leite, L.; Lukevics, E. Theoretical Study of the Ketonization Reaction Mechanism of Acetic Acid. *Latvian J. Chem.* **2009**, *1*, 61–67.

38. Mehlana, G.; Bourne, S.A.; Ramon, G. A New Class of Thermo- and Solvatochromic Metal–Organic Frameworks Based on 4-(Pyridin-4-Yl)Benzoic Acid. *Dalton Trans.* **2012**, *41*, 4224. [CrossRef] [PubMed]

39. Hu, Z.; Deibert, B.J.; Li, J. Luminescent Metal–Organic Frameworks for Chemical Sensing and Explosive Detection. *Chem. Soc. Rev.* **2014**, *43*, 5815–5840. [CrossRef] [PubMed]

40. Gao, Q.; Xu, J.; Cao, D.; Chang, Z.; Bu, X.H. A Rigid Nested Metal–Organic Framework Featuring a Thermoresponsive Gating Effect Dominated by Counterions. *Angew. Chem. Int. Ed.* **2016**, *55*, 15027–15030. [CrossRef]

41. Macrae, C.F.; Bruno, I.J.; Chisholm, J.A.; Edgington, P.R.; Mccabe, P.; Pidcock, E.; Rodriguez-monge, L.; Taylor, R.; Van De Streek, J.; Wood, P.A. Mercury CSD 2.0—New Features for the Visualization and Investigation of Crystal Structures. *J. Appl. Crystallogr.* **2008**, *41*, 466–470. [CrossRef]

42. Hwang, Y.K.; Hong, D.Y.; Chang, J.S.; Jhung, S.H.; Seo, Y.K.; Kim, J.; Vimont, A.; Daturi, M.; Serre, C.; Férey, G. Amine Grafting on Coordinatively Unsaturated Metal Centers of MOFs: Consequences for Catalysis and Metal Encapsulation. *Angew. Chem. Int. Ed.* **2008**, *47*, 4144–4148. [CrossRef] [PubMed]

43. Prodi, L.; Bolletta, F.; Montalti, M.; Zaccheroni, N. Luminescent Chemosensors for Transition Metal Ions. *Coord. Chem. Rev.* **2000**, *205*, 59–83. [CrossRef]

44. Britt, D.; Tranchemontagne, D.; Yaghi, O.M. Metal-Organic Frameworks with High Capacity and Selectivity for Harmful Gases. *Proc. Natl. Acad. Sci. USA* **2008**, *105*, 11623–11627. [CrossRef] [PubMed]

45. Dybtsev, D.N.; Chun, H.; Kim, K. Rigid and Flexible: A Highly Porous Metal–Organic Framework with Unusual Guest-Dependent Dynamic Behavior. *Angew. Chem. Int. Ed.* **2004**, *116*, 5143–5146. [CrossRef]

46. Kim, H.; Kim, S.; Kim, J.; Ahn, W. Liquid Phase Adsorption of Selected Chloroaromatic Compounds over Metal Organic Frameworks. *Mater. Res. Bull.* **2013**, *48*, 4499–4505. [CrossRef]

47. Mehlana, G.; Bourne, S.A.; Ramon, G.; Öhrström, L. Concomitant Metal Organic Frameworks of Cobalt(II) and 3-(4-Pyridyl) Benzoate: Optimized Synthetic Conditions of Solvatochromic and Thermochromic Systems. *Cryst. Growth Des.* **2013**, *13*, 633–644. [CrossRef]

48. Ozawa, T. A New Method of Analyzing Thermogravimetric Data. *Bull. Chem. Soc. Jpn.* **1965**, *38*, 1881–1886. [CrossRef]

49. Khuong, T.; Ramsahye, N.A.; Trens, P.; Tanchoux, N.; Serre, C.; Fajula, F.; Férey, G. Microporous and Mesoporous Materials Adsorption of C5–C9 Hydrocarbons in Microporous MOFs MIL-100 (Cr) and MIL-101 (Cr): A Manometric Study. *Microporous Mesoporous Mater.* **2010**, *134*, 134–140.

50. Xian, S.; Yu, Y.; Xiao, J.; Zhang, Z.; Xia, Q.; Wang, H.; Li, Z. RSC Advances Competitive Adsorption of Water Vapor with VOCs Dichloroethane, Ethyl Acetate and Benzene on MIL-101(Cr) in Humid Atmosphere. *RSC Adv.* **2015**, *5*, 1827–1834. [CrossRef]

51. Costa, C.; Dzikh, I.P.; Lopes, M.; Lemos, F. Activity–Acidity Relationship in Zeolite ZSM-5. Application of Bronsted-Type Equations. *Mol. Catal. A* **2000**, *154*, 193–201. [CrossRef]

 *chemistry*

Article

Controlled Stepwise Synthesis and Characterization of a Ternary Multicomponent Crystal with 2-Methylresorcinol

Daniel Komisarek [1], Carsten Schauerte [2] and Klaus Merz [3,*]

[1] Universitaet Duesseldorf, Inorganic Chemistry, Universitaetsstrasse 1, 40225 Düsseldorf, Germany; Daniel.komisarek@hhu.com or Daniel.Komisarek@rub.de

[2] Solid-Chem GmbH, Universitätstrasse 136, 44799 Bochum, Germany; schauerte@solid-chem.com

[3] Inorganic Chemistry, Ruhr-University Bochum, Inorganic Chemistry 1, Universitaetstrasse 150, 44801 Bochum, Germany

* Correspondence: klaus.merz@rub.de; Tel.: +49-234-3224187

Received: 6 February 2020; Accepted: 28 February 2020; Published: 4 March 2020

Abstract: A typical approach of a multicomponent crystal design starts with a retrosynthetic analysis of the target molecule followed by a one-pot reaction of all components. To develop protocols for multicomponent crystal syntheses, controlled stepwise syntheses of a selected crystalline ternary multicomponent system **1** involving 2-methylresorcinol (MRS), tetramethyl-pyrazine (TMP), and 1,2-bis(4-pyridyl)ethane (BPE) are presented. The obtained binary cocrystals **2** (involving MRS and TMP) and **3** (involving MRS and BPE) as well as the final resulting ternary multicomponent system **1** were characterized by X-ray analysis.

Keywords: multicomponent cocrystal; cocrystallization mechanism; cocrystal synthesis

---

1. Introduction

Polymorphism and cocrystallization are tools to add or enhance desirable properties of crystalline products [1–3]. The developments in this field in recent years have enabled more carefully tunable properties [4]. The tunability of the properties of (co)crystalline systems is of much importance, especially in the fields of pharmacy and medicine [5–7]. By introducing various coformers, factors like solubility, stability, bioavailability, or tolerability of active pharmaceutical ingredients can be influenced [8–10]. A better understanding of the driving force behind the genesis of molecular aggregation processes in the solid state can therefore be helpful to increase the efficiency of the synthesis of such crystalline products. Steed et al. have explained that asymmetry of crystalline structures can be exploited to purposefully produce medically relevant cocrystals [11]. While binary cocrystals have been focused on more commonly, less attention has been imposed on higher multicomponent systems like ternary [12], or even quaternary and quintenary [13] structures. In contrast to a well thought out stepwise multistage organic synthesis, a typical approach of a multicomponent crystal design starts with a retrosynthetic analysis of the target molecule followed by a one-pot reaction of all components in solution or solid state. To develop sophisticated and reliable protocols for multicomponent crystal syntheses, the complex modular assembly processes must be investigated by analysis of a hierarchy of intermolecular interactions and the molecular environment of involved components on each aggregation step [14].

In recent years, the group of Desiraju proposed that differing structural environments in the (n − 1)-multicomponent systems could lead to a state in which the incorporation of a new coformer would be more favourable for the overall structure [13]. Furthermore, Nangia and Bolla have confirmed that when using geometrically similar (in regards to size, planarity, etc.) heterosynthons, the formation of multicomponent crystals could simply be predicted by the strongest hydrogen bond former interactions, showing the importance of the kinetic effect [15]. Results presented by Aakeröy and Gunawardana also lay an emphasis on the geometric environment of (n − 1)-crystal systems as crucial for the formation of higher component products [16].

Hence, in this contribution, in contrast to a one-pot synthesis, the controlled stepwise syntheses of a selected crystalline ternary multicomponent system is presented. The mechanism of the formation of multicomponent cocrystals could be figured out based on the analysis of molecular arrangement in solid state of the selected components. The binary cocrystals as well as the final resulting ternary multicomponent system were characterized by X-ray analysis.

For this, 2-methylresorcinol (MRS) formation capabilities in conjunction with the N-bases tetramethylpyrazine (TMP) and 1,2-bis(4-pyridyl)ethane (BPE) (Scheme 1) were investigated, using neat, drop-, or liquid-assisted grinding techniques. Additionally, sequential and one-pot reactions were performed. The obtained crystalline products were characterized using Powder X-ray Diffraction (PXRD) and single crystal structure analysis.

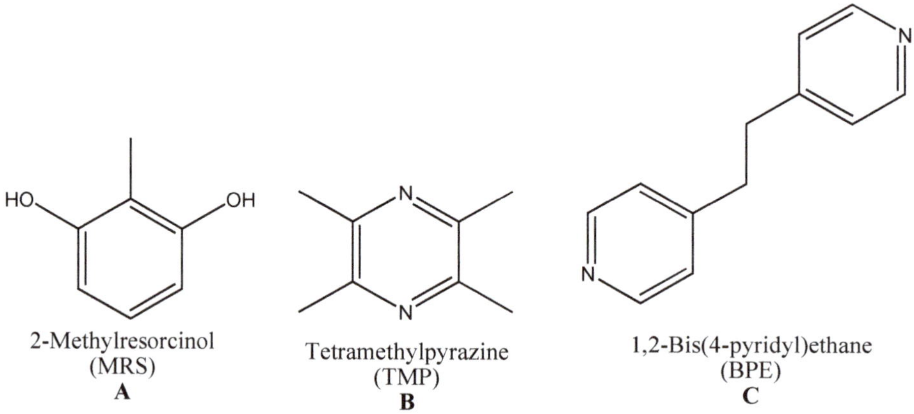

Scheme 1. Selected compounds.

2. Materials and Methods

By varying the component ratios or the solvent during the grinding experiments, it was possible to access cocrystals of A:2B **3** and A:C **2** as well as the ternary cocrystal 2A:2B:C **1**. Modified parameters include the ratio of the used components; grinding conditions like neat, drop-, or liquid-assisted grinding; and the choice of the solvents tetrahydrofuran vs. *n*-hexane. Only liquid-assisted grinding experiments in the presence of tetrahydrofuran were successful. Modification of the grinding conditions or the usage of *n*-hexane resulted in the observation of physical component mixtures. The experiments were either done as one-pot reactions of three components or by adding a component to sequentially synthesized and characterized binary cocrystals (Scheme 2).

PXRD was performed on each of the received crystalline products. Furthermore, single crystalline entities suitable for X-ray structure analysis of these multicomponent crystals was grown from solution (Table 1).

Preparation of **2**: 1 mmol of A and 2 mmol of B were placed in a mortar. Ten drops of tetrahydrofuran (THF) were added, and the resulting mixture was ground for 5 min. The resulting powder was characterized by PXRD.

Preparation of **3**: 1 mmol of A and C, respectively, were placed in a mortar. Ten drops of THF were added, and the resulting mixture was ground for 5 min. The resulting powder was characterized by PXRD.

Preparation of **1**:

a.   One-pot synthesis: 2 mmol of A, 2 mmol B, and 1 mmol C were placed in a mortar and pestle. Ten drops of THF were added, and the resulting mixture was ground for 5 min. The resulting powder was characterized by PXRD.

b.   From **2**: The latter was prepared in the previously mentioned way and identified as **2** by PXRD. The powder of **2** was placed in a mortar, and 1 mmol of A and 1 mmol of C and 10 drops of THF were added to the mixture. It was then ground for 5 min. The resulting powder was characterized by PXRD.

c.   From **3**: The latter was prepared in the previously mentioned way and identified as **3** by PXRD. The powder of **3** was placed in a mortar, and 1 mmol of A and 2 mmol of B and 10 drops of THF were added to the mixture. It was then ground for 5 min. The resulting powder was characterized by PXRD.

d.   From the physical mixture of B and C: 1 mmol of each were put in a mortar. 10 drops of THF were added, and the mixture was ground for 5 min. The resulting powder was characterized by PXRD. Only reflections of the physical mixture (B and C) were observed in this PXRD. Thereafter, the powder was placed in a mortar and 2 mmol of A and 1 mmol of B, and 10 drops of THF were added. The mixture was ground for 5 min, and the resulting powder was characterized by PXRD.

All obtained crystalline materials were characterized by powder X-ray diffraction (PXRD) (Figure 1), measured on a Bruker D2 PHASER diffractometer in flat mode and Bragg–Brentano geometry using filtered CuKα and CuKβ radiation.

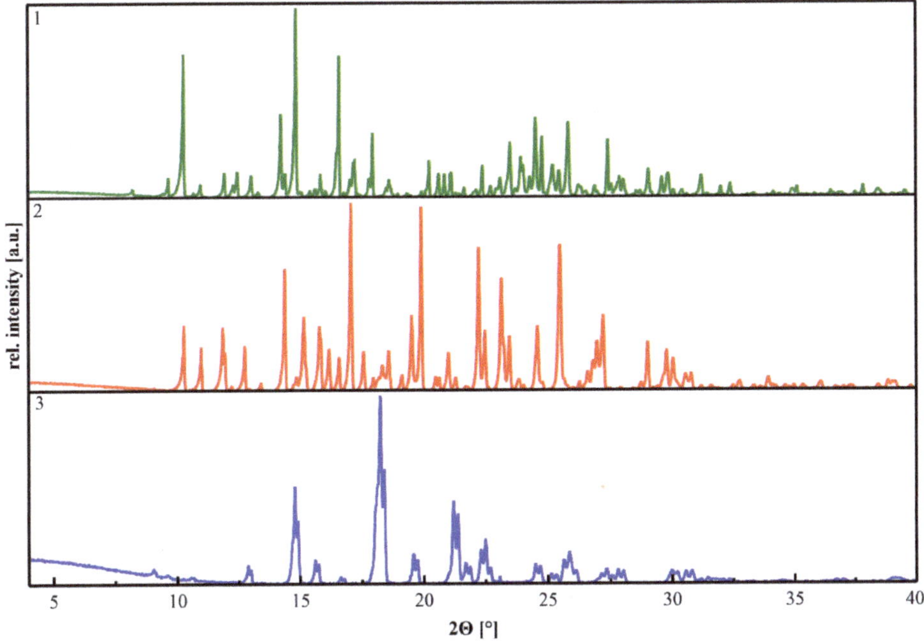

Figure 1. The diffractograms show the diffraction patterns of the obtained multicomponent crystals **1** (2A:2B:C) green graph, **2** (A:2B) red graph, and **3** (A:C) blue graph.

Single-crystal X-ray diffraction measurements of **2**, **3** were carried out on a Rigaku Synergy diffractometer using monochromated Mo Kα radiation ($\lambda = 0.71073$). Single-crystal X-ray diffraction measurement of **1** was carried out on a Rigaku Supernova using monochromated Cu Kα radiation (λ = 1.54184). Structures were solved by direct methods, and all nonhydrogen atoms were refined anisotropically on F^2 (program SHELXTL-97, G.M. Sheldrick, University of Göttingen, Göttingen, Germany). Crystallographic crystal data and processing parameters are shown in Table 1. CIF-files giving X-ray data with details of refinement procedures for **1, 2,** and **3** CCDC Nr. 1981507-1981509 are available free of charge via the Internet at http://pubs.acs.org.

The UNI Force Field Calculations of Mercury crystallography package were used to calculate inter-molecular potentials of the crystal structure **2** with normalized hydrogens (kJ/mol).

Table 1. Summary of crystal data, data collection, and refinement parameters for **1, 2,** and **3**.

	1 **2A:2B:C**	**2** **A:2B**	**3** **A:C**
formula	$2(C_7H_8O_2)$: $2(C_8H_{12}N_2)$: $C_{12}H_{12}N_2$	$C_7H_8O_2$: $2(C_8H_{12}N_2)$	$C_7H_8O_2$: $C_{12}H_{12}N_2$
formula weight	704.90	396.53	308.37
temperature [K]	173	173	173
crystal system	triclinic	monoclinic	monoclinic
space group	P-1	$P2_1/c$	$P2_1/c$
a	7.5145(2)	7.2801(2)	8.7006(9)
b	8.6750(2)	24.9768(8)	12.026(1)
c	15.0243(4)	12.0692(4)	17.180(2)
α	100.440(2)	90	90
β	101.969(2)	102.504(3)	103.78(1)
γ	92.396(2)	90	90
V [Å^3]	939.04(4)	2142.5(1)	1745.9(3)
Z	1	4	4
F (000)	378	856	656
no. of rflns. measured	16,811	16,224	23,798
no. of unique rflns.	3699	5440	3074
μ [mm^{-1}]	0.646	0.080	0.077
parameters	242	270	291
S (F^2)	1.035	1.086	1.022
$R1$ [$I > 2\sigma(I)$]	0.0547	0.0440	0.0545
$wR2$ (all rflns.)	0.1363	0.1280	0.1713

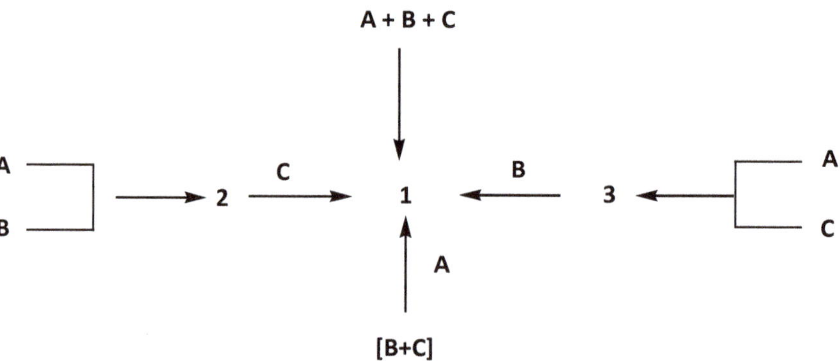

Scheme 2. Performed synthesis routes for the multicomponent crystal 1 (2A:2B:C).

3. Results

The crystalline product obtained by liquid-assisted grinding of a mixture of A, B, and C in the presence of THF shows reflections in the PXRD that cannot be attributed to the educts. The crystalline product was dissolved in nitromethane, and by cooling crystallization a single crystal was obtained (Table 1). The X-ray structure analysis confirms the successful crystallization of a ternary 2:2:1 A:B:C multicomponent system **1** (Figure 2).

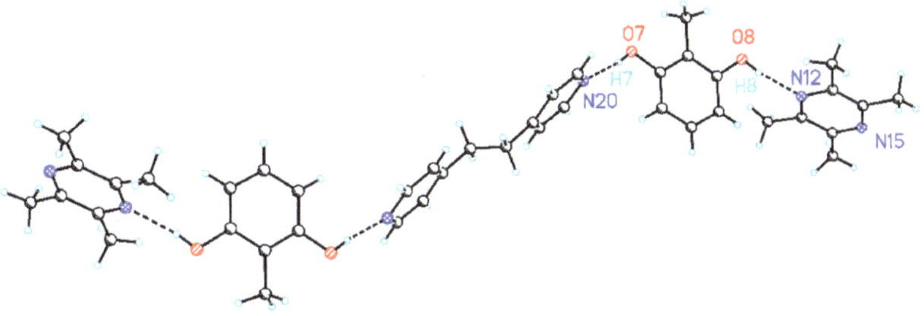

Figure 2. Packing view of ternary 2A:2B:C multicomponent system **1**.

The central 1,2-bis(4-pyridyl)ethane building block C forms hydrogen bonds via its two aromatic nitrogen centers to hydroxy groups from each of the two neighbouring 2-methylresorcinol molecules A. The remaining hydroxy groups of A coordinate via hydrogen bonds to one nitrogen center each of the two terminal tetramethylprazine molecules. Surprisingly, the second nitrogen centers of B are not involved in any further intermolecular interactions. Consequently, the 2:2:1 A:B:C multicomponent system **1** does not form infinite chains. This behavior is observed for B in only few selected multicomponent systems. Examples are cocrystalline systems with Br-C6F4-OH12 or Br-C6F4-COOH12 [16]. The analysis of the coordination behavior of A shows that the hydrogen bonds from A to B, in comparison to A to C, have no significant differences in the distances O $\cdots$ N 2750 Å and O $\cdots$ N 2755 Å as well as in the angles of the ring planes A:B 75° and A:C 69°.

As mentioned, the geometric environment of the involved building blocks and their intermolecular interactions are crucial parameters for the formation of higher multicomponent crystals. The one-pot reaction of all components (A, B, and C) leads to the ternary 2A:2B:C system **1**. The 2A:2B:C motif is discretely isolated by the unusual coordination behavior of the tailored B molecules. The hypothesis is that the geometric molecular environment is responsible in each aggregation step in the design of **1**. Based on differing structural environments in the crystalline products obtained by first aggregation steps, A with B and A with C, the crystal packing of 2A:2B:C—starting from A—can be built up by successive substitution of B by C or C by B. To investigate this hypothesis and to gain a deeper understanding of the formation of the multicomponent system, the cocrystals involving A and B as well A and C were crystallized and the crystal packing analyzed.

Cocrystals of **2** (A:2B) and **3** (A:C) could be obtained by grinding a 1:2 ratio (2) and 1:1 ratio (3). After that, suitable single crystals for XRD were formed after recrystallization of the crystalline powder from THF (**2**) and nitromethane (**3**). In contrast to the known crystal structure of the A:B cocrystal [8], the crystal structure of **2** represents a 2:1 cocrystal with one A and two B molecules (Figure 3). Due to the fact that the two molecules of B in **2** show explicit different structural environments, the description A:B:B′ seems more appropriate for this system. The central A unit forms chains via hydrogen bonds (O1 $\cdots$ N1 2.84 Å, O2 $\cdots$ N2 2.85 Å) to two neighbouring B molecules. While the D distances of the hydrogen bonds are practically identical, the dihedral angles between the planes of the arene ring of A and the neighboring pyrazine rings of B show slight variations (62°/68°). Additionally, present B′ molecules form columns with B by π-π stacking interactions. A comparison of the intermolecular potentials in the molecular arrangement of **2** pointed out the dominating role of the π-π stacking motif (−42 and −43 kJ/mol) in contrast to the described hydrogen bonds (−32 and −33 kJ/mol).

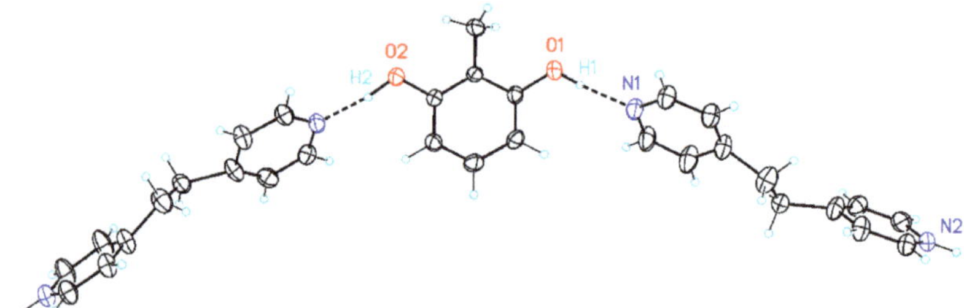

Figure 3. Packing view of (a) **2** (A:2B) O1-H1: 0.900 Å, O1 ⋯ N3 2.84 Å, O1-H1 ⋯ N3 161°, O2-H2: 0.883 Å, O2 ⋯ N2 2.85 Å, and O2-H2 ⋯ N4 167°·

The crystal structure **3** is built up by chains of linked 1:1 A:C cocrystals (Figure 4). Both hydroxyl groups of the central A molecule form hydrogen bonds to nitrogen centers of the two neighbouring C molecules. A comparison of both hydrogen bonds that indicates slight differences is be remarkable. In contrast to the hydrogen bond O1-H1 ⋯ N1, O2-H2 ⋯ N2 shows enlarged covalent O1-H2 and O2 ⋯ N2 hydrogen bonds. Besides the slightly different hydrogen bonds to **2**, differences in dihedral angles between the planes of the arene ring of A and the neighboring pyridyl rings of B are observed (68°/75°).

Figure 4. Packing view of (b) **3** (A:C). O1-H1: 0.908 Å, O1 ⋯ N1 2.716 Å, O1-H1 ⋯ N1 175°, O2-H2: 1.041 Å, O2 ⋯ N2 2.743 Å, and O2-H2 ⋯ N2 171°.

4. Discussion

If the formation of the described ternary multicomponent crystal **1** in a one-pot-synthesis is due to the different environments during the molecular recognition of A, B, and C in the crystallization process, **1** should also be synthesizable from the cocrystals **2** and **3**. In both cocrystals (**2** and **3**), the building block A shows significant differences in the molecular environment (Figure 5). In order to verify the assumption, appropriate crystallization experiments were carried out. Cocrystal **2** was grinded with C, and cocrystal **3** was grinded with B. Additionally, a physical mixture of B and C was grinded with A. The powder diffractograms of all three crystallization experiments show exclusively the reflections of the expected ternary multicomponent system **1**. It seems, when attempting to design a multicomponent crystal of nth degree, it is helpful to take a closer look at the (n − 1)th degree predecessors.

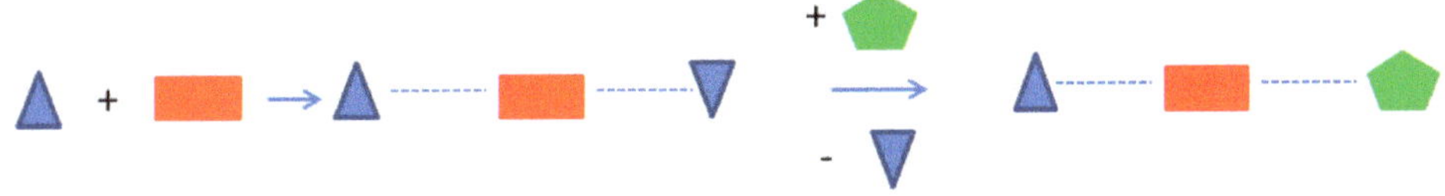

Figure 5. General proposed formation process of a ternary multicomponent crystal.

The hypothesis that differences in the molecular environment during the molecular recognition in the crystallization process are responsible for the formation of multicomponent crystal has been demonstrated in this study by using the building blocks 2-methylresorcinol (A), tetramethylpyrazine B and 1,2-bis(4- pyridyl)ethane C. The carefully designed cocrystals **2** (A:2B) and **3** (A:C) with differences in their molecular environment are suitable starting materials for successfully synthesizing the ternary multicomponent crystal **1** (2A:2B:C). Furthermore, the results indicate that Ostwald's rule of stages can be applied to solid state synthesis just as well as it can be applied to crystallization from solution.

Author Contributions: Project administration, supervision, and single crystal structure determination, K.M.; investigation and original draft preparation, D.K.; X-ray powder diffraction and solid state analysis, C.S. All authors have read and agreed to the published version of the manuscript.

Funding: This research was supported by the DFG Open Access Publication Funds of the Ruhr-Universität Bochum.

Conflicts of Interest: The authors declare no conflict of interest.

References

1. Bhowal, R.; Biswas, S.; Tumbarathil, A.; Koner, A.L.; Chopra, D. Exploring the Relationship between Intermolecular Interactions and Solid-State Photophysical Properties of Organic CoCrystals. *J. Phys. Chem. C* **2019**, *123*, 9311–9322. [CrossRef]
2. Soliman, I.I.; Kandil, S.M.; Abdou, E.M. Gabapentin-saccharin cocrystals with enhanced physicochemical properties and in vivo absorption formulated as oro-dispersible tablets. *Pharm. Dev. Technol.* **2020**, *2*, 227–236. [CrossRef] [PubMed]
3. Fondren, N.S.; Fondren, Z.T.; Unruh, D.K.; Maiti, A.; Cozzolino, A.; Lee, Y.J.; Hope-Weeks, L.; Weeks, B.L. Study of Physicochemical and Explosive Properties of a 2,4,6-Trinitrotoluene/Aniline Cocrystal Solvate. *Cryst. Growth Des.* **2020**, *1*, 116–129. [CrossRef]
4. Desiraju, G.R. Crystal engineering: Structure, property and beyond. *IUCr* **2017**, *4*, 710–711. [CrossRef] [PubMed]
5. Sathisaran, I.; Dalvi, S.V. Engineering Cocrystals of Poorly Water-Soluble Drugs to Enhance Dissolution in Aqueous Medium. *Pharmaceutics* **2018**, *10*, 108. [CrossRef] [PubMed]
6. Nangia, A.K.; Desiraju, G.R. Crystal Engineering: An Outlook for the Future. *Angew. Chem. Int. Ed.* **2019**, *58*, 4100–4107. [CrossRef] [PubMed]
7. Cains, P.W. Classical methods of preparation of polymorphs and alternative solid forms. In *Drugs and the Pharmaceutical Sciences*, 2nd ed.; Brittain, H.G., Ed.; CRC Press: New York, NY, USA, 2009; Volume 192, pp. 76–138.
8. Berry, D.J.; Steed, J.W. Pharmaceutical cocrystals, salts and multicomponent systems; intermolecular interactions and property based design. *Adv. Drug Deliv. Rev.* **2017**, *117*, 3–24. [CrossRef] [PubMed]
9. Duggirala, N.K.; Perry, M.L.; Almarsson, O.; Zaworotko, M.J. Pharmaceutical cocrystals: Along the path to improved medicines. *Chem. Commun.* **2016**, *52*, 640–655. [CrossRef] [PubMed]
10. Bolla, G.; Nangia, A. Pharmaceutical cocrystals: Walking the talk. *Chem. Commun.* **2016**, *52*, 8342–8360. [CrossRef] [PubMed]
11. Anderson, K.M.; Probert, M.R.; Whiteley, C.N.; Rowland, A.M.; Goeta, A.E.; Steed, J.W. Designing CoCrystals of Pharmaceutically Relevant Compounds That Crystallize with Z′ > 1. *Cryst. Growth Des.* **2008**, *9*, 1082–1087. [CrossRef]
12. Mir, N.A.; Dubey, R.; Tothadi, S.; Desiraju, G.R. Combinatorial crystal synthesis of ternary solids based on 2-methylresorcinol. *CrystEngComm* **2015**, *17*, 7866–7869. [CrossRef]

13. Mir, N.A.; Dubey, R.; Desiraju, G.R. Four- and five-component molecular solids: Crystal engineering strategies based on structural inequivalence. *IUCr* **2016**, *3*, 96–101. [CrossRef] [PubMed]
14. Gunawardana, C.A.; Aakeröy, C.B. Cocrystal synthesis: Fact, fancy, and great expectations. *Chem. Commun.* **2018**, *54*, 14047–14060. [CrossRef]
15. Bolla, G.; Nangia, A. Multicomponent ternary cocrystals of the sulfonamide group with pyridine-amides and lactams. *Chem. Commun.* **2015**, *51*, 15578–15581. [CrossRef] [PubMed]
16. Aakeroy, C.B.; Spartz, C.L.; Dembowski, S.; Dwyre, S.; Desper, J. A systematic structural study of halogen bonding versus hydrogen bonding within competitive supramolecular systems. *IUCrJ* **2015**, *2*, 498–510. [CrossRef] [PubMed]

chemistry

Article

Understanding Conformational Polymorphism in Ganciclovir: A Holistic Approach

Lorella Spiteri, Ulrich Baisch * and Liana Vella-Zarb *

Department of Chemistry, University of Malta, MSD, 2080 Msida, Malta; lorella.spiteri@um.edu.mt
* Correspondence: ulrich.baisch@um.edu.mt (U.B.); liana.vella-zarb@um.edu.mt (L.V.-Z.)

Abstract: We present a holistic crystallographic study of the antiviral ganciclovir, including insights into its solid-state behavior, which could prove useful during drug development, making the process more sustainable. A newly developed methodology was used incorporating a combination of statistical and thermodynamic approaches, which can be applied to various crystalline materials. We demonstrate how the chemical environment and orientation of a functional group can affect its accessibility for participation in hydrogen bonding. The difference in the nature and strength of intermolecular contacts between the two anhydrous forms, exposed through full interaction maps and Hirshfeld surfaces, leads to the manifestation of conformational polymorphism. Variations in the intramolecular geometry and intermolecular interactions of both forms of ganciclovir were identified as possible predictors for their relative thermodynamic stability. It was shown through energy frameworks how the extensive supramolecular network of contacts in form I causes a higher level of compactness and lower enthalpy relative to form II. The likelihood of the material to exhibit polymorphism was assessed through a hydrogen bond propensity model, which predicted a high probability associated with the formation of other relatively stable forms. However, this model failed to classify the stability of form I appropriately, suggesting that it might not have fully captured the collective impacts which govern polymorphic stability.

Keywords: supramolecular structure; conformational polymorphism; intermolecular contacts

Citation: Spiteri, L.; Baisch, U.; Vella-Zarb, L. Understanding Conformational Polymorphism in Ganciclovir: A Holistic Approach. *Chemistry* **2021**, *3*, 126–137. https://doi.org/10.3390/chemistry3010010

Received: 22 December 2020
Accepted: 20 January 2021
Published: 26 January 2021

Publisher's Note: MDPI stays neutral with regard to jurisdictional claims in published maps and institutional affiliations.

1. Introduction

In the last decades, there has been considerable investment into creating sustainable chemical technologies and reactions, specifically within the pharmaceutical industry, in order to maximize the efficiency of the traditionally lengthy and expensive process of drug development [1–3]. Computational advancements have been developed in parallel to such efforts, namely to unravel the intrinsic entanglement between the crystalline material and its solid-state properties, both for the possibility of exploiting such knowledge to engineer a pharmaceutical drug with the desired properties, and also to ensure its stability and reliability [3,4]. Given that the orientation and interactions between molecules within a crystal determine a material's physiochemical characteristics such as solubility, any alterations in this arrangement could have substantial implications, especially in a pharmaceutical context in relation to bioavailability and shelf life. Numerous publications illustrate examples of how different computational techniques are being applied to investigate and predict such features [5–9].

This case study revolves around ganciclovir (Figure 1), which is an acyclic guanine nucleoside analogue, that acts as a DNA synthesis inhibitor to various forms of herpes viruses, thereby reducing the rate of viral growth within the host [10]. In general, ganciclovir is formulated as a sodium salt and is administered through an intravenous route (Cytovene®) or also as an implant (Vitrasert®) [11]. It is also used as an ophthalmic gel which contains ganciclovir in its pure form, and a very small percentage of water (Virgan®) [11,12]. In spite of its importance as a BCS class III antiviral, and most especially as a treatment against cytomegalovirus, ganciclovir's physicochemical properties have still not been optimized [13,14]. They include poor permeability and limited bioavailability, all of which hinder the drug's performance. Due to such properties, this antiviral is administered in frequent and high doses, exposing the patient to higher risks of toxicity.

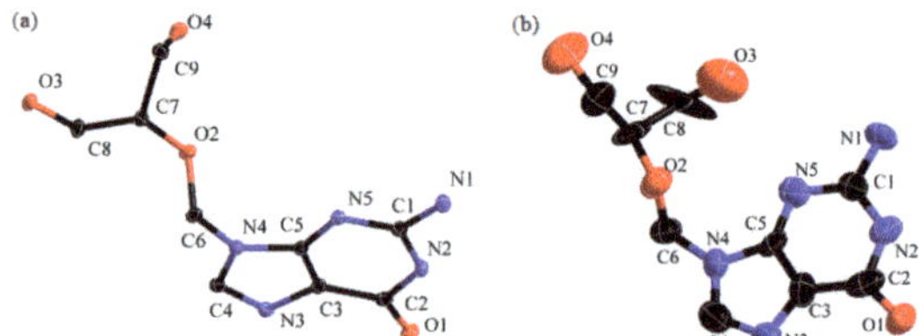

Figure 1. Chemical structure of ganciclovir.

In view of all the above, the need for a complete study arose, entailing a detailed analysis at intramolecular, intermolecular and supramolecular levels using computational methods. This investigation is a comparative study between two anhydrous, enantiotropically related conformational polymorphs I (USP reference standard) and II (Figure 2) in an attempt to identify the factors which influence polymorph stability and assess the possibility of forming other polymorphs [15,16]. Published experimental data characterizing ganciclovir polymorphs are used to complement the observations extracted from the developed methodology [15,17,18], which incorporates a combination of different statistical and thermodynamic approaches, namely the full set of programs and interfaces available through the Cambridge Structural Database (*Mercury, Mogul, Isostar*) and *CrystalExplorer*. This approach can be applied not only to the API (Active Pharmaceutical Ingredient) discussed in this article but also to other crystalline materials.

Figure 2. The molecular structure of: (**a**) anhydrous form I (CCDC refcode: UGIVAI01, 50% probability level); (**b**) anhydrous form II (CCDC refcode: UGIVAI, 50% probability level) [19]. Hydrogen atoms were omitted for clarity. The different temperatures at which diffraction data were collected (form I: 100 K, form II: 293 K) had no effect on the occurrence of the two crystalline forms themselves.

2. Computational Methods

2.1. Non-Bonding Interactions Analysis

The intramolecular geometry analysis was conducted using V1.7.5 *Mogul*, as described in a paper by Galek et al., together with some adjustments to optimize the results [20]. During this investigation, all organometallic entries were excluded, as their geometric parameters may have introduced a bias in the histograms provided in *Mogul* when analyzing a purely organic molecule. In cases where few fragments were available, the relevance threshold was reduced from 1.00 to 0.75.

The electrostatic potential was calculated through *Mercury* by means of the MOPAC (Molecular Orbital Package) interface, using the RM1 semi-empirical Hamiltonian method, and was mapped onto the VdW molecular surface [21–23].

IsoStar V2.2.5 was utilized for initial examination of the interactions within ganciclovir, with functional groups acting as the central groups [24]. The analysis of possible intermolecular interactions was taken further by means of the construction of full interaction maps in

Mercury, taking into consideration the environmental effects of collective factors and steric exclusion to produce maps that are unique to each form and conformation [25,26].

The detection of any intramolecular and intermolecular hydrogen bonding in the asymmetric unit was investigated using *Mercury* for both anhydrous forms individually. The definition of a hydrogen bond was modified to have an angle of more than 120°, and the distance range was slightly extended further to be less or equal to the sum of VdW radii + 1.00 Å, to ensure that all potential interactions were included [27].

Hirshfeld surfaces were plotted for each of the anhydrous forms using *CrystalExplorer*, so as to gain a better understanding of the network of non-bonding contacts, beyond conventional hydrogen bonds [28]. The participants of such contacts were identified through fingerprint scatterplots, which map the distance from a point on the surface to the closest nucleus inside the surface, d_i, against the distance outside the surface, d_e. Further details about the calculation of each descriptive variable used can be found elsewhere [29].

Crystal packing similarity was investigated through *Mercury*, using both anhydrous forms as reference, and the whole database was explored for any entry having sufficient packing similarity. Default selection options were retained, including a molecular cluster size of 15, 20% distance tolerance and 20° angle tolerance.

2.2. Energy Framework Analysis

Energy frameworks were constructed in *CrystalExplorer* using the CE-B3LYP model (basis set: 6-31G(d,p)), taking into account only intermolecular interactions within the radius of 3.8 Å from the centroid on the central molecule [30]. All frameworks were scaled equally so as to facilitate comparisons.

2.3. Polymorph Assessment through Hydrogen Bond Propensity Models

Initially, the definition of a hydrogen bond was set as established earlier; however, this resulted in an excessive number of contacts within each form, some of which were chemically incorrect. Therefore, the bond angle was modified to be larger than 133° (the lower limit of the hydrogen bonds detected in both forms). All hydrogen bond acceptors and donors were selected, with the latter list including carbon. Systems with errors or disorder were excluded from the study, together with organometallic compounds and entries with an R-factor > 0.075.

A set of data containing entries whose chemistry was relevant to ganciclovir was generated so as to build the statistical model for prediction. Potential bias or error was decreased by ensuring that each functional group (5 substructures; Please refer to Supplementary Materials) was represented by an adequate number of hits. After analysis, donor or acceptor candidates with a low number of relevant hits were omitted to avoid regression failures. A logistic regression model was fitted on the data, producing its corresponding area under the receiver operating characteristic (ROC) curve value as a measure of the extent of correct predictions. This procedure was performed separately for both polymorphs.

3. Results and Discussion

3.1. Intramolecular Level

In order to initiate the study, *Mogul* was used to predict the geometric preferences of every bond length, angle, torsion angle and ring within the molecules present in both forms, by accessing a depository of CSD-based libraries [31]. The results for geometric characteristics of form I showed that there were no unusual parameters, as opposed to some properties of form II that were not within the normal CSD distributions. Although these outlying parameters might be a product of poor data quality, which could be inferred from the atomic displacement parameters of form II (Figure 2), some of these unusual geometric characteristics (Please refer to Supplementary Materials) were also observed in other systems involving the ganciclovir molecule, such as the HCl salt and the monohydrate [32].

The electrostatic map of form I (Figure 3) clearly displays the large variety of hydrogen bond participants, with maxima (dark blue) around the hydrogen atoms of the amine group, and minima (dark red) around the oxygen of the carbonyl group and N3 of the imidazole ring, due to the presence of lone pairs of electrons. In addition, the electrostatic map also proves the presence of other potential hydrogen bond contributors, such as the hydroxyl groups, the ether oxygen atom and also carbon atoms. The electrostatic map of form II was very similar to that of form I (Please refer to Supplementary Materials).

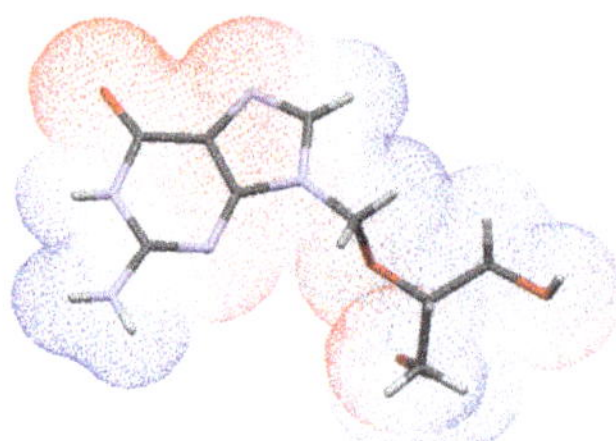

Figure 3. Electrostatic map of form I, mapped on the molecular vdW surface. Double bonds are omitted only for visualization purposes.

3.2. Intermolecular Level

The guanine ring system was one of the substructures identified via *IsoStar*, illustrated in Figure 4a, with nitrogen and oxygen atoms acting as probes. The distribution of the data points is very close to what was expected, with the majority concentrated around polar contributors. Most of the structures have bifurcated hydrogen bonds, each forming a contact with both O1 and N3. However, the presence of very few turquoise data points indicates that only a minority of hits seems to be at a close distance to the guanine rings (see frequency distribution plot in Supplementary Materials). This observation was unexpected due to the high electronegative character of both nitrogen and oxygen atoms, which in general enables their involvement in relatively strong contacts.

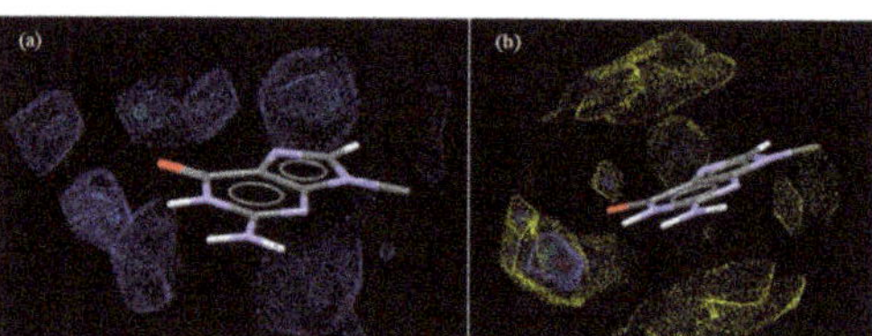

Figure 4. *IsoStar* contour plot with guanine as the central group having (**a**) nitrogen and oxygen as the contact groups (internal scaling: level 25—dark blue, level 50—lighter blue, level 75—turquoise) and (**b**) only carbon as the contact group (internal scaling: Level 10—Yellow, Level 25—Blue, Level 50—Green, Level 75—Red).

The contour plot in Figure 4b highlights the ability of carbon to act as a hydrogen bond donor, as well as the dominant presence of possible C–H . . . π interactions or π . . . π stacking, created due to the delocalized electron density in the aromatic guanine ring system. Other contour plots were constructed using an aliphatic ether as hydrogen bond acceptor and hydroxyl group as the central group. The distributions of the resultant hits of both plots were as expected (Please refer to Supplementary Materials).

Full interaction maps are sensitive to the specific conformation, meaning that each of the forms will have a different map [25]. Comparison of the maps (Figure 5) reveals that the maps associated with form I have a larger area and higher intensity than those pertaining to form II, indicating that the conformation of form I is more accessible for hydrogen bonding. The fully extended conformation of form I enables it to form more hydrogen bonds, which ultimately might be a major contributor to its thermodynamic stability at ambient conditions. Such stability is also evidenced through the fact that form III (hydrate form) converts to form I at temperatures above 180 °C, an exothermal transition which proves the monotropic relationship between the two forms [18]. The lack of accessibility in relation to the ether oxygen in form II, due to the proximity of the hydroxyl group, was earlier highlighted by the orientation of the torsion angle in *Mogul* results, which clearly demonstrated a degree of folding.

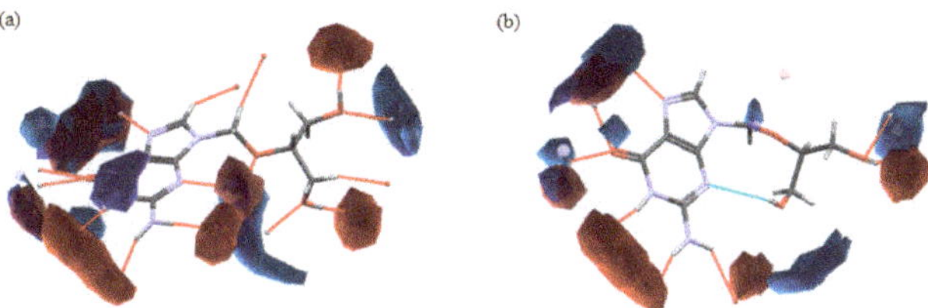

Figure 5. Full interaction map of (**a**) form I and (**b**) form II, with uncharged NH nitrogen (blue), RNH3 nitrogen (light blue), alcohol oxygen (light red) and carbonyl oxygen (red), at level 6.0. The hydrogen bonds predicted by *Mercury* are also shown (intermolecular, red, intramolecular, turquoise).

The geometric features of the hydrogen bonds of both forms were compared with those in the literature, and all results were compatible except those involving a C–H donor, which to the best of our knowledge were not included in any published list (Please refer to Supplementary Materials) [17]. The full interaction maps could not predict the involvement of the carbon donors for form I unless the distance levels were significantly increased. Even though the geometric parameters of the contacts with C–H donors are associated with characteristics of weaker interactions, they have a collective effect on the crystal packing and physicochemical properties of form I [33,34]. The addition of methyl carbon and aromatic carbon as separate probes did not alter maps significantly, suggesting that the C–H . . . π contacts and π . . . π stacking are less influential in form I, as opposed to in form II (Please refer to Supplementary Materials).

3.3. Hirshfeld Surface Analysis

Hirshfeld surface analysis was performed [9,35,36] in order to gain a better understanding of the differences in hydrogen bond networks and the contribution of weaker contacts within these conformational polymorphs. Since the Hirshfeld surface depends on the spherical atomic electron densities of a particular molecule within its crystal structure, surface differ even between polymorphs [29]. Examination of the Hirshfeld surfaces in Figure 6 shows the distinguishable features of each form, particularly the position and intensity of some of the red areas, which represent close contacts. The Hirshfeld surface of form I is characterized by more red spots relative to that of form II, which as remarked earlier through the full interaction maps, has a folding feature that hinders its accessibility to hydrogen bonds.

The nature of all non-bonding contacts in both polymorphs is presented through the fingerprint plots in Figure 7, by plotting d_i against d_e and hence translating the information provided by the Hirshfeld surface into a 2D format. The plots are overall strongly related, probably due to the similarities between these conformational polymorphs. The greatest proportion of interactions H . . . H are with a higher percentage contribution (Figure 8) and smaller minimum distance $d_i \approx d_e \approx 1.1$ Å associated with form II. Although such contacts are believed to be repulsive in nature, Matta et al. demonstrated how the net result of their presence has a stabilizing effect of up to a 10 kcal/mol decrease in the total energy of that particular structure [37].

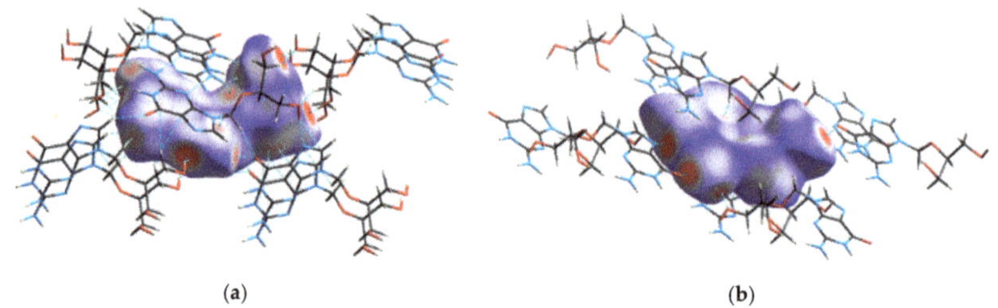

(a)   (b)

Figure 6. Hirshfeld surfaces with d_{norm} as mapped function, constructed for (**a**) form I and (**b**) form II, with neighboring molecules to illustrate the network of non-bonding contacts, represented by dotted lines (not only hydrogen bonds).

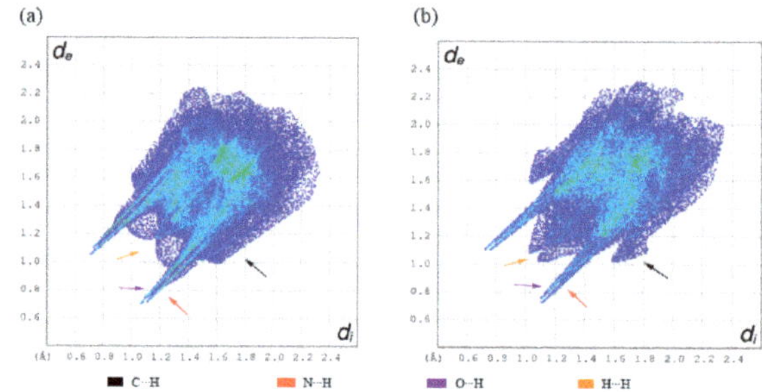

Figure 7. Fingerprint plots of (**a**) form I and (**b**) form II. The arrows represent interactions between different atoms, as indicated in the legend below. Given that the plots are approximately symmetrical, the arrows can be mirrored through the x–y diagonal.

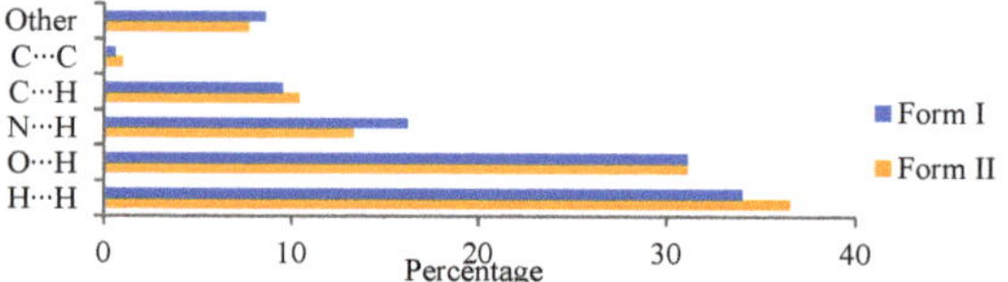

Figure 8. Bar-chart representing the percentage contribution of the main contacts in the Hirshfeld surfaces of forms I and II. The category of "other" includes C . . . N and C . . . O interactions.

The spikes at the bottom left of both plots are very prominent, exhibiting the dominance of N . . . H and O . . . H contacts, due to the highly polar functional groups in ganciclovir. The presence of these contacts is in agreement with the full interaction maps and the hydrogen bonds predicted by *Mercury*. The greener shades on the spikes of the form I surface plot indicate a higher frequency of contacts having relatively shorter distances. The percentage contribution of O . . . H interactions is equivalent for both forms, but the global minimum distance $d_i + d_e$ of around 1.7 Å pertains to form I.

The C–H . . . π contribution is evident in both polymorphs, as shown in the Hirshfeld surface maps; however, it is more significant in form II (Figure 8). Although these interactions are weaker and less directional than the conventional hydrogen bonds, they still contribute to self-assembly and molecular recognition processes, as they reinforce the stability within the supramolecular structure [38]. On the other hand, the absence of large maxima around the area at $d_i + d_e \approx 3.8$ Å and of the typical bold red and blue pairs of triangles on the shape index surfaces, indicate the less prominent π . . . π contacts [39]. Examination of the network of interactions in the two forms, reveals how the side chain prevents close direct stacking of the aromatic systems, hence minimizing the effect of π ... π contacts. Any C ... C contact seems to be distant and concentrated on the periphery of the guanine ring system, which hints that the aromatic systems are either far apart and/or shifted relative to one another.

3.4. Supramolecular Analysis

Form I is centrosymmetric ($P2_1/c$) with a unit cell containing two pairs of molecules that are identical within the pair and inversely-related between pairs. Form II is non-centrosymmetric ($P2_1$), thus lacking an inversion center.

The extensive hydrogen bonding network in form I is highly visible along its packing pattern, which seems to follow zigzag lines when viewed along *c* (Figure 9a). This type of pattern allows the layers of molecules to be very close to one another, hence enabling the optimal use of space available [40]. While the molecular volume of form I is greater than that of form II, packing in the former is more compact as evidenced by the volume of its Hirshfeld surface, packing coefficient and density value (Table 1). Figure 9b shows where the purine backbones seem to be parallel to one another, while the side chains are "out of plane". Layering of the guanine rings creates an off-center parallel stacking, but distance and angular parameters of the guanine ring centroids are at the upper limits for effective π ... π interactions, suggested by the literature [41,42]. This property was also highlighted by a low percentage contribution of C ... C contacts in the fingerprint plots.

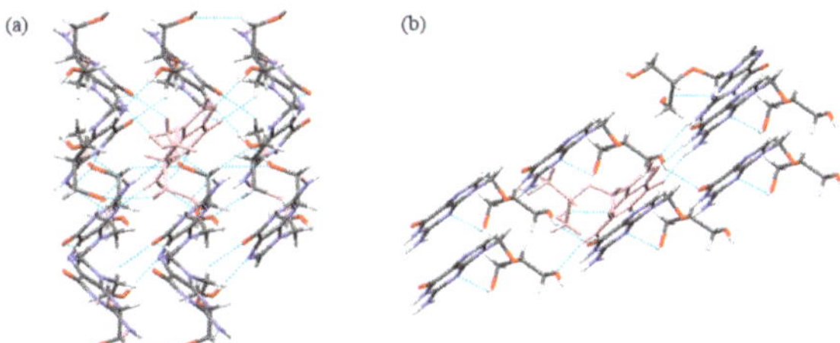

Figure 9. Packing in (**a**) form I and (**b**) form II, with a central molecule (pink) each having the same back-bone orientation. Hanging contacts were removed for better visualization.

Table 1. Numerical Hirshfeld surface packing data for forms I and II.

	Form I	Form II
Volume (Å^3)	250.18	268.07
Packing coefficient	0.770978	0.725452
Density (g/cm^3)	1.653	1.549

The packing arrangements in both forms were compared to other crystal structures with similar molecular arrangements. However, even though there were five more entries in the CSD which involved ganciclovir as a component, only the two hydrates (mono- and tri-), resulted in having one out of a cluster of 15 molecules, with a degree of similarity when compared to both anhydrates [32] demonstrating the uniqueness of the packing patterns of form I and II. This test also verified the significant differences between the way in which both anhydrous forms were arranged three-dimensionally resulting in an RMS of 1.017 (see Figure 8).

3.5. Energy Frameworks

The thermodynamic stability profile of each of the polymorphs is a direct implication of the nature and strength of their non-bonding interactions. Energy frameworks were calculated in an attempt to understand the topological dissimilarities of the energy components of the two polymorphs, and subsequently potentially link these characteristics to their respective packing and thermal behavior [35]. The total energy is divided into Coulomb forces/electrostatic potential forces and dispersion energy.

The negative energy components in form I are distributed along both parallel and non-parallel molecules (Figure 10, top row), hence stabilizing the energy architecture throughout the supramolecular structure in multiple directions. The dispersion forces can mainly be traced along the off-centered parallel stacking (along the axis *a*), as well as between adjacent molecules whose aromatic systems lie along the same plane. The strongest energy contribution (-164 kJ/mol) is attributed to N1–H . . . O2 and N2–H . . . O4 interactions, located between parallel molecules. The strength of these stabilizing interactions is due to both electrostatic and dispersion forces, the former being the dominant one. The O4–H . . . O3 contact contributes towards a lowering of energy (-56.4 kJ/mol), while the combination of O3–H . . . N3 and C6–H . . . O1 interactions induces a further stabilizing effect of -63.7 kJ/mol.

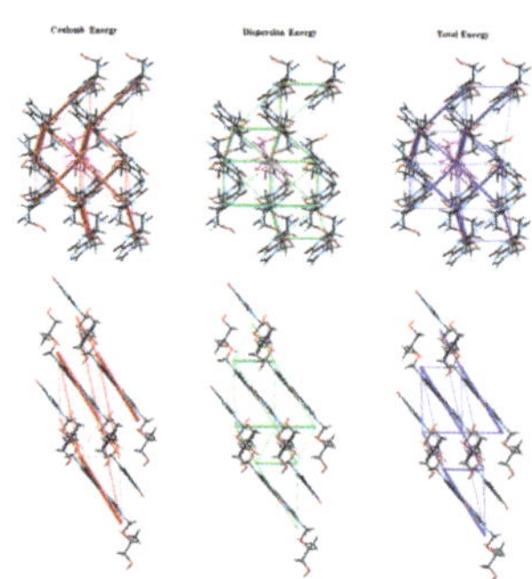

Figure 10. Energy framework diagrams, using the CE-B3LYP model, a scale tube size of 20 and cut-off value of 0 kJ/mol. The top row represents form I (viewed down the *c* axis), and the bottom row is associated with form II (viewed down the *b* axis). Molecules are clustered around the central molecule (pink) with a 3.8 Å radius.

In form II, N1–H . . . O1 and N2–H . . . N3 are the strongest interactions both of which contribute towards the dominant electrostatic energy (-73.6 kJ/mol) along the planes of aromatic rings. A secondary energy contribution (-45.2 kJ/mol) in the same direction is attributable to the N1–H . . . O4 contact. The out-of-plane interactions are mainly a result of dispersion contacts and the O4–H . . . O1 electrostatic interaction. However, such a contribution is considerably low (-18.3 kJ/mol) relative to the other energy components along the aromatic systems plane.

The less exhaustive network of intermolecular bonding in form II, in conjunction with the presence of an intramolecular bond, seems to enhance stability within the molecule. In contrast, form I entails a very complex extensive framework of intermolecular contacts, with 16 hydrogen bonds per molecule (and eight different neighboring molecules). Such a multidirectional framework creates a greater stabilizing effect in form I, as illustrated by the relatively thicker cylindrical radius of the total energy tubes (Figure 10). Such an influence, coupled with a more compact packing arrangement, might contribute towards a lower enthalpy in form I relative to form II, which according to L. Yu must be accompanied by a lower entropy in order to conserve the enantiotropic relationship between the two polymorphs [18,43]. With an increase in temperature, in the range of 222 °C to 228 °C, enough energy is absorbed to break the intermolecular framework in form I, which then transitions to form II (as confirmed by variable temperature X-ray diffraction and differential scanning calorimetry) [15,17,18].

The relatively loose packing and the intermolecular framework in form II seems to be constructed in such a way as to cater for higher energy environments. The energy framework of this polymorph reveals how the stabilizing effect is expanded along planes, whereas much fewer interactions are observed between layers. This lower extent of intermolecular forces between parallel planes in form II might create an environment that can accommodate higher entropy within the crystal structure, thereby allowing it to be thermodynamically stable at higher temperatures.

3.6. Polymorph Assessment

The objective of this analysis revolved around stability assessment of the hydrogen bond network present in both forms, given the complexity of the chemical environment of ganciclovir. The reliability and predictive ability of the fitted logistic regression model were mirrored by the value of the area under the ROC curve, which was equal to 0.878, and by the reduction from the null to the residual deviance [44,45]. The same procedure was performed for each form individually, and since they are polymorphs of the anhydrous form of ganciclovir, very similar coefficients were obtained (Please refer to Supplementary Materials). Minor differences can be attributed to the non-deterministic nature of the process involving the fitting of the model. This outcome confirms the robustness of this method, which is capable of assessing the stability of various polymorphs having different hydrogen bonds simultaneously.

The final model was used to calculate the propensities of all possible intermolecular hydrogen bonds, with those in form II having the highest probabilities (more conventional contacts) (Please refer to Supplementary Materials). The overall low likelihood of the intermolecular bonds in form I was illustrated in the putative structure landscape, which categorized this polymorph as having the least stable hypothetical forms (Please refer to Supplementary Materials). The utilization of every functional group in form I, for inter-molecular bonding, might have been prioritized over the formation of the fewer and more probable contacts.

It is understandable that due to low frequencies, the contacts in form I were ranked as having lower probability, and hence low stability was predicted. However, this outcome was not in agreement with the fact that form I is the thermodynamically stable polymorph, under ambient conditions. At this point, it is essential to recall the statistical mechanism behind the construction of the Hydrogen Bond Propensity (HBP) model, which is based on the occurrence of hydrogen bonds in similar chemical environments. Therefore, in some examples, the resultant model might not be sufficient to explain the complexity of hydro-gen bonding and to capture the collective effects of multiple factors that determine the polymorph stability [6]. In his research paper, Abramov commented how in general, a HBP model cannot account for enantiotropic relationships between polymorphs, such as the one between form I and II [6]. Moreover, one has to take into account that the directional features and geometric parameters of the contacts are not being considered in the model, and these characteristics have a significant effect on stability.

In an attempt to overcome these limitations, the HBP model was constructed using a much larger training set, but similar results were obtained. Despite such limitations, there was still valuable information that could be extracted from these results. The putative struc-ture landscape in Figure 11 portrays the presence of data points located very close to form II, representing reasonable hypothetical structures having very strong hydrogen bond interactions. The viability of the formation of such forms is highly encouraging in view of further research dedicated to the exploration of other possible ganciclovir polymorphs.

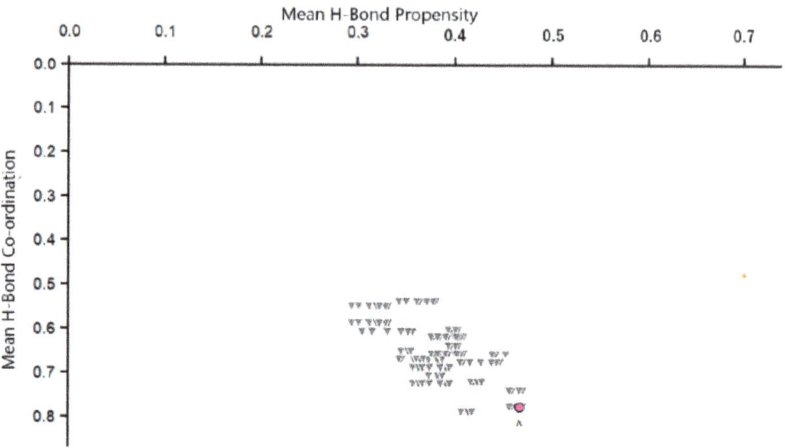

Figure 11. The propensity participation chart output showing form II (violet dot), which is found at the location with the optimal conditions (large values for both axes) that are usually associated with thermodynamically stable crystal forms.

4. Conclusions

This investigation presents an extensive study of ganciclovir from a crystallographic point of view, ranging from a structural to energy framework analysis. The methodology developed has proven how the use of multiple computational tools which target the examination of different characteristics, can be successful in providing a reasonable insight into the mechanism of polymorphism. Furthermore, it can be implemented to extract information about the influence of specific bonds and substituents on the formation of a particular form, regardless of the complexity or nature of the molecule.

Such a dominant impact was observed through the folding of the chain moiety in form II, which intrinsically affects its intramolecular and intermolecular characteristics. On the other hand, full interaction maps and Hirshfeld surfaces demonstrated how the extended chain moiety in form I enables the molecule to have an extensive network of non-covalent contacts, with a significant degree of directionality. These properties were identified as significant factors responsible for the known thermodynamic behavior of both forms, ultimately resulting in the conformational polymorphism present between them. The collective analysis of results from all molecular levels and energy frameworks provides a plausible explanation for the stability of both polymorphs at different temperature ranges.

The nature of this methodology makes it applicable to a wide range of crystal forms, even beyond active pharmaceutical ingredients. Such knowledge would facilitate the selection of co-formers or environmental conditions to selectively target the formation of a particular co-crystal. It would also aid navigation through the possibility of forming other relatively stable polymorphs, as suggested by the putative landscape in the case of ganciclovir. Therefore, the utilization of such data could lead to a more sustainable process of drug development, as well as possibly improving the pharmacokinetic properties of this active pharmaceutical ingredient.

Supplementary Materials: The following are available online at https://www.mdpi.com/2624-8549/3/1/10/s1. Detailed information about the computational methods applied as well as a list of all results.

Author Contributions: Conceptualization, L.V.-Z. and L.S.; methodology, L.S.; software, L.S.; validation, L.V.-Z., L.S. and U.B.; formal analysis, L.S.; investigation, L.S.; resources, L.V.-Z.; data curation, L.S.; writing—original draft preparation, L.S.; writing—review and editing, U.B. and L.V.-Z.; visualization, L.S.; supervision, L.V.-Z. and U.B.; project administration, L.V.-Z.; funding acquisition, L.S. and L.V.-Z. All authors have read and agreed to the published version of the manuscript.

Funding: This research was funded by the Tertiary Education Scholarships Scheme (TESS) (L.S.) and by the project: "Setting up of transdisciplinary research and knowledge exchange (TRAKE) complex at the University of Malta (ERDF.01.124)", which is being co-financed through the European Union through the European Regional Development Fund 2014–2020 (L.V.-Z. and U.B.).

Institutional Review Board Statement: Not applicable.

Informed Consent Statement: Not applicable.

Data Availability Statement: The data presented in this study are available in the Supplementary Materials.

Conflicts of Interest: The authors declare no conflict of interest.

References

1. Salazar, D.E.; Gormley, G. Modern Drug Discovery and Development. In *Clinical and Translational Science*, 2nd ed.; Robertson, D., Williams, G.H., Eds.; Academic Press: Nashville, TN, USA, 2017; pp. 719–743.
2. Koenig, S.G.; Leahy, D.K.; Wells, A.S. Evaluating the Impact of a Decade of Funding from the Green Chemistry Institute Pharmaceutical Roundtable. *Org. Process Res. Dev.* **2018**, *22*, 1344–1359. [CrossRef]
3. Berdigaliyev, N.; Aljofan, M. An overview of drug discovery and development. *Future Med. Chem.* **2020**, *12*, 939–947. [CrossRef] [PubMed]
4. Datta, S.; Grant, D.J.W. Crystal structures of drugs: Advances in determination, prediction and engineering. *Nature Rev. Drug Discov.* **2004**, *3*, 42. [CrossRef] [PubMed]
5. Price, S.L. Computed Crystal Energy Landscapes for Understanding and Predicting Organic Crystal Structures and Polymorphism. *Acc. Chem. Res.* **2009**, *42*, 117–126. [CrossRef]
6. Abramov, Y.A. Current Computational Approaches to Support Pharmaceutical Solid Form Selection. *Org. Process Res. Dev.* **2013**, *17*, 472–485. [CrossRef]
7. Sandhu, B.; McLean, A.; Sinha, A.S.; Desper, J.; Sarjeant, A.A.; Vyas, S.; Reutzel-Edens, S.M.; Aakeröy, C.B. Evaluating Competing Intermolecular Interactions through Molecular Electrostatic Potentials and Hydrogen-Bond Propensities. *Cryst. Growth Des.* **2018**, *18*, 466–478. [CrossRef]
8. Spiteri, L.; Baisch, U.; Vella-Zarb, L. Correlations and statistical analysis of solvent molecule hydrogen bonding—A case study of dimethyl sulfoxide (DMSO). *CrystEngComm* **2018**, *20*, 1291–1303. [CrossRef]
9. Thomas, S.P.; Spackman, M. The Polymorphs of ROY: A Computational Study of Lattice Energies and Conformational Energy Differences*. *Aust. J. Chem.* **2018**, *71*. [CrossRef]
10. Grinde, B. Herpesviruses: Latency and reactivation—Viral strategies and host response. *J. Oral Microbiol.* **2013**, *5*. [CrossRef]
11. Verheyden, J.P.; Martin, J.C. 9-(1,3-dihydroxy-2-propoxymethyl)-guanine as an Antiviral. U.S. Patent No. 4355032, 19 October 1982.
12. Mallipeddi, S.M.; Sterling, J.E.; Koneru, P.B. Ganciclovir Compositions and Related Methods. U.S. Patent No. 9486530, 8 November 2016.
13. Parr, A.; Hidalgo, I.J.; Bode, C.; Brown, W.; Yazdanian, M.; Gonzalez, M.A.; Sagawa, K.; Miller, K.; Jiang, W.; Stippler, E.S. The Effect of Excipients on the Permeability of BCS Class III Compounds and Implications for Biowaivers. *Pharm. Res.* **2016**, *33*, 167–176. [CrossRef]
14. World Health Organization. WHO Drug Information 2017. *WHO Drug Inf.* **2017**, *31*, 385–492. Available online: https://apps.who.int/iris/handle/10665/330952 (accessed on 26 January 2021).
15. Sarbajna, R.; Anil, P.; Sivalakshmi Devi, A.; Suryanarayana, M.V.; Sethi, M.; Dutta, D. *Studies on Crystal Modifications of Ganciclovir*; Taylor & Francis: London, UK, 2011; Volume 537, pp. 141–154.
16. Kawamura, T.; Hirayama, N. Crystal Structure of Ganciclovir. *X-Ray Struct. Anal. Online* **2009**, *25*, 51–52. [CrossRef]
17. Roque-Flores, R.L.; Guzei, I.A.; Matos, J.d.R.; Yu, L. Polymorphs of the antiviral drug ganciclovir. *Acta Crystallogr. Sect. C* **2017**, *73*, 1116–1120. [CrossRef] [PubMed]
18. Roque-Flores, R.L.; Matos, J.d.R. Simultaneous measurements of X-ray diffraction–differential scanning calorimetry. *J. Therm. Anal. Calorim.* **2019**, *137*, 1347–1358. [CrossRef]
19. Putz, H.; Brandenburg, K. *Diamond—Crystal and Molecular Structure Visualisation, 4.6.4*; Crystal Impact: Bonn, Germany, 2020.
20. Galek, P.T.A.; Pidcock, E.; Wood, P.A.; Bruno, I.J.; Groom, C.R. One in half a million: A solid form informatics study of a pharmaceutical crystal structure. *CrystEngComm* **2012**, *14*, 2391–2403. [CrossRef]
21. Lewars, E.G. Semiempirical calculations. In *Computational Chemistry: Introduction to the Theory and Applications of Molecular and Quantum Mechanics*; Kluwer Academic Publishers: Hingham, MA, USA, 2016.
22. Rocha Gerd, B.; Freire Ricardo, O.; Simas Alfredo, M.; Stewart James, J.P. RM1: A reparameterization of AM1 for H, C, N, O, P, S, F, Cl, Br, and I. *J. Comput. Chem.* **2006**, *27*, 1101–1111. [CrossRef]
23. Macrae, C.F.; Sovago, I.; Cottrell, S.J.; Galek, P.T.A.; McCabe, P.; Pidcock, E.; Platings, M.; Shields, G.P.; Stevens, J.S.; Towler, M.; et al. Mercury 4.0: From visualization to analysis, design and prediction. *J. Appl. Crystallogr.* **2020**, *53*, 226–235. [CrossRef]
24. Bruno, I.; Cole, J.; Lommerse, J.; Rowland, R.; Taylor, R.; Verdonk, M. IsoStar: A library of information about nonbonded interactions. *J. Comput. Aided Mol. Design* **1997**, *11*, 525–537. [CrossRef]
25. Wood, P.A.; Olsson, T.S.G.; Cole, J.C.; Cottrell, S.J.; Feeder, N.; Galek, P.T.A.; Groom, C.R.; Pidcock, E. Evaluation of molecular crystal structures using Full Interaction Maps. *CrystEngComm* **2013**, *15*, 65–72. [CrossRef]

26. Bruno Ian, J.; Cole Jason, C.; Edgington Paul, R.; Kessler, M.; Macrae Clare, F.; McCabe, P.; Pearson, J.; Taylor, R. New software for searching the Cambridge Structural Database and visualizing crystal structures. *Acta Crystallogr. Sect. B* **2008**, *58*, 389–397. [CrossRef]

27. Wood, P.A.; Allen, F.H.; Pidcock, E. Hydrogen-bond directionality at the donor H atom—Analysis of interaction energies and database statistics. *CrystEngComm* **2009**, *11*, 1563–1571. [CrossRef]

28. Wolff, S.K.; Grimwood, D.J.; McKinnon, J.J.; Turner, M.J.; Jayatilaka, D.; Spackman, M.A. *CrystalExplorer*; University of Western Australia: Perth, Australia, 2017.

29. Spackman, M.A.; Jayatilaka, D. Hirshfeld surface analysis. *CrystEngComm* **2009**, *11*, 19–32. [CrossRef]

30. Mackenzie, C.F.; Spackman, P.R.; Jayatilaka, D.; Spackman, M.A. CrystalExplorer model energies and energy frameworks: Extension to metal coordination compounds, organic salts, solvates and open-shell systems. *IUCrJ* **2017**, *4*, 575–587. [CrossRef]

31. Taylor, R.; Cole, J.; Korb, O.; McCabe, P. Knowledge-Based Libraries for Predicting the Geometric Preferences of Druglike Molecules. *J. Chem. Inf. Modeling* **2014**, *54*, 2500–2514. [CrossRef]

32. Fernandes, J.A.; Galli, S.; Palmisano, G.; Volante, P.; Mendes, R.F.; Paz, F.A.A.; Masciocchi, N. Reviewing the Manifold Aspects of Ganciclovir Crystal Forms. *Cryst. Growth Design* **2016**, *16*, 4108–4118. [CrossRef]

33. Scheffer, J.; Wong, Y.F.; Patil, A.O.; Curtin, D.Y.; Paul, I.C. CPMAS (cross-polarization magic angle spinning) carbon-13 NMR spectra of quinones, hydroquinones, and their complexes. Use of CMR to follow a reaction in the solid state. *J. Am. Chem. Soc.* **1985**, *107*, 4898–4904. [CrossRef]

34. Steiner, T. C–H⋯O hydrogen bonding in crystals. *Crystallogr. Rev.* **2003**, *9*, 177–228. [CrossRef]

35. Tan, S.L.; Jotani, M.M.; Tiekink, E.R.T. Utilizing Hirshfeld surface calculations, non-covalent inter-action (NCI) plots and the calculation of inter-action energies in the analysis of mol-ecular packing. *Acta Cryst. E Cryst. Commun.* **2019**, *75*, 308–318. [CrossRef]

36. Vella-Zarb, L.; Dinnebier, R.E.; Baisch, U. The Devil is in the Detail: A Rare H-Bonding Motif in New Forms of Docetaxel. *Cryst. Growth Design* **2013**, *13*, 4402–4410. [CrossRef]

37. Matta, C.F.; Hernández-Trujillo, J.; Tang, T.-H.; Bader, R.F.W. Hydrogen–Hydrogen Bonding: A Stabilizing Interaction in Molecules and Crystals. *Chem. A Eur. J.* **2003**, *9*, 1940–1951. [CrossRef]

38. Nishio, M. The CH/π hydrogen bond in chemistry. Conformation, supramolecules, optical resolution and interactions involving carbohydrates. *Phys. Chem. Chem. Phys.* **2011**, *13*, 13873–13900. [CrossRef]

39. McKinnon, J.J.; Fabbiani, F.P.A.; Spackman, M.A. Comparison of Polymorphic Molecular Crystal Structures through Hirshfeld Surface Analysis. *Cryst. Growth Design* **2007**, *7*, 755–769. [CrossRef]

40. Carugo, O.; Blatova, O.A.; Medrish, E.O.; Blatov, V.A.; Proserpio, D.M. Packing topology in crystals of proteins and small molecules: A comparison. *Sci. Rep.* **2017**, *7*, 13209. [CrossRef] [PubMed]

41. Alvarez, S. A cartography of the van der Waals territories. *Dalton Trans.* **2013**, *42*, 8617–8636. [CrossRef] [PubMed]

42. Yao, Z.-F.; Wang, J.-Y.; Pei, J. Control of π–π Stacking via Crystal Engineering in Organic Conjugated Small Molecule Crystals. *Cryst. Growth Design* **2018**, *18*, 7–15. [CrossRef]

43. Yu, L. Inferring thermodynamic stability relationship of polymorphs from melting data. *J. Pharm. Sci.* **1995**, *84*, 966–974. [CrossRef] [PubMed]

44. Le, C.T. *Applied Categorical Data Analysis and Translational Research*; Wiley: Hoboken, NJ, USA, 2010.

45. Forte, R.M. *Mastering Predictive Analytics with R*; Packt Publishing: Birmingham, UK, 2015.

Article

Exploring the Structural Chemistry of Pyrophosphoramides: N,N′,N″,N‴-Tetraisopropylpyrophosphoramide

Duncan Micallef, Liana Vella-Zarb and Ulrich Baisch *

Department of Chemistry, Faculty of Science, University of Malta, Msida MSD2080, Malta;
duncan.micallef.11@um.edu.mt (D.M.); liana.vella-zarb@um.edu.mt (L.V.-Z.)
* Correspondence: ulrich.baisch@um.edu.mt; Tel.: +356-2340-3425

Abstract: $N,N′,N″,N‴$-Tetraisopropylpyrophosphoramide **1** is a pyrophosphoramide with documented butyrylcholinesterase inhibition, a property shared with the more widely studied octamethylphosphoramide (Schradan). Unlike Schradan, **1** is a solid at room temperature making it one of a few known pyrophosphoramide solids. The crystal structure of **1** was determined by single-crystal X-ray diffraction and compared with that of other previously described solid pyrophosphoramides. The pyrophosphoramide discussed in this study was synthesised by reacting iso-propyl amine with pyrophosphoryl tetrachloride under anhydrous conditions. A unique supramolecular motif was observed when compared with previously published pyrophosphoramide structures having two different intermolecular hydrogen bonding synthons. Furthermore, the potential of a wider variety of supramolecular structures in which similar pyrophosphoramides can crystallise was recognised. Proton (^{1}H) and Phosphorus 31 (^{31}P) Nuclear Magnetic Resonance (NMR) spectroscopy, infrared (IR) spectroscopy, mass spectrometry (MS) were carried out to complete the analysis of the compound.

Keywords: $N,N′,N″,N‴$-Tetraisopropylpyrophosphoramide; pyrophosphoramide; synthons; supramolecular motifs; X-ray crystallography

Citation: Micallef, D.; Vella-Zarb, L.; Baisch, U. Exploring the Structural Chemistry of Pyrophosphoramides: $N,N′,N″,N‴$-Tetraisopropylpyrophosphoramide. *Chemistry* **2021**, *3*, 149–163. https://doi.org/10.3390/chemistry3010013

Received: 20 December 2020
Accepted: 18 January 2021
Published: 28 January 2021

Publisher's Note: MDPI stays neutral with regard to jurisdictional claims in published maps and institutional affiliations.

1. Introduction

$N,N′,N″,N‴$-Tetraisopropylpyrophosphoramide **1** (O((iPrNH)$_2$PO)$_2$) is a commercially available pyrophosphoramide known to be a butyrylcholinesterase inhibitor [1,2]. Although the molecular structure of the compound is known, crystallographic support for this structure has not been published. Another pyrophosphoramide that has shown cholinesterase inhibition is octamethylpyrophosphoramide (O(Me$_2$N)$_2$PO)$_2$), which is commonly known as Schradan. This compound, unlike **1**, was used as an insecticide for sucking and chewing insects, which are agricultural pests but it has fallen out of use since the 1950's [3,4]. Schradan is a liquid at room temperature making its storage, use, and structural characterisation more problematic than **1**. In the mid-twentieth century, apart from its use as a pesticide, Schradan was also studied for its chelation ability with numerous metal centres. These studies mainly dealt with the complexation of Schradan with alkali earth, transition and actinide metals, as well as Sn^{4+} [5–9]. Although the complexation ability was proven through numerous analytical methods, including X-ray diffraction (XRD), no industrial or chemical use for these complexes was discussed. Only six other analogues of pyrophosphoramides were found in the literature, namely $N,N′,N″,N‴$-Tetrakis(2-methylphenyl)- oxybis(phosphonic diamide) (O((2-MePh)NH)$_2$PO)$_2$), $N,N′,N″,N‴$-tetra-tert-butoxybis(phosphonic diamide) (O((tBuNH)$_2$PO)$_2$), $N,N′,N″,N‴$-tetrakis(4-methylphenyl)-oxybis(phosphonic diamide) (O((4-MePh)NH)$_2$PO)$_2$), $N,N′,N″,N‴$-tetrakis(benzyl)-$N,N′,N″,N‴$-tetrakis(methyl)oxybis-(phosphonicdiamide) (O((BzMeN)$_2$PO)$_2$), 2,2′-Oxybis(1,3-bis(2,6-diisopropylphenyl)-1,3,2-diaza-phospholidine) 2,2′-dioxide (O((C$_2$H$_4$(2,5-iPrPhN)$_2$)$_2$PO)$_2$) and 2,2′-oxybis(1,3-bis- [(naphthalen-1-yl)methyl]octahydro-1H-1,3,2λ^5-benzodiaza phosphol-2-one) monohydrate (O((1,2-Cy(NaphN)$_2$)$_2$PO)$_2$·H$_2$O) [10–15].

These solids were characterised by single-crystal X-ray diffraction (SXRD) and published as novel structures with little detailed discussion on any similarities in structure and chemistry as opposed to the research carried out on Schradan. The latter two of the four were compared as part of a Hirshfeld analysis for phosphoramides. $O((t\text{BuNH})_2PO)_2$ was complexed with manganese(II) to give the complex $[Mn(O(t\text{BuNH})_2PO)_2)_2DMF_2][Cl]_2 \cdot 2H_2O$ [16]. This was the only published non-Schradan complex of this group of analogous pyrophosphoramides.

The main objective of this work was to study the supramolecular features of **1** and other, already published, pyrophosphoramides in light of known intermolecular interactions and arrangements [17–21] and to understand their influence on the physical properties of these compounds.

2. Materials and Methods

2.1. General Considerations

All reactions were carried out under an argon atmosphere using Schlenk line techniques. Chloroform was dried over a P_2O_5 still, while diethyl ether, THF, petroleum ether and chloroform used in the work-up procedures of the product were dried over 4 Å molecular sieves. FT-IR (Fourier Transform Infrared) spectra were recorded using KBr pellet samples and a Shimadzu IRAffinity-1 FTIR spectrophotometer. ^{1}H NMR and ^{31}P NMR (Nuclear Magnetic Resonance) spectra were collected on a Bruker Ascend NMR spectrometer with a probe having a set frequency of 500.13 MHz for ^{1}H NMR and 202.457 MHz for ^{31}P NMR. Gas chromatography Mass spectroscopy (GC MS) data was collected using a Thermo DSQ II GC/MS spectrometer with samples being prepared by dissolution of the products in chloroform. Single-crystal X-ray diffraction data was collected on a STOE STADIVARI diffractometer.

2.2. Synthesis of N,N′,N″,N‴-Tetraisopropylpyrophosphoramide (O((iPrNH)₂PO)₂)

The synthesis of $O((i\text{PrNH})_2PO)_2$ **1** was performed through a modification of the synthesis of the analogous $O((\text{Me}_2N)_2PO)_2$ (Schradan) as described by Goehring, M. and Niedenzu, K. in 1956 [22]. Iso-propyl amine (5 mL, 0.058 mol) was dissolved in 10 mL of chloroform as a solvent. The mixture was cooled to −78 °C and pyrophosphoryl tetrachloride (1 mL, 0.007 mol) was added dropwise to the reaction solution using a glass syringe. Upon addition of pyrophosphoryl tetrachloride, the formation of white fumes was noted, along with the formation of a solid crystalline mass. The reaction was left at a temperature of -78 °C until the white fumes dissipated and was subsequently allowed to reach room temperature. The reaction mixture was left to react at room temperature overnight. The white suspension thus formed was subsequently heated to 60 °C for 3 h to complete the reaction. The yellow solution obtained was left overnight to form a clear colourless crystalline mass ($i\text{PrNH}_3Cl$). The mixture was then filtered to collect a clear colourless crystalline mass ($i\text{PrNH}_3Cl$) and a yellow solution. The latter was layered with diethyl ether to yield a white solid. Yield, 41% (crude product with respect to pyrophosphoryl tetrachloride), ^{1}H NMR (CDCl₃): 8.31 ppm (s, 1H, NH3), 3.64 ppm (td, 1H, CH), 3.40 ppm (m, 1H, CH), 2.26 ppm (s, 1H, NH), 1.39 ppm (d, 2H, CH3) and at 1.14 ppm (t, 6H, CH3), ^{31}P {^{1}H} NMR (CDCl₃): 14.33 ppm (s), FT-IR (KBr, cm^{-1}): 3400 (w), 3252 (sb), 2965 (s), 2870 (m), 1637 (m), 1527 (m), 1466 (m), 1432 (m), 1396 (m), 1367 (m), 1257 (s), 1229 (s), 1167 (m), 1137 (w), 1055 (s), 1026 (s), 945 (m), 914 (m), 886 (s), 804 (m), 750 (m), GC MS (EI; 70 eV) m/z: 44.12, 58.08, 79.01, 93.97, 137.07, 179.10 $[(i\text{PrNH})_2PO_2]^+$, 195.16.

Crystals suitable for SXRD studies were obtained through liquid–liquid diffusion crystallisation: A sample of the white solid product was dissolved in a minimum volume of chloroform to yield a saturated solution. This was layered with 30–40 °C petroleum ether in a 1:5 volume ratio of chloroform/petroleum ether. This produced a liquid-liquid diffusion crystallisation set up which yielded a white crystalline solid over the course of two days. The solid was collected by cannula filtration and crystals suitable for SXRD studies were collected from this solid.

2.3. Purification by Column Chromatography

Although single crystals of O((*i*PrNH)$_2$PO)$_2$ were obtained from the product, it is evident from NMR and GC MS data that this solid was not pure and therefore a sample of O((*i*PrNH)$_2$PO)$_2$ was also purified by column chromatography to determine whether this was a possible method of purification. The sample was dissolved in dichloromethane and activated Keiselgel 60 was used as the stationary phase, while THF/acetonitrile (1:1 *v/v*) was used as a mobile phase. ^{1}H NMR (CDCl$_3$): 3.37 (m, 1H, CH), 2.29 (bt, 1H, NH), 1.16 (t, 6H, CH$_3$), ^{31}P {^{1}H} NMR (CDCl$_3$): 14.33 ppm (s), FT-IR (KBr, cm^{-1}): 3414 (w), 3255 (bm), 2965 (s), 2870 (m), 1465 (m), 1428 (m), 1367 (m), 1257 (s), 1205 (s), 1167 (m), 1137 (m), 1051 (s), 1021 (s), 916 (m), 886 (m), 833 (sh), 800 (vw), 771 (m), GC MS (EI; 70 eV) *m/z*: 44.12, 58.11, 79.01, 93.93, 137.09, 179.10 [(^{i}PrNH)$_2$PO$_2$]$^+$, 195.16.

2.4. Single-Crystal X-ray Diffraction

Crystals were collected under oil and mounted in oil. Data was collected using Cu radiation on a STOE STADIVARI diffractometer, using a 200 K Pilatus detector at 293 K. The ShelXT intrinsic phasing method and ShelXL least squares method were used for structure solution and refinement, respectively (Table 1, Figures 1 and 2). Details regarding the structure solution and refinement procedures are described in Appendix A, while further detailed structural information for this structure is given in Tables S5–S9 in the Supplementary Materials.

Table 1. Crystal data and structure refinement for O(*i*PrNH)$_2$PO)$_2$.

Empirical formula	C$_{12}$H$_{32}$N$_4$O$_3$P$_2$	μ/mm^{-1}	2.133
Formula weight/gmol^{-1}	342.35	F(000)	744.0
Temperature/K	293	Crystal size/mm^3	0.8 × 0.6 × 0.3
Crystal system	orthorhombic	Radiation	CuKα (λ = 1.54186)
Spacegroup	*Pca*2$_1$	2θ range for data collection/°	8.696 to 137.268
a/Å	20.342(2)	Index ranges	$-24 \leq h \leq 23, -3 \leq k \leq 6, -20 \leq l \leq 23$
b/Å	5.0495(6)	Reflections collected	43032
c/Å	19.103(3)	Independent reflections	3317 [Rint = 0.0557, Rsigma = 0.0226]
α/°	90	Data/restraints/parameters	3317/1/214
β/°	90	Goodness-of-fit on F2	1.044
γ/°	90	Final R indexes [I > = 2σ (I)]	R1 = 0.0402, wR2 = 0.1071
Volume/Å^3	1962.2(4)	Final R indexes [all data]	R1 = 0.0408, wR2 = 0.1075
Z	4	Largest diff. peak/hole/eÅ^{-3}	0.28/−0.29
ρ calc/gcm^{-3}	1.159	Flack parameter	0.01(3)

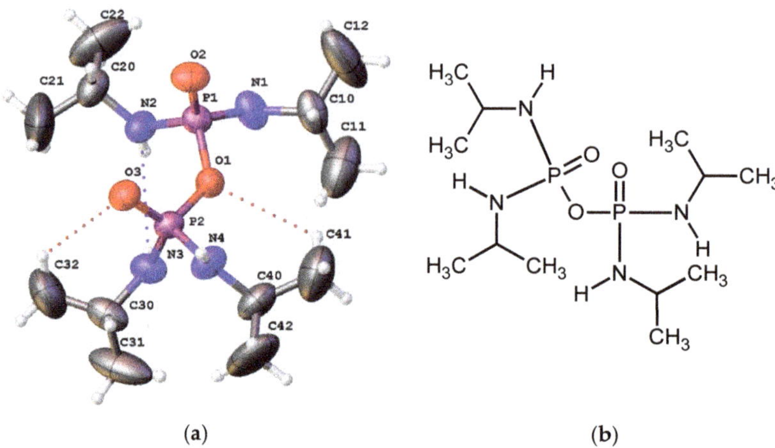

(a)   (b)

Figure 1. (**a**) Molecular structure obtained for O((*i*PrNH)₂PO)₂ the hydrogen bonding N3–H1···N2 (blue) and the weak hydrogen bonding observed as C32–H32A···O3 and C41–H41A···O1 (red)(generated using Olex2 [23]); (**b**) molecular diagram for O((*i*PrNH)₂PO)₂ (generated using ChemSketch [24]).

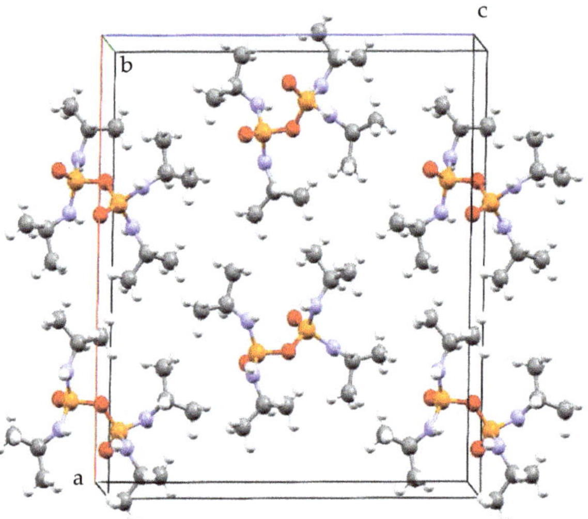

Figure 2. Unit cell for O((*i*PrNH)₂PO)₂ viewed along the b-axis (generated in Mercury [25]).

3. Results and Discussion

3.1. Structural Comparison with Previously Described Pyrophosphoramides

Most intermolecular motifs that build up the supramolecular structure of a compound are due to the moieties within the molecule in question. Therefore, compounds with similar moieties and chemical structures may be expected to have similar intermolecular motifs and in turn, similar crystal structures [17,18]. In this regard, the various pyrophosphoramides discussed in this study can be divided into five main groups, dependent on their chemical structure (Table 2).

Table 2. Categorisation of pyrophosphoramides discussed in this publication including crystal structures and space groups published from data obtained by single-crystal X-ray diffraction for the solid products, along with the liquid $O((Me_2N)_2PO)_2$.

Pyropshophoramides				
Di-N-Substituted Pyrophosphoramides			**Mono-N-Substituted Pyrophosphoramides**	
$O((R_2N)_2PO)_2$ (Symmetric secondary amine)	$O((R^1R^2N)_2PO)_2$ (Asymmetric secondary amine)	$O((R^1(NR^2)_2)_2PO)_2$ (*N,N'*-substituted diamine derivatives)	$O((AlkylNH)_2PO)_2$ (Primary alkyl amine derivatives)	$O((ArlyNH)_2PO)_2$ (Primary aryl amine derivatives)
$O((Me_2N)_2PO)_2$ Liquid [3,4]	$O((BzMeN)_2PO)_2$ *C2/c* [13]	$O((C_2H_4(2,5\text{-}iPrPhN)_2)_2PO)_2$ *P2_1/n* [14]	$O((tBuNH)_2PO)_2$ *P2_1/c* [12]	$O((2\text{-}MePhNH)_2PO)_2$ *P2_1/c* [10,11]
		$O((1,2\text{-}Cy(NaphN)_2)_2PO)_2$ *C2* [15]	$O((iPrNH)_2PO)_2$ *Pca2_1* (current)	$O((4\text{-}MePhNH)_2PO)_2$ *Pccn* [13]

On comparison of the crystal structures of the compounds classified in each group, the most important molecular difference that seemed to affect the supramolecular structure was the presence of the N–H moiety. Its presence caused a significant difference between the structure of the mono-N-substituted pyrophosphoramides and the di-N-substituted pyrophosphoramides, which lack this moiety. In the crystal structures of mono-N-substituted pyrophosphoramides P=O···H–N hydrogen bonds were the most prominent and common intermolecular bonds in the structure. These formed numerous intermolecular and intramolecular synthons, which were the primary cause for the formation of the various supramolecular motifs observed, and which dictated both structure and symmetry [17–19]. The di-N-substituted pyrophosphoramides were shown to form crystalline structures wherein the pyrophosphoramide moieties did not interact with each other directly. Thus, no strong intermolecular interactions were found. The organic moieties therefore had a significant impact on the packing of molecules in the crystal structures obtained in contrast to the mono-N-substituted pyrophosphoramides (Table 3).

Table 3. Hydrogen bond distances in the crystal structure of **1**.

D	H	A	d(D-A)/Å	d(H-A)/Å	d(D-A)/Å	D-H-A/°
C41	H41A	O1	0.96	2.99	3.601(11)	122.9
C42	H42B	O2 [1]	0.96	2.58	3.505(8)	161.4
C32	H32A	O3	0.96	2.85	3.511(10)	126.4
N3	H3	O3 [2]	0.75(5)	2.34(5)	3.063(4)	164(5)
N3	H3	N2	0.75(5)	2.98(5)	3.491(5)	128(4)
N1	H1	O2 [2]	0.70(4)	2.22(4)	2.907(4)	170(5)
N4	H4	N3 [3]	0.92(6)	2.79(6)	2.979(4)	161(5)
N2	H2	O3 [2]	0.85(6)	2.16(6)	2.979(4)	161(5)

[1] $3/2 - X, +Y, \frac{1}{2} + Z$; [2] $+X, -1 + Y, +Z$; [3] $+X, 1 + Y, +Z$.

3.1.1. Structural Comparison of Mono-N-Substituted Pyrophosphoramides.

Given that the compound characterised in this current study, **1**, was a mono-N-substituted pyrophosphoramide its novel structure is best discussed in relation to other mono-N-substituted pyrophosphoramides. Despite its chemical similarities to $O((tBuNH)_2PO)_2$, it not only crystallised in a different spacegroup, namely *Pca2_1*, but it also showed a different supramolecular motif. First, the already published structures will be discussed and then compared with **1**.

The structures of Mono-N-substituted pyrophosphoramides $O((RNH)_2PO)_2$ were found to crystallise in two space groups: either *P2_1/c* or *Pccn*. Common supramolecular arrangements form part of the crystal structures of $O((tBuNH)_2PO)_2$ and the two $O((2\text{-}MePhNH)_2PO)_2$ polymorphs [10–12] (Figure 3). The first one is a ring synthon with a $R_2^2(8)$ graph set binding molecules through intermolecular hydrogen bonding, while in the other [17,19] a partially eclipsed conformations is reached (Figure 3). The second synthon constricts the molecules from taking on other conformations and therefore minimises the number of possible supramolecular structures available. These two synthons together create the same supramolecular structure for all of the above crystal structures, namely a chain like packing as given in Figure 4. The respective structures for the three mono-N substituted pyrophosphoramides are shown in Figure 5. This structure motif is repeated through translational symmetry to form infinite chains. The two molecules that form the actual supramolecular units are related to each other by the glide plane denoted in the $P2_1/c$ spacegroup.

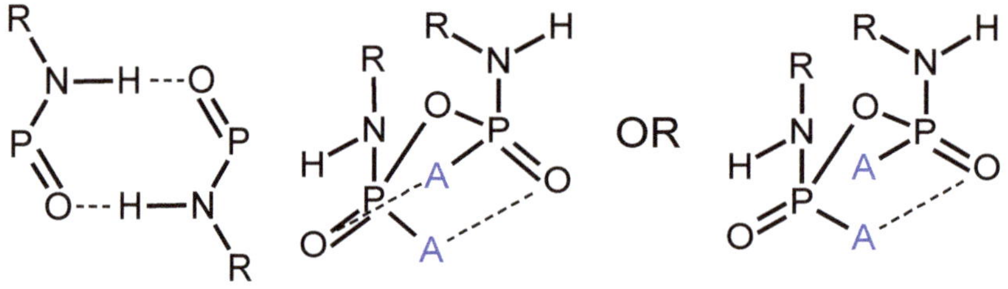

Figure 3. H-bonding synthons in $P2_1/c$ structures: (**a**) intermolecular ring synthon; (**b**) the two variants of the intramolecular synthon where A is the non-intermolecular bonding amine (all generated using ChemSketch [24]).

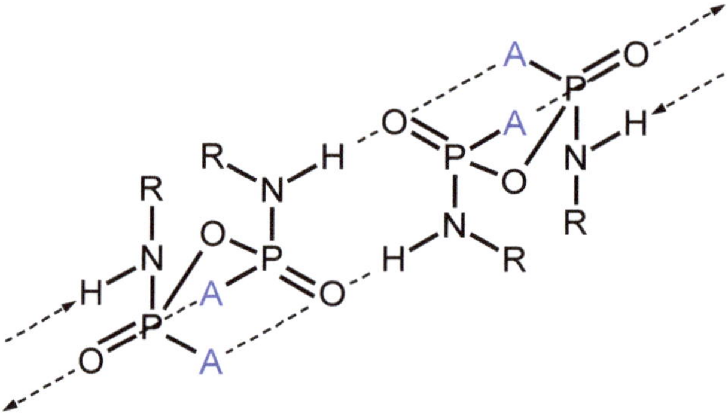

Figure 4. Basic unit for the supramolecular H-bonding motif that is common to all three mono-N-substituted pyrophosphoramides crystallising in $P2_1/c$ (generated using ChemSketch [24]).

Formation of this hydrogen bonding motif seems to be independent of the nature of the organic substituent on the amide nitrogen, as it occurs for both the alkyl tert-butyl and aryl 2-methylphenyl analogues. The packing of these chains is, however, influenced by the organic substituents. The tert-butyl groups in O((tBuNH)$_2$PO)$_2$ pack in a staggered formation resulting in the closest possible packing of the chains [12]. No additional intermolecular interactions are detected within the usual limits.

The packing effects of the 2-methylphenyl groups are more complex than those of the tert-butyl groups. They give rise to two polymorphs of this compound. The two molecular components of these supramolecular units are related to each other by the glide plane imposed by the $P2_1/c$ spacegroup. In the two O((2-MePhNH)$_2$PO)$_2$ polymorphs however, the direction of the motif is different. The motif in the polymorph reported by Pourayoubi, M. et al. is built along the c axis while the same motif in the polymorph described by Cameron, S.T. et al. is perpendicular to the c axis. Given that the hydrogen bond interaction is identical in both polymorphs, the difference lies in the way the supramolecular chains pack. The polymorph discussed by Pourayoubi et al. [11] showed closer packing between the chains. In both polymorphs the 2-methylphenyl groups are oriented antiparallel to each other (Figures 6 and 7). The stacking distances between the intramolecular 2-methylphenyl moieties are 3.539 Å and 3.933 Å [10,11]. In one polymorph [11] the position of the chains seems mainly influenced by the supramolecular structure given in the diagram shown in Figure 6. A square like motif is formed wherein each side is composed of the Ph(π)$\cdots$H–C(meta) interactions [17]. Two contact distances of 3.159 Å and 3.188 Å are present, which are shorter than any contact distance reported in the polymorph described by Cameron et al. [10]. The latter forms similar inter- and intramolecular interactions with one much larger contact distances of 4.717 Å and a very oblique interaction between neighbouring molecules, not bound by hydrogen bonding, with a distance of 3.626 Å [10]. The former interaction is longer than any interactions described for the other polymorph and indicates a less efficient packing. Thus, the different modes of packing of the organic moieties are most probably the cause of the formation of the two polymorphs.

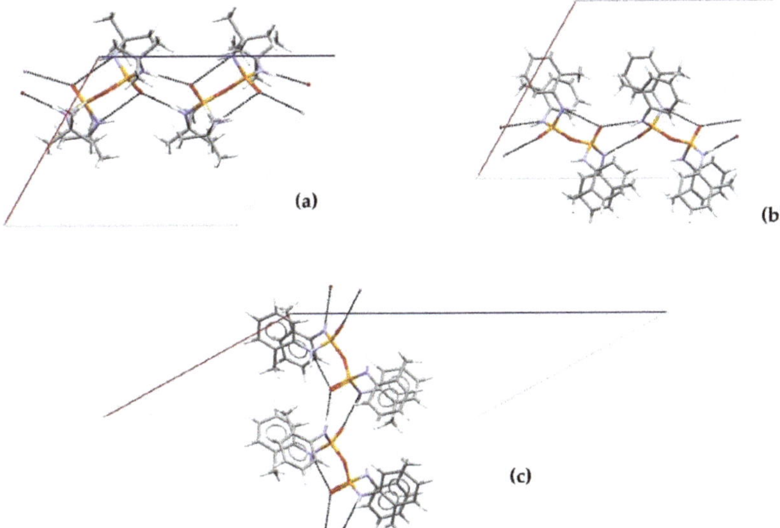

Figure 5. The hydrogen bonding motif common to all $P2_1/c$ as noted in the published structures: (**a**) O((tBuNH)$_2$PO)$_2$; (**b**) O((2-MePhNH)$_2$PO)$_2$ published by Pourayoubi et al. [11]; (**c**) O((2-MePhNH)$_2$PO)$_2$ published by Cameron et al. [10] (all generated in Mercury [25]).

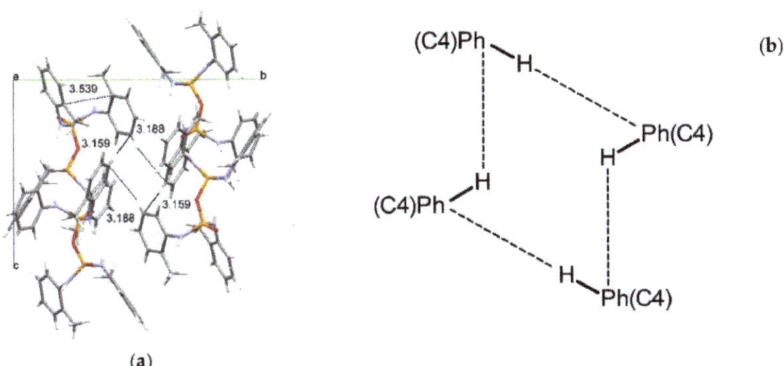

Figure 6. (**a**) O((2-MePhNH)$_2$PO)$_2$ polymorph published by Pourayoubi et al. showing both intermolecular and intramolecular C–H$\cdots$Ph interactions (generated in Mercury [25]); (**b**) a diagram of the intermolecular C–H$\cdots$Ph interactions for clarity (generated using ChemSketch [24]).

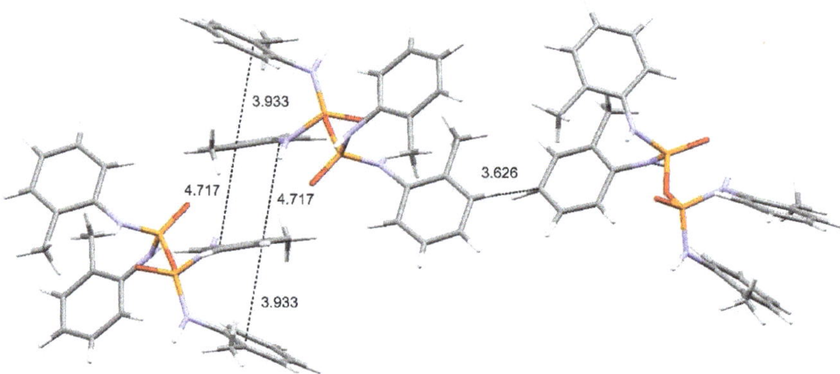

Figure 7. O((2-MePhNH)$_2$PO)$_2$ polymorph published by Cameron et al. showing both intermolecular and intramolecular C–H···Ph interactions (generated in Mercury [25]).

Experimental procedures on how to obtain only one of two polymorphs are not reported in the literature. The O((2-MePhNH)$_2$PO)$_2$ structure published by Cameron et al. was obtained by the reaction of phosphenyl chloride with ortho-methylaniline followed by recrystallisation in methanol. The synthesis of the polymorph obtained by Pourayoubi et al. was not described in the literature to the best of our knowledge. Interestingly, the crystallographic data of the two structures was collected at different temperatures (150 K for [11] and 295 K for [10,11]). It is, therefore, possible that a polymorphic transition occurs at lower temperatures. This is also in agreement with the difference in the corresponding unit cell volumes.

The mono-N-substituted pyrophosphoramides, which do not crystallise in the $P2_1/c$ spacegroup, show different inter- and intramolecular hydrogen bonding motifs. Given that the compounds crystallising in the same spacegroup ($P2_1/c$) contain both alkyl and aryl moieties and show a different packing of these moieties, a similar observation is expected for the 4-methylphenyl and the iso-propyl analogues. However, a different supramolecular structure was formed by the 4-methylphenyl derivative O((4-MePhNH)$_2$PO)$_2$, crystallising in spacegroup *Pccn* [13]. There is no intramolecular synthon present (Figure 3b), resulting in a different pattern of supramolecular interactions. The phosphoryl oxygen and the amide nitrogen, which typically form the intramolecular P=O···H–N motif in the $P2_1/c$ structures, do not show any hydrogen bonding and therefore, the molecules are not limited to an eclipsed conformation (*vide supra*). They form a staggered conformation that enables the formation of the basic supramolecular building block for this structure. Two molecules are connected via two intermolecular P=O···H–N hydrogen bonds, forming a ring synthon with graph set R$_2^2$(12) (Figure 8) [13,19]. Further N–H···Ph and P=O···H–C(Ph(C2)) interactions are observed. The N–H···Ph interactions are formed through the remaining amide moieties which do not interact in the R$_2^2$(12) ring described prior and this additional set of interactions seems to stabilise the motif by increasing the packing efficiency and stopping this moiety from forming the previously mentioned intramolecular P=O···H–N bonding. The P=O···H–C(Ph(C2)) interactions also seem to add stability by further aiding the P=O oxygen atoms to obtain the orientation necessary to form this synthon [17,21].

The two molecules are related by a glide plane on the *a–c* plane along the *c* axis (Figure 9) which results in the formation of infinite chains along cell axis *c*. Neighbouring chains positioned anti-parallel to each other along the *b* axis and related to each other by a second glide plane along the diagonal of the *ab* plane. Chains neighbouring each other along the *a* axis are parallel and again related by translation. The main interaction responsible for arrangement in antiparallel chains is the Ph(π)···H–C(para-methyl) interaction between the closely situated 4-methyphenyl groups bound to different molecules in different chains [21], as noted in Figure 9 [13]. This guarantees the closest possible packing of the various phenyl groups in the molecule. Thus, the 4-methylphenyl groups seem to play a structure forming role in both the formation of a different hydrogen bonding motif and a different packing of the molecules in this structure compared to other pyrophosphoramides.

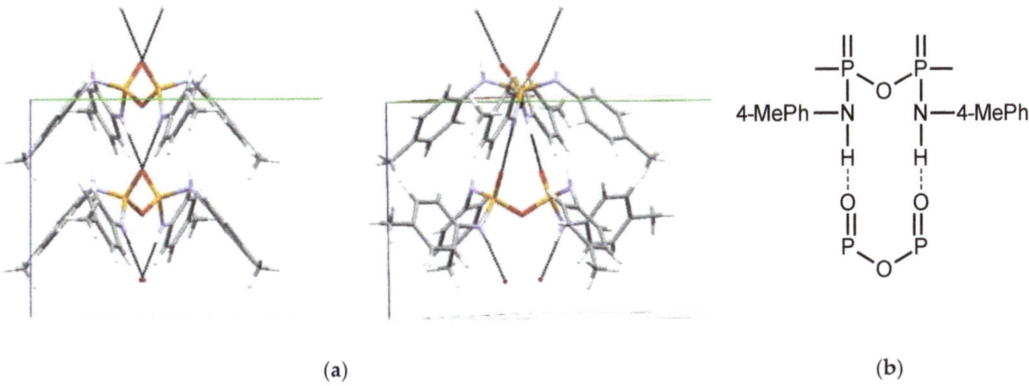

(**a**)   (**b**)

Figure 8. Hydrogen bonding motif for O((4-MePhNH)$_2$PO)$_2$ (**a**) as viewed along the *a*-axis (**left**) and offset to show the staggered conformation of the pyrophosphoramide (**right**) (generated in Mercury [25]); (**b**) The ring synthon responsible for the motif given in a simplified diagram (generated using ChemSketch [24]).

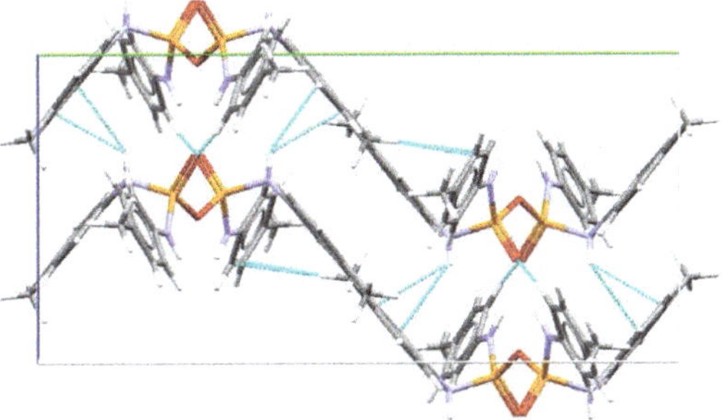

Figure 9. Intermolecular and intramolecular non-hydrogen bonding interactions in the structure of O((4-MePhNH)$_2$PO)$_2$ (generated in Mercury [25]).

The 2-methylphenyl and 4-methylphenyl derivatives both crystallise in different supramolecular structures and space groups. However, the 2-methylphenyl pyrophosphoramide crystallises in the same space group and shows the same supramolecular bonding as that for the tert-butyl analogue. O((4-MePhNH)$_2$PO)$_2$. The latter was collected at 90 K, whereas the O((2-MePhNH)$_2$PO)$_2$ polymorphs were collected at 150 K and 295 K [10,11,13]. Thus, it is unclear whether the occurrence of different structural motifs is mainly caused by the different nature of the side groups or simply because the single crystal data was collected for each compound at different temperatures

No intramolecular P=O···H–N hydrogen bonding is present in the crystal structure of **1**. This leads to the formation of a staggered arrangement of the pyrophosphoramide group similar to that observed for O((4-MePhNH)$_2$PO)$_2$. Thus again, the lack of intramolecular bonding between amide and phosphoroxide seems to hinder an eclipsed conformation (*vide supra*, Figure 3b) and related supramolecular motifs. The most important supramolecular motif in the structure of **1** is therefore the P=O···H–N interaction. This is in agreement with what was observed also in all other mono-N-substituted pyrophosphoramides, where intermolecular hydrogen bonding was a major factor in the formation of the relevant supramolecular motifs [10–13]. The actual synthon is shown in Figure 10. It is formed by two molecules of **1** related to each other by translation symmetry via two different P=O···H–N hydrogen bonding synthons. To the best of our knowledge this is the only mono-N-substituted pyrophosphoramide structure to exhibit multiple intermolecular hydrogen bonding synthons. The first synthon is a ring with a $R_2^1(8)$ graph set connecting one P=O group with two neighbouring N–H groups, which are bound to the two different phosphorus centres of the second molecule. The second synthon consists of a P=O···H–N interaction between the other P=O group of the first molecule and a neighbouring N–H group. The latter is not part of the previously discussed synthon; it shares, however, a phosphorus centre with one of the previously discussed amide groups. Neighbouring chains along the *a* axis pack anti-parallel to each other via a 2_1 screw axis (Figure 11). The remaining amide does not seem to participate in any type of intermolecular or intramolecular bonding, even though, theoretically, infinite chains along the *b* axis could be formed in a similar manner as observed in other mono-N-substituted pyrophosphoramides (*vide supra*). Therefore, the main mode of intermolecular interactions is the Van der Waals forces that caused the hydrophobic iso-propyl groups to stack.

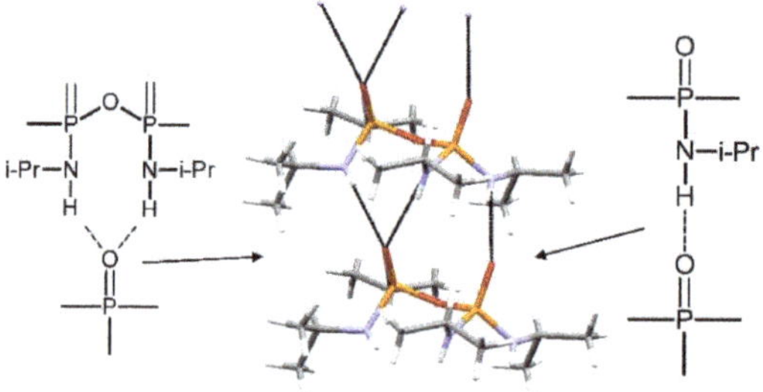

Figure 10. Hydrogen bonding synthons present in the structure of O((*i*PrNH)$_2$PO)$_2$ (generated in Mercury [25] and ChemSketch [24]).

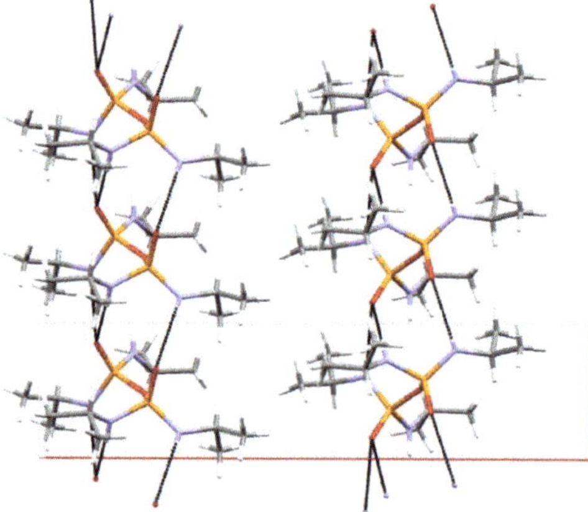

Figure 11. Antiparallel chains of O((*i*PrNH)$_2$PO)$_2$ noted along the a-axis (generated in Mercury [25]).

The cause for the formation of this different supramolecular motif is difficult to determine as the different molecular structure, synthesis, crystallisation techniques, and solvents used can all affect crystallization process and lead to this crystal structure. The temperature at which single crystal data was obtained is unlikely to be the cause of this as it falls into the same range as for the other compounds crystallising in spacegroup

$P2_1/c$ [10–12]. The synthesis and crystallisation approach used to obtain single crystals of **1** differs from published procedures to obtain crystalline material for the related compounds. Given the chemical similarities between the iso-propyl and tert-butyl moiety the difference in supramolecular motifs between the two is unexpected.

3.1.2. Structural Comparison of di-N-Substituted Pyrophosphoramides

Because of the lack of a N-H donor in di-N-substituted pyrophosphoramides the main supramolecular interactions present in the crystal structure derive from the organic substituents. A very good and well-known example of a di-N-substituted pyrophosphoramide $O((R_2N)_2PO)_2$ is Schradan, $O((Me_2N)_2PO)_2$ [3–9,22]. Schradan is a liquid at room temperature. No crystal structures collected of crystals below the melting point temperature are reported in the literature. Structural information from experimental data is only available for $O((R^1R^2N)_2PO)_2$ and the *N,N'*-substituted diamine derivatives $O((R^1(NR^2)_2)_2PO)_2$, i.e., $O((BzMeN)_2PO)_2$, $O((C_2H_4(2,5\text{-}iPrPhN)_2)_2PO)_2$ and $O((1,2\text{-}Cy(NaphN)_2)_2PO)_2$ [13–15]. No significant classical intermolecular interactions between the pyrophosphoramide or organic moieties are present in the corresponding published crystal structures. The pyrophosphoramide backbone forms a similar staggered conformation in all three compounds. In $O((BzMeN)_2PO)_2$, the main intermolecular interactions effecting the supramolecular structure are weak C–H···Ph interactions [13]. These form in two different synthons, namely N–Me···Ph(C2) and CH_2(benzyl)···Ph. The molecules pack in layers linked along the axis *c* (Figure 12) [17,21].

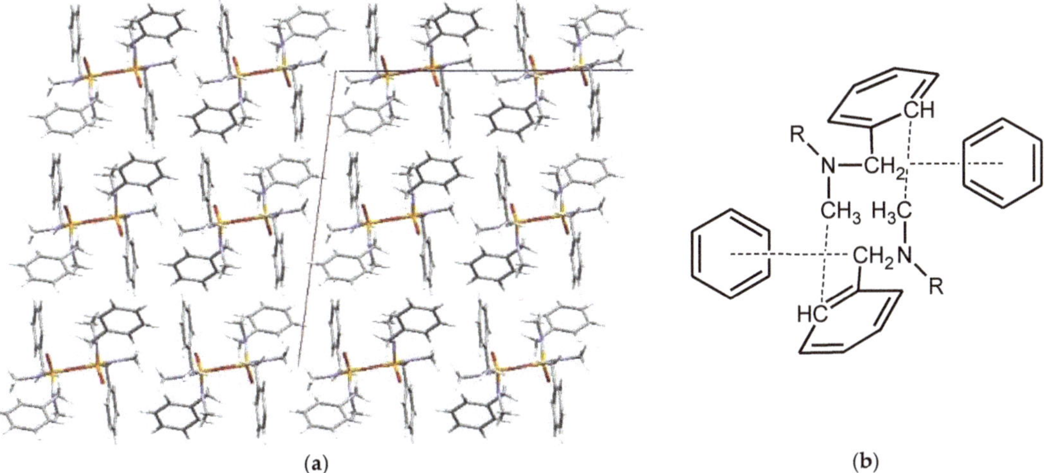

Figure 12. Packing of $O((BzMeN)_2PO)_2$: (**a**) C–H···Ph interactions which act as the main supramolecular building block (generated in Mercury [25]); (**b**) simplified diagram of the supramolecular motif (generated using ChemSketch [24]).

The crystal structure of $O((C_2H_4(2,5\text{-}iPrPhN)_2)_2PO)_2$ shows only very weak intermolecular interactions [14]. The pyrophosphoramide moieties seem to be isolated from each other by the bulky hydrophobic organic moieties (Figure 13).

In the crystal structure of $O((1,2\text{-}Cy(NaphN)_2)_2PO)_2$ intermolecular motifs (C–H···Ph interactions) are present similar to those observed in the structure of $O((BzMeN)_2PO)_2$. Intermolecular interactions are noted between the benzyl methylene, the naphthyl groups and the cyclohexyl –CH_2– groups. Three distinct C–H···Ph interactions are present namely CH_2(benzyl methylene)···naphthyl(C4), C–H(Naphthyl(C8))···naphthyl(C10), and CH_2(cyclohexyl)···napthyl(C3) [15]. These interactions are unidirectional, with the proton donors binding to the closest naphthalene along the *c* axis, which is typically the molecule diagonal to the proton donor molecule in the layers running along axis *a* (Figure 14). The pyrophosphoramide group is connected to hydrate water by hydrogen bonding forming a $R_2^2(8)$ ring [15]. This ring motif separates the pyrophosphoramide molecule structurally from any further interaction with typical hydrogen bonding electron acceptors.

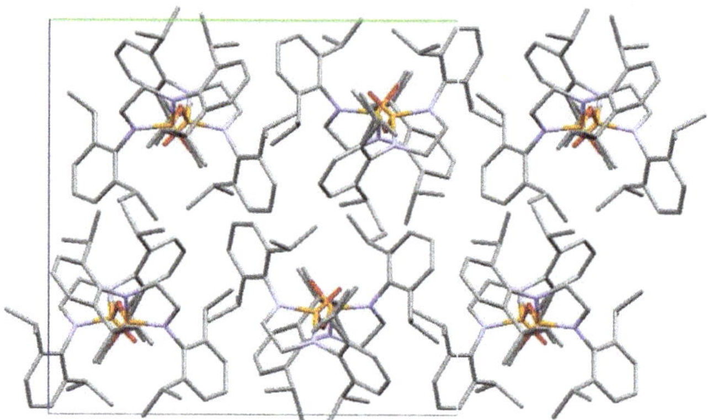

Figure 13. Structure of $O((C_2H_4(2,5\text{-}i\text{PrPhN})_2)_2PO)_2$ as viewed along the a-axis (generated in Mercury [25]).

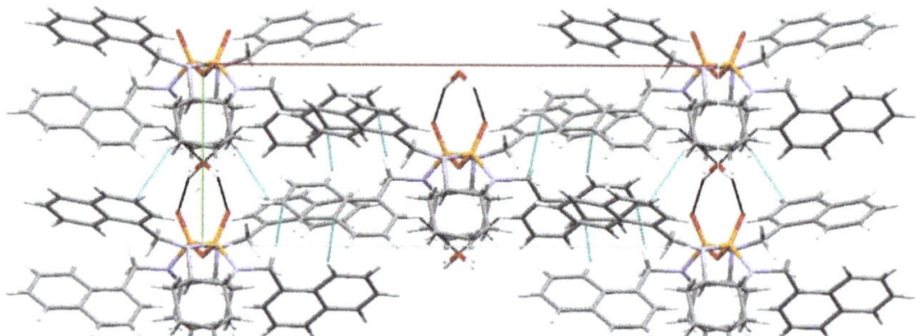

Figure 14. Structure of $O((1,2\text{-}Cy(\text{NaphN})_2)_2PO)_2$ viewed along the c-axis showing hydrogen bonding (black contacts) and C–H···Ph interactions (light blue contacts) (generated in Mercury [25]).

In all three cases the supramolecular structure is dominated by the organic substituents. There is complete lack of π–π stacking interactions. A possible reason for this might be that the pyrophosphoramide backbone always takes on a more staggered conformation forcing the organic substituents into unfavorable positions to form π–π stacking. The staggered conformation itself is most likely caused by the lack of the amine N–H bonds, which usually force the molecule in a more eclipsed conformation through inter/intramolecular bonding as observed in some of the mono-N-substituted pyrophosphoramides.

4. Conclusions

The crystal structure of the mono-N-substituted pyrophosphoramide, $O((i\text{PrNH})_2PO)_2$ **1** has been determined from single-crystal X-ray diffraction data while the chemical identity of the species was supported by IR, ^{1}H NMR, ^{31}P NMR, and GC MS data (see SI). A thorough structural comparison of **1** with other pyrophosphoramides for which the crystal structures have been published previously was carried out. **1** forms a novel supramolecular motif previously unattested for mono-N-substituted pyrophosphoramides. This motif was composed of two different synthons with P=O···H–N interactions. The great impact of this type of hydrogen bonding on the supramolecular motifs in all previously published mono-N-substituted pyrophosphoramides could be confirmed also for **1**. Trends regarding the effects of the various organic moieties within the different compounds were difficult to describe due to the lack of systematic data. No comparison of **1** with the di-N-substituted pyrophosphoramides was undertaken, given that the trends in packing are significantly different between this group and the mono-N-substituted pyrophosphoramides.

The differences observed between the supramolecular motifs present in **1** and the supramolecular features of other mono-N-substituted pyrophosphoramides indicate that there are possibly other supramolecular motifs that have not yet been discovered yet. The crystal structure of **1** further expands the diversity of possible supramolecular synthons. The different synthon in **1** ($R_2^1(8)$ring), not known in previously described mono-N-substituted pyrophosphoramides, adds a new strong structural motif to the viable synthons known experimentally for mono-N-substituted pyrophosphoramides. It can also be considered to be a viable option for the discovery of new co-crystals of these species [20,21]. Different solid-state forms of pyrophosphoramides as well as co-crystals formed with other organic and inorganic species will have different physical properties and, at times different chemical properties compared to the currently marketed compounds without changing the actual molecule [26–30]. This is important for mono-N-substituted pyrophosphoramides given their possible use as a pesticide. The formation of different forms with different thermodynamic and kinetic stabilities can aid in complexation reactions and e.g., diminish decomposition on storage, a property of great importance for use in agriculture [28,29,31,32].

However, given the disparity between the various experimental techniques used in both the structures described in the literature and the current study further work at different temperatures, including X-ray data from powders, must be undertaken to obtain a more comprehensive understanding of the relationship between the chemical and physical properties and the crystal structures of various pyrophosphoramides.

Supplementary Materials: The following are available online at https://www.mdpi.com/2624-8549/3/1/13/s1. Interpretation of IR, NMR and mass spectra. Figure S1: IR spectrum of $O((iPrNH)_2PO)_2$. Figure S2: IR spectrum of $O((iPrNH)_2PO)_2$ obtained from column chromatography. Figure S3: ^1H NMR spectrum of $O((iPrNH)_2PO)_2$. Figure S4: ^1H NMR spectrum of $O((iPrNH)_2PO)_2$ from column chromatography. Figure S5: $^{31}P\{^1H\}$ NMR spectrum of $O((iPrNH)_2PO)_2$. Figure S6: $^{31}P\{^1H\}$ NMR spectrum of $O((iPrNH)_2PO)_2$ from column chromatography. Figure S7: Gas chromatograph of $O((iPrNH)_2PO)_2$. Figure S8: Gas chromatograph of $O((iPrNH)_2PO)_2$ from column chromatography. Figure S9: Mass spectrum of $O((iPrNH)_2PO)_2$ from column chromatography. Table S1: ^1H NMR experimental data and assignment for proton peaks of $O((iPrNH)_2PO)_2$ in $CDCl_3$. Table S2: ^1H NMR experimental data and assignment for proton peaks of $O((iPrNH)_2PO)_2$ from column chromatography in $CDCl_3$. Table S3: Mass spectra peak data for the $O((iPrNH)_2PO)_2$ Gas chromatography peaks at 6.618 and 7.313 min.

Author Contributions: Conceptualization, D.M., U.B. and L.V.-Z.; Methodology and Practical Chemical and Spectroscopic studies, D.M.; XRD data collection and structure solution, U.B.; Writing—original draft preparation, D.M.; Writing—review and editing, U.B. and L.V.-Z.; Project administration, U.B. and L.V.-Z.; funding acquisition, L.V.-Z. and U.B. All authors have read and agreed to the published version of the manuscript.

Funding: The research work disclosed in this publication is partially funded by the Endeavour Scholarship Scheme (Malta). Scholarships are part-financed by the European Union—European Social Fund (ESF)—Operational Programme II—Cohesion Policy 2014–2020 "Investing in human capital to create more opportunities and promote the well-being of society". It was also funded by the project: "Setting up of transdisciplinary research and knowledge exchange (TRAKE) complex at the University of Malta (ERDF.01.124)", which is being co-financed through the European Union through the European Regional Development Fund 2014–2020 (L.V.-Z. and U.B.).

Data Availability Statement: The data presented in this study are available in the supplementary materials.

Acknowledgments: The authors would like to thank Robert M. Borg for his contribution regarding NMR data collection and analysis and Godwin Sammut for his contribution in GC MS data collection both of the Department of Chemistry, Faculty of Science, University of Malta. We would also like to thank Jens Meyer from STOE & Cie GmbH for his contribution in SXRD data collection.

Conflicts of Interest: The authors declare no conflict of interest.

Appendix A

Appendix A.1. Single Crystal X-ray Diffraction Experimental Description

Single crystals of $C_{12}H_{32}N_4O_3P_2$ [O(iPrNH)$_2$PO)$_2$] were [Layering CHCl$_3$ solution with 30–40 petroleum ether]. A suitable crystal was selected and [Collected in oil and frozen] on a STOE STADIVARI diffractometer. The crystal was kept at 293 K during data collection. Using Olex2 [1], the structure was solved with the ShelXT [2] structure solution program using Intrinsic Phasing and refined with the ShelXL [3] refinement package using Least Squares minimisation. CCDC 2055906 contains the supplementary crystallographic data for this paper. The data can be obtained free of charge from The Cambridge Crystallographic Data Centre via www.ccdc.cam.ac.uk/structures.

1. Dolomanov, O.V., Bourhis, L.J., Gildea, R.J, Howard, J.A.K. & Puschmann, H. (2009), J Appl. Cryst. 42, 339–341.
2. Sheldrick, G.M. (2015). Acta Cryst. A71, 3–8.
3. Sheldrick, G.M. (2015). Acta Cryst. C71, 3–8.

Crystal structure determination of [O(iPrNH)$_2$PO)$_2$]

Crystal Data for $C_{12}H_{32}N_4O_3P_2$ (M = 342.35 g/mol): orthorhombic, space group $Pca2_1$ (no. 29), a = 20.342(2) Å, b = 5.0495(6) Å, c = 19.103(3) Å, V = 1962.2(4) Å^3, Z = 4, T = 295 K, μ(CuKα) = 2.133 mm^{-1}, $Dcalc$ = 1.159 g/cm^3, 43032 reflections measured (8.696° $\leq$ 2Θ $\leq$ 137.27°), 3317 unique (R_{int} = 0.0557, R_{sigma} = 0.0226) which were used in all calculations. The final R_1 was 0.0408 (I > 2σ(I)) and wR_2 was 0.1075 (all data).

Appendix A.2. Refinement Model Description

Number of restraints—1, number of constraints—unknown.
Details:

1. Fixed Uiso

- At 1.2 times of all C(H) groups
- At 1.5 times of all CH$_3$ groups

2a. Ternary CH refined with riding coordinates: C10(H10), C20(H20), C30(H30), C40(H40)
2b. Idealised Me refined as rotating group: C31(H31A,H31B,H31C), C11(H11A,H11B,H11C), C41(H41A,H41B,H41C), C42(H42A,H42B,H42C), C12(H12A,H12B,H12C), C22(H22A, H22B,H22C), C32(H32A,H32B,H32C), C21(H21A, H21B,H21C)

References

1. Li, B.; Stribley, J.A.; Ticu, A.; Xie, W.; Schopfer, L.M.; Hammond, P.; Brimijoin, S.; Hinrichs, S.H.; Lockridge, O. Abundant Tissue Butyrylcholinesterase and Its Possible Function in the Acetylcholinesterase Knockout Mouse. *J. Neurochem.* **2002**, *75*, 1320–1331. [CrossRef]
2. Pohanka, M. Cholinesterases, a target of pharmacology and toxicology. *Biomed. Pap.* **2011**, *155*, 219–223. [CrossRef] [PubMed]
3. Lickerish, L.A. Studies on commercial octamethylpyrophosphor- amide (schradan) V.*-Insecticidal Comparisons of the Two Main Constituents. *J. Sci. Food Agric.* **1953**, *4*, 24–28. [CrossRef]
4. Rediske, J.H.; Lawrence, W.H. Octamethylpyrophosphoramide (OMPA) As a Systemic Animal Repellent for Douglas-Fir Seedlings. *For. Sci.* **1964**, *10*, 93–103.
5. Joesten, M.D. The donor properties of pyrophosphate derivatives. V. Complexes of molybdenum (V) oxychloride, uranyl ion, and thorium (IV) ion with octamethylpyrophosphoramide. *J. Inorg. Nucl. Chem.* **1967**, *6*, 1598–1599. [CrossRef]
6. du Preez, J.G.H.; Sadie, F.G. Hexamethylphosphoramide and octamethylpyrophosphoramide complexes of tetravalent metal chlorides. *J. S. Afr. Chem. Inst.* **1966**, *19*, 73–84.

7. Joesten, M.D.; Hussain, M.S.; Lenhert, P.G. Structure studies of pyrophosphate chelate rings. I. Crystal structures of tris-octamethylpyrophosphoramide complexes of cobalt (II), magnesium (II), and copper (II)perchlorates. *Inorg. Chem.* **1970**, *9*, 151–161. [CrossRef]

8. Hussain, M.S.; Joesten, M.D.; Lenhert, P.G. Structuee Studies of Pyrophosphate Chelate Rings. II. The Crystal Structuee of Bis(Perchlorato)Bis(Octamethylpyrophosphoramide)Copper(II). *Inorg. Chem.* **1970**, *9*, 162–168. [CrossRef]

9. Kepert, D.L.; Patrick, J.M.; White, A.H. Structure and Stereochemistry in f-Block Complexes of High Co-Ordination Number. Part 3.* The [M(Bidentate Ligand)2(Unidentate Ligand)4] System: Crystal Structures of Tetrakis (Isothiocyanato)-Bis(Octamethylpyrophosphoramide. *Dalton Trans.* **1983**, *0*, 559–566.

10. Cameron, T.S.; Cordes, R.E.; Jackman, F.A. Synthesis and Crystal Structure of μ-Oxo-bis(phosphenyl-ortho-toluidide). *Z. Nat. B* **1978**, *33*, 728–730. [CrossRef]

11. Pourayoubi, M.; Padělková, Z.; Rostami Chaijan, M.; Růžička, A. *N*,*N'*,*N''*,*N'''*-Tetrakis(2-Methylphenyl)-Oxybis(Phosphonic Diamide): A Redetermination at 150 K with Mo K Radiation. *Acta Crystallogr. Sect. E Crystallogr. Commun.* **2011**, *67*, 466–470. [CrossRef] [PubMed]

12. Pourayoubi, M.; Tarahhomi, A.; Karimi Ahmadabad, F.; Fejfarová, K.; Van Der Lee, A.; Dušek, M. Two New XP(O)[NHC(CH$_3$)$_3$]$_2$ Phosphor-Amidates, with X = (CH$_3$)$_2$N and [(CH$_3$)$_3$CNH]$_2$P(O)(O). *Acta Crystallogr. Sect. C Cryst. Struct. Commun.* **2012**, *68*, 164–169. [CrossRef] [PubMed]

13. Tarahhomi, A.; Pourayoubi, M.; Golen, J.A.; Zargaran, P.; Elahi, B.; Rheingold, A.L.; Ramírez, M.A.L.; Percino, T.M. Hirshfeld surface analysis of new phosphoramidates. *Acta Crystallogr. Sect. B Struct. Sci. Cryst. Eng. Mater.* **2013**, *69*, 260–270. [CrossRef]

14. Giffin, N.A.; Hendsbee, A.D.; Roemmele, T.L.; Lumsden, M.D.; Pye, C.C.; Masuda, J.D. Preparation of a Diphosphine with Persistent Phosphinyl Radical Character in Solution: Characterization, Reactivity with O2, S8, Se, Te, and P4, and Electronic Structure Calculations. *Inorg. Chem.* **2012**, *51*, 11837–11850. [CrossRef] [PubMed]

15. Yerramsetti, N.; Unruh, D.K.; Li, G. CCDC 1559159: Experimental Crystal Structure Determination. Available online: https://www.ccdc.cam.ac.uk/structures/search?id=doi:10.5517/ccdc.csd.cc1pbfgx&sid=DataCite (accessed on 28 June 2017).

16. Tarahhomi, A.; Pourayoubi, M.; Fejfarová, K.; Dusek, M. A Novel Amido-pyrophosphate MnII Chelate Complex with the Synthetic Ligand O{P(O)[NHC(CH$_3$)$_3$]$_2$}2(L): [Mn(L)2{OC(H)N(CH$_3$)$_2$}2]Cl$_2$.2H$_2$O. *Acta Crystallogr. Sect. C Struct. Chem.* **2013**, *69*, 225–229. [CrossRef]

17. Desiraju, G.R. Supramolecular Synthons in Crystal Engineering—A New Organic Synthesis. *Angew. Chem. Int. Ed.* **1995**, *34*, 2311–2327. [CrossRef]

18. Desiraju, G.R. Designer Crystals: Intermolecular Interactions, Network Structures and Supramolecular Synthons. *Chem. Commun.* **1997**, *16*, 1475–1482. [CrossRef]

19. Desiraju, G.R. Hydrogen Bridges in Crystal Engineering: Interactions without Borders. *Acc. Chem. Res.* **2002**, *35*, 565–573. [CrossRef]

20. Desiraju, G.R. Crystal Engineering: A Holistic View. *Angew. Chem. Int. Ed.* **2007**, *46*, 8342–8356. [CrossRef]

21. Desiraju, G.R. Crystal engineering: A brief overview. *J. Chem. Sci.* **2010**, *122*, 667–675. [CrossRef]

22. Goehring, M.; Niedenzu, K. Diphosphorsäure-tetrakis-dimethylamid. *Angew. Chem.* **1956**, *68*, 704. [CrossRef]

23. Dolomanov, O.V.; Bourhis, L.J.; Gildea, R.J.; Howard, J.A.K.; Puschmann, H. OLEX2: A complete structure solution, refinement and analysis program. *J. Appl. Crystallogr.* **2009**, *42*, 339–341. [CrossRef]

24. *ACD/ChemSketch*; Advanced Chemistry Development: Toronto, ON, Canada, 2013.

25. Macrae, C.F.; Bruno, I.J.; Chisholm, J.A.; Edgington, P.R.; McCabe, P.; Pidcock, E.; Rodriguez-Monge, L.; Taylor, R.J.; Van De Streek, J.; Wood, P.A. Mercury CSD 2.0—New features for the visualization and investigation of crystal structures. *J. Appl. Crystallogr.* **2008**, *41*, 466–470. [CrossRef]

26. Bernstein, J.; Sheva, B. Pinching Polymorphs. *Nat. Mater.* **2005**, *4*, 427–429. [CrossRef]

27. Haleblian, J.; McCrone, W. Pharmaceutical Applications of Polymorphism. *J. Pharm. Sci.* **1969**, *58*, 911–929. [CrossRef]

28. De Villiers, M.; Van Der Watt, J.; Lötter, A. Kinetic study of the solid-state photolytic degradation of two polymorphic forms of furosemide. *Int. J. Pharm.* **1992**, *88*, 275–283. [CrossRef]

29. Foltz, M.F.; Coon, C.L.; Garcia, F.; Nicholas, A.L., III. The Thermal Stability of the Polymorphs of Hexanitrohexaazaisowurtzitane. *Part I Propellants Explos. Pyrotech.* **1994**, *19*, 19–25. [CrossRef]

30. Hilfiker, R.; Blatter, F.; Von Raumer, M. Relevance of Solid-state Properties for Pharmaceutical Products. In *Polymorphism*; Wiley: Hoboken, NJ, USA, 2006; pp. 1–19.

31. Haywood, A.; Mangan, M.; Grant, G.; Glass, B. Extemporaneous Isoniazid Mixture: Stability Implications. *J. Pharm. Pract. Res.* **2005**, *35*, 181–182. [CrossRef]

32. Sarcevica, I.; Orola, L.; Veidis, M.V.; Podjava, A.; Belyakov, S. Crystal and Molecular Structure and Stability of Isoniazid Cocrystals with Selected Carboxylic Acids. *Cryst. Growth Des.* **2013**, *13*, 1082–1090. [CrossRef]

MDPI

St. Alban-Anlage 66

4052 Basel

Switzerland

Tel. +41 61 683 77 34

Fax +41 61 302 89 18

www.mdpi.com

Chemistry Editorial Office

E-mail: chemistry@mdpi.com

www.mdpi.com/journal/chemistry